This helpful and practical compendium provides clinical scientists with an essential guide to the basic techniques of molecular medicine. It serves as a laboratory manual and a source of reference, and helps bridge the gap from protocol to publication. It is suitable for those wishing to perform basic semi-quantitative experiments such as Northern or Southern blots and also those wishing to undertake more specialized genetic manipulations such as gene cloning, expression and creation of cDNA libraries. It will give clinical scientists a unique insight into the potential of these techniques, their interpretation, problems and pitfalls, and advice on troubleshooting and quality control.

As stated by Sir David Weatherall in the foreword to the volume, 'It should be of great value to both established research workers and young scientists coming into the field for the first time. It deserves every success'.

Clinical gene analysis and manipulation

Tools, techniques and troubleshooting

POSTGRADUATE MEDICAL SCIENCE

This important new series is based on the successful and internationally well-regarded specialist training programme at the Royal Postgraduate Medical School in London. Each volume provides an integrated and self-contained account of a key area of medical science, developed in conjunction with the course organisers and including contributions from specially invited authorities.

The aim of the series is to provide biomedical and clinical scientists with a reliable introduction to the theory and to the technical and clinical applications of each topic.

The volumes will be a valuable resource and guide for trainees in the medical and biomedical sciences, and for laboratory-based scientists.

Titles in the series

Radiation protection of patients edited by Richard Wootton

Monoclonal antibodies edited by Mary A. Ritter and Heather M. Ladyman

Image analysis in histology: conventional and confocal microscopy edited by R. Wootton, D. R. Springall and J. M. Polak

Molecular neuropathology edited by Gareth W. Roberts and Julia M. Polak.

Clinical gene analysis and manipulation

Tools, techniques and troubleshooting

BY

JANUSZ A. Z. JANKOWSKI, MD, PhD, MRCP (UK) (Editor)

Senior lecturer in Medicine, Queen Elizabeth University Hospital, Birmingham, UK
Formerly Assistant Professor in Medicine, University of California, San Francisco, USA

and

JULIA M. POLAK, PhD, FRCPath (Deputy Editor)

Professor and Honorary Consultant, Department of Histochemistry, Royal Postgraduate Medical School, London, UK

FOREWORD:

Professor Sir David Weatherall MD, FRCP, FRS

Institute of Molecular Medicine, Oxford University, UK

Published in association with
The Royal Postgraduate Medical School
University of London by

CAMBRIDGE
UNIVERSITY PRESS

1996

Published by the Press Syndicate of the University of Cambridge
The Pitt Building, Trumpington Street, Cambridge CB2 1RP
40 West 20th Street, New York, NY 10011-4211, USA
10 Stamford Road, Oakleigh, Melbourne 3166, Australia

First published 1996

Printed in Great Britain at the University Press, Cambridge

A catalogue record for this book is available from the British Library

Library of Congress cataloguing in publication data

Clinical gene analysis and manipulation : tools, techniques, and
    troubleshooting / by Janusz A. Z. Jankowski, editor ; Julia Polak,
    deputy editor ; foreword, Sir David Weatherall.
        p.   cm. — (Postgraduate medical science)
    Includes bibliographical references.
    ISBN 0 521 47326 8 (hardcover). – ISBN 0 521 47896 0 (pbk.)
    1. Genetic engineering – Laboratory manuals.   2. Molecular
genetics – Laboratory manuals.   I. Jankowski, Janusz A. Z.
II. Polak, Julia M.   III. Series.
    [DNLM: 1. Genetics, Biochemical.   2. Genetic Techniques.   3. Gene
Expression Regulation.   QH 450 C641 1996]
QH442.C565 1996
574.87'3224 – dc20
DNLM/DLC
for Library of Congress   95-47219 CIP

ISBN 0 521 47326 8 hardback
ISBN 0 521 47896 0 paperback

KT

Contents

Contributors

Janusz A. Z. Jankowski PhD, MD	(Editor), Senior Lecturer and Director of the Gastrointestinal Gene Group, Department of Medicine, Queen Elizabeth University Hospital, Edgebaston, Birmingham, B15 2TH, UK
Julia M. Polak PhD FRCPath	(Deputy Editor) Professor and Honorary Consultant, Department of Histochemistry, Royal Postgraduate Medical School, Du Cane Rd, London, W12 0HS
Fiona Bedford PhD	(Coordinator of Molecular Biology), NARSAD Young Investigator, Box 0556, Department of Physiology, University of California, 513 Parnassus St, San Francisco, USA
Karen Henderson PhD	(Coordinator of Cell Biology), ICRF Scientist, Department of Histopathology, Imperial Cancer Research Fund, Royal Postgraduate Medical School, DuCane Road, London, W12 0HS
Nicholas Bates MB, ChB PhD	Clinical Scientist, ICRF Gene Transcription Laboratory, Cyclotron Building, Royal Postgraduate Medical School, London, W12 0HS
Eliel Bayever MB, MRCP	Assistant Professor, Department of Paediatrics and Pharmacology, University of Nebraska, Omaha, USA
Anne Bishop PhD	Senior Scientist, Department of Histochemistry, Royal Postgraduate Medical School, Du Cane Rd, London, W12 0HS
Brian Boullier PhD	Lecturer, Department of Biomedical Sciences, Bradford University, West Yorkshire, BD7 1DP, UK
Paul Buss MD	Consultant Paediatrician, University of Wales College of Medicine, Royal Gwent Hospital, Newport, Gwent, NP9 2UB
Simon Castleden PhD	ICRF Research Fellow, Gene Therapy for Cancer Laboratory, ICRF Richard Dimbleby Unit, St Thomas' Hospital, SE1 7EH
Rebecca Chinery PhD	Postdoctoral Fellow, Imperial Cancer Research Fund Laboratories, 44 Lincolns Inn Rd, London, WC2A 3PN

Robin Dover PhD   Clinical Scientist, Imperial Cancer Research Fund Laboratories, 44
                       Lincoln's Inn Rd, London, WC2A 3PN

Ian Freshney PhD   Director of Molecular Cell Laboratories, Cancer Research Cam-
                       paign, Beatson Laboratories, Department of Medical Oncology,
                       Alexander Stone Building, Bearseden, Glasgow G61 1BD

James Goldenring   Professor, Department of Medicine, Institute for Molecular Medi-
MD PhD   cine and Genetics, Medical College of Georgia, Augusta, Georgia,
                       30912-3175, USA

Lee Gordon PhD   Research Fellow, Department of Histochemistry, Royal Post-
                       graduate Medical School, London, W12 0HS

William Gullick   Senior Clinical Scientist and Honorary Professor, Department of
PhD   Molecular Oncology, Royal Postgraduate Medical School, Lon-
                       don, W12 0HS

Helen Hurst PhD   Director of the ICRF Gene Transcription Laboratory, Cyclotron
                       Building, Royal Postgraduate Medical School, London, W12 0HS

P. L. Iverson MD   Clinical Fellow, Department of Paediatrics and Pharmacology,
                       University of Nebraska, Omaha, USA

Suresh C. Jhanwar   Associate Member and Head, Laboratory of Solid Tumor Genet-
PhD   ics, Department of Human Genetics, Memorial Sloan-Kettering
                       Cancer Centre, 1275 York Ave, New York, New York 10021,
                       USA

Michael Jones PhD   Wellcome Lecturer, Department of Virology, Royal Postgraduate
                       Medical School, London, W12 0HS

Susan Kirkland   Senior Scientist, Department of Histopathology, Royal Post-
PhD   graduate Medical School, London, W12 0HS

Steve Legon PhD   Lecturer in Molecular Biology, Department of Chemical Patho-
                       logy, Royal Postgraduate Medical School, London, W12 0HS

Irvin Modlin MD,   Professor, GI Surgical Pathobiology Research Group, Department
PhD FACS   of Surgery, Yale University, New Haven, Connecticut 06520-8062

Jane Prosser PhD   Senior Clinical Scientist, MRC Human Genetics Unit, Western
                       General Hospital, Edinburgh, EH4 2XU

T. Rajkumar PhD   ICRF Research Fellow, Department of Molecular Oncology,
                       Royal Postgraduate Medical School, London, W12 0HS

Roger Phillips PhD   Research Fellow, Department of Biomedical Sciences, Bradford
                       University, West Yorkshire, BD7 1DP, UK

Susan Ridge PhD   Postdoctoral Fellow, Leukaemia Research Fund, Chester Beatty
                       Institute, Fulham Rd, London, SW3 6JB

Karol Sikora   Professor and Honorary Consultant, Department of Clinical Onco-
MD MRCP   logy, Royal Postgraduate Medical School, London, W12 0HS

Richard Vile PhD   Senior Clinical Scientist, Gene Therapy for Cancer Laboratory,
                      ICRF Richard Dimbleby Unit, St Thomas' Medical School, SE1
                      7EH, London

Professor Sir David   Regius Professor of Medicine, Institute of Molecular Medicine,
Weatherall FRS   Oxford, UK

Alan Wright MD   Clinical Scientist, MRC Human Genetics Unit, Edinburgh, EH4
                      2XU

Foreword

The past four decades have witnessed a revolution in the biological sciences, the equal of that in physics earlier this century. It is already clear that the remarkable achievements of molecular and cell biology have great potential for medical research and practice in the future and that the medical sciences have entered the most exciting phase of their long evolution.

It is much too early to try to assess the extent to which what we like to call 'molecular medicine' will alter medical practice in the next millennium, though we now know enough to suggest that its impact will be very considerable. It has already shown us some of the mysteries of the relationship between phenotype and genotype in monogenic disease and, as an added benefit, has helped us to devise better strategies for carrier identification, prenatal detection and, in some cases, management of these conditions. It is starting to throw up some tantalizing clues about the genetic basis for susceptibility or resistance to some of the environmental agents that are involved in the common, chronic intractible diseases of western society, such as heart disease, stroke, diabetes, psychiatric disease, and the rest. As we come to understand the functions of the genes involved we shall no doubt learn more about the underlying causes of our problem diseases and how to prevent or treat them more effectively. The field has also provided us with some remarkable insights into the pathogenesis of cancer. It has already shown that many malignancies result from acquired mutations of sets of genes that are involved in the regulation of cell growth and differentiation and is starting to build a picture of how these genetic changes are orchestrated to produce a malignant phenotype, and how they may be mediated by external or endogenous carcinogens.

Many of these advances in our understanding of the molecular pathology of disease are starting to have practical applications in the clinic. They are revolutionizing our ability to avoid monogenic disease, producing new strategies for engineering diagnostic agents for a wide variety of infections, and changing the face of forensic medicine. And a tentative start has been made at gene transfer therapy, both for monogenic disease and for the management of different forms of cancer.

As we start to understand the pathogenesis of disease at a molecular level there is the expectation that the pharmaceutical industry may change completely from the era of medicinal chemistry to the evolution of designer drugs and, as a bonus, the molecular sciences are offering new approaches to understanding fundamental issues of human biology including development, ageing, and behaviour. Much of this will take many years before it has direct clinical application but it now seems certain that sooner or later much of it will.

It follows that this field of medical research, to be exploited to be its maximum, requires the interaction of complex technologies derived from the fields of molecular and cell biology. Indeed, molecular medicine is very much a technique-driven enterprise; many of the concepts are not new but the availability of a remarkable new technology often allows them to be explored in a completely novel way. Because many of the tools of the trade are derived from a wide background of disciplines they are particularly bewildering to those coming into the field for the first time. This book attempts to provide a birds-eye view of the basic techniques of molecular medicine and hence has presented a particularly daunting project for its editors; they appear to have done a remarkably thorough job in bringing all this disparate information together. It should be of great value to both established research workers and young scientists coming into the field for the first time. It deserves every success.

D. J. Weatherall
Oxford

SECTION I

Introduction, background and essential theory

1

Introduction and applications

JANUSZ A. Z. JANKOWSKI

1.1 Introduction

As the title suggests, this book is composed of four main sections: tools – equipment, core knowledge and clinical applications required for gene analysis and manipulation; techniques – choices and preferences are given for each experiment; trouble-shooting – guide for rapid resolution of experimental failure; and gene manipulation.

> There are no such things as applied sciences, only the applications of science
>
> Louis Pasteur, 1872

The science of molecular cell biology has advanced dramatically over the last century and in the last decade this has had important applications to medical research, particularly to oncology, microbiology, immunology and inherited genetic diseases. Globally, oncological sciences have the most resources devoted to them in terms of specialist scientific journals, research funding, national institutions and number of scientists. The increasing use of molecular cell biological techniques has led to considerable diversity in the methodological practice and interpretation of genetic analysis. It is particularly important, therefore, that the calibre of clinical scientific endeavour keeps pace with the scope and potential of new developments in basic science.

The clinical scientist has a unique place in medical research, however, clinically related and administrative duties frequently take precedence over other scientific priorities. Furthermore, the rapid and exponential increase in knowledge and potential clinical applications of the new molecular techniques requires that scientists are abreast of many developments so that the strengths and weaknesses of techniques can be fundamentally understood. In this regard clinical research must also be based upon high technical proficiency and the appropriate application of selected techniques.

Clinical scientists have such diverse backgrounds that it is difficult for new researchers to learn comprehensive and fundamental aspects of molecular cell techniques from their colleagues. In some clinical research institutes training is sporadic and unstructured or confined to narrow areas. In many other academic centres training is either rudimentary or learnt by 'trial and error', making research in this area unnecessarily expensive in time and resources. As a result, it is imperative that a standard text written by clinical scientists is available both as a reference and also as a laboratory manual.

The aim of this book is to enable the clinical researcher to undertake molecular studies and utilize the current powerful tools of recombinant DNA technology in an attempt to improve clinical assessment and subsequent therapy. We recognize, however, that researchers must gain a profound understanding of molecular experiments in order to appreciate sound laboratory practice and gain confidence to tackle other appropriate techniques. This book is directed at those wishing to perform basic semi-quantitative experiments such as Northern or Southern blots and also wishing to undertake more specialized genetic manipulation such as gene cloning, expression and creation of cDNA libraries.

The various contributors to this book are involved daily in the supervision and training of clinical scientists and the contents therefore give a unique insight into the learning requirements and problems encountered both by advanced and inexperienced researchers. In addition, our contributors are affiliated to major clinical research organizations and therefore they have experience in many recent innovations in gene cloning, gene expression and use of new commercial products.

We conclude the book by providing a synopsis of contemporary and avant-garde applications of gene analysis and manipulation to molecular medicine. This will enable readers to understand the versatility of these techniques in clinical medicine and it will also serve as a guide to key references in this area.

Louis Pasteur, a founding father of contemporary medicine, recognized that in order to achieve excellence in science there are no easy options. In this regard we offer an organized strategy of laboratory training, a wide scope of tested techniques and a background of relevant knowledge to apply to the research of disease. This volume is particularly designed as a companion throughout clinical studies, from protocol to publication.

We believe the format of the book will enable researchers of mixed ability to equip a molecular cell laboratory and practice new methods whilst improving their technical expertise and knowledge in other areas. In addition, this book will enable them to interpret, troubleshoot and apply quality control measures to their laboratory. The potential readership of this volume includes post-graduate scientists working in clinical research centres and clinicians working in molecular medicine; undergraduate biological scientists and veterinary, medical and dental students.

This volume has several novel aspects that enable the aspiring clinical scientist to plan experiments independently and is designed as a companion for clinical studies before, during and after molecular experimentation. In this regard the volume can be used to organize laboratory design beforehand by consulting the 'tools' and 'appendices' sections. As experimentation starts, consultation of the comprehensive list of available techniques and the trouble-shooting guide will aid the researcher through the major obstacles. Once experimentation is completed, the researcher has a ready source of related published work in the applications of molecular medicine section.

1.2 Clinical applications of modern molecular biology

Most molecular techniques can be arbitrarily divided into those that assess gene expression at the protein or RNA level and those that allow detailed manipulation of genetic material between cells.

1.2.1 Gene expression

The advent of PCR has dramatically improved the analysis of genes because it efficiently and quickly enables amplification of small quantities of genes. Genes can be studied by coarse methods, which enable approximate sizes of genes or their products to be determined by gel blotting. Fine analysis of the genetic structure is permitted by the technique of genetic sequencing.

1.2.2 Genetic manipulation and gene cloning

Gene cloning is the formation of multiple clones of identical single genes to allow analysis and manipulation of single genes and their products. There are many applications including structure, expression and mutational analysis, yet the procedure is simple and involves four basic steps: purification of DNA, modification of DNA, transfection of cells and identification of clones. This technique can be used with single genes or with entire genomes, such as is commonly performed in the construction of cDNA libraries.

2

The history of molecular genetics

REBECCA CHINERY

The late twentieth century has witnessed the golden age of scientific discovery: the identification of DNA as the genetic material and the establishment of its double helical structure can be considered as starting points in the rapid growth of the new discipline of 'molecular genetics'. The structure proposed by Watson and Crick readily indicated how DNA could be duplicated to ensure hereditary continuity during cell division. However, a decade elapsed before it could be established how DNA could also control the cell's metabolic processes by providing information for the synthesis of cellular enzymes. This relationship, namely the genetic code (Figure 2.1), is in a sense the key to molecular

UUU Phe	UCU Ser	UAU Tyr	UGU Cys
UUC Phe	UCC Ser	UAC Tyr	UGC Cys
UUA Leu	UCA Ser	UAA Stop	UGA Stop
UUG Leu	UCG Ser	UAG Stop	UGG Trp
CUU Leu	CCU Pro	CAU His	CGU Arg
CUC Leu	CCC Pro	CAC His	CGC Arg
CUA Leu	CCA Pro	CAA Gln	CGA Arg
CUG Leu	CCG Pro	CAG Gln	CGG Arg
AUU Ile	ACU Thr	AAU Asn	AGU Ser
AUC Ile	ACC Thr	AAC Asn	AGC Ser
AUA Ile	ACA Thr	AAA Lys	AGA Arg
AUG Met*	ACG Thr	AAG Lys	AGG Arg
GUU Val	GCU Ala	GAU Asp	GGU Gly
GUC Val	GCC Ala	GAC Asp	GGC Gly
GUA Val	GCA Ala	GAA Glu	GGA Gly
GUG Val	GCG Ala	GAG Glu	GGG Gly

* Transcription start site

Figure 2.1. The Universal Genetic Code (nuclear genes) for amino acids.

genetics. Its elucidation is one of the major achievements of genetics, which has developed with such astonishing speed that it has already become difficult to describe molecular genetics from a strictly historical standpoint. All scientists should, however, have some idea of the chronological order in which certain events have had a bearing upon this science. The following chronology will fill this need, even though many 'experts' will argue about the inclusion of some events and the omission of others. Whatever the criticisms, by the end of this decade, many of the recent 'molecular' discoveries may well be relegated to less prominent positions.

1944 O. T. Avery, C. M. MacLeod and M. McCarty describe the *pneumococcus* transforming principle, the foundations of which was laid by F. Griffith (1928). The fact that bacteria are rich in DNA suggests that DNA and not protein is the hereditary chemical.

1948 J. Lederberg and N. Zinder and independently, B. D. Davis develop the penicillin selection technique for isolating biochemically deficient bacterial strains.

1952 A. D. Hershey and M. Chase demonstrate that the DNA is the genetic material of bacteriophage.

F. Sanger and colleagues describe the complete amino acid sequence for the protein hormone 'insulin' and show that it contains two polypeptide chains held together by disulfide bonds.

J. Lederberg and E. M. Lederberg invent the replica plating technique.

1953 J. D. Watson and F. H. C. Crick propose the 'double helical' model for DNA.

1957 F. H. C. Crick suggests that during protein formation the amino acid is carried to the template by an 'adaptor molecule' containing nucleotides and that the 'adaptor' is the part that actually fits on the RNA template. Crick was predicting the discovery of transfer RNA.

1958 M. B. Hoagland discovers that transfer RNA acts as the 'adaptor' molecule.

1959 J. D. Watson and A. Tissieres demonstrate that ribosomes act as the template for protein synthesis.

1960 F. Jacob and J. Monod postulate the existence of messenger RNA.

1961 M. W. Nirenberg and H. Matthaei use a cell-free system for translation.

F. H. C. Crick and colleagues show that the genetic language (code) is made up of three-letter words.

1965 R. B. Merrifield and J. Stewart develop an automated method for synthesizing polypeptides on a solid support.

1969 Laemmli, O'Farrell and Rousseau develop the method of electrophoresis to separate proteins and nucleic acids through acrylamide gels.

1970 M. Mandel and A. Higa develop a general method for introducing DNA into *E. coli*.

1972 W. Arber, H. O. Smith and D. Nathans develop techniques utilizing restriction endonucleases to study the organization of genetic systems.

1973 S. N. Cohen, A. C. Y. Chang, H. W. Boyer and R. B. Helling construct the first biologically functional, hybrid bacterial plasmids.

1975 E. M. Southern describes a method for transferring DNA fragments from agarose gels to nitrocellulose filters and detecting bands by hybridization to radioactive RNA (Southern blotting).

M. Grunstein and D. S. Hogness develop the colony hybridization method for the isolation of cloned DNAs containing specific DNA segments or genes.

1976 H. R. B. Pelham and R. J. Jackson describe a simple and efficient mRNA-dependent *in vitro* translation system using rabbit reticulocyte lysates.

W. Y. Kan, M. S. Golbus and A. M. Dozy are the first to use recombinant DNA technology in a clinical setting. They develop a prenatal test for alpha thalassaemia utilizing molecular hybridization techniques.

1977 J. C. Alwine, D. J. Kemp and G. R. Stark describe methods for transferring electrophoretically separated RNA bands from an agarose gel to diazobenzyloxymethyl paper. Specific RNA bands are then detected by hybridization with radioactive DNA probes, followed by autoradiography. This method is the reverse of that described by Southern (1975) in that RNA rather than DNA is transferred to a solid support; thus the method is known as 'Northern blotting'.

J. Collins and B. Holm develop cosmids for cloning large DNA fragments.

A. M. Maxam and W. Gilbert publish the 'chemical method' of DNA sequencing.

1978 T. Maniatis and colleagues develop a procedure for gene isolation, which involves construction of cloned libraries of eukaryotic DNA and screening of these libraries for individual sequences by hybridization to specific nucleic acid probes.

1980 M. R. Capecchi describes a technique for efficient transformation of cultured mammalian cells by direct microinjection of DNA into cells with glass micropipettes.

1981 P. R. Langer, A. A. Waldrop and D. C. Ward develop a procedure for sythesizing biotinylated DNA probes that hybridize with complementary DNA, providing an anchor for streptavidin-linked, color-generating systems.

1982 Eli Lilly International Co. is the first to market a drug (insulin: 'Humulin') made by recombinant DNA techniques.

1984 D. C. Schwartz and C. R. Cantor demonstrate that pulsed-field gradient electrophoresis can be used to separate large (up to 2 000 kbp) DNA fragments. This method overcomes the limitation of agarose gel electrophoresis, which can only separate molecules of much smaller sizes (up to 50 kbp).

1985 K. B. Mullis and colleagues describe the use of the polymerase chain reaction (PCR) to allow enzymatic amplification *in vitro* of specific sequences from DNA.

A. J. Jeffries, V. Wilson and S. L. Thien develop the DNA fingerprint technique and describe its potential use in forensic science.

J. R. Miller, A. D. McLachlan and A. Klug report the isolation and characterization of a zinc finger protein from *Xenopus* oocytes. This protein binds to the 5S RNA gene and controls its transcription.

1987 M. R. Kuehn and colleagues introduce a human gene into the mouse to allow its study in a conventional laboratory rodent. They employ a mutant allele of the gene encoding HPRT and use a retrovirus as a vector to insert it into cultured mouse embryonic stem cells. These cells are then implanted into mouse embryos to form chimeras.

1988 S. L. Mansour, K. R. Thomas and M. R. Capecchi describe a general strategy for gene targeting in the laboratory mouse (gene 'knock-out').

1992 P. Liang and A. B. Pardee describe the differential display method.

3

Molecules of gene expression

REBECCA CHINERY

3.1 Introduction

The basic unit of biological material is the cell. This unit is made up from inanimate molecules, of which nearly 30% consists of *macromolecules* that contain a series of subunits in their structures. These macromolecules function as structural components, energy stores, repositories of genetic information and as molecules for controlling the chemical processes within the cell.

Four major types of macromolecule have been identified in all cell species, namely *nucleic acids, proteins, carbohydrates*, and *lipids*. Nucleic acids and proteins are considered to be the *informational macromolecules* because of their structure. Most nucleic acids contain four different types of subunit and proteins contain 20. The arrangement of these subunits in linear sequence contains genetic information, but whereas nucleic acids can be genetic material, proteins are gene products. Many proteins have a specific catalytic activity as functional *enzymes* but others serve as structural elements. In contrast to the informational macromolecules, the *non-informational* carbohydrates and lipids contain repeats of identical subunits, and serve the cell as fuel stores and as structural components. Since molecular biology primarily focuses on the former molecules, descriptions of carbohydrates and lipids will be omitted from this section.

Macromolecules organize themselves into supramolecular assemblies, such as, for example, ribosomes, which are nucleoproteins. An important aspect of these assemblies is that they are not attached by covalent bonds but are held together by a combination of weakly electrostatic hydrogen bonds, or non-polar (hydrophobic) interactions, together with special stereochemical 'spatial' relationships. This means that they can be dissociated into their respective components fairly easily.

3.2 Structure of nucleic acids

3.2.1 Components and primary structure of nucleic acids

In order to appreciate the ways in which nucleic acids can be analysed and manipulated in the laboratory, it is essential to have a basic understanding of their structure and functions. Despite their ultimate complexity, ribonucleic acid (RNA) and deoxyribonucleic acid (DNA) are made up of relatively few components and are defined by their constituent sugar: D-ribose (RNA) and D-2′-deoxyribose (DNA). Each pentose sugar is attached to a heterocyclic (i.e. containing more than one type of atom) ring base, forming a *nucleoside*. Only four different heterocyclic ring bases are found in nucleic acids. They fall into two types: the double ring containing *purines*, which includes adenine (A) and guanine (G) and the single ring containing *pyrimidines*, which includes thymine (T) and cytosine (C) in the case of DNA but uracil (U) and cytosine in the case of RNA. In some DNA molecules other bases, that are derivatives of the standard bases, are occasionally found but these are special cases. Furthermore, some classes of RNA such as transfer RNA and ribosomal RNA contain several different types of bases called *rare* or *minor bases*.

A *nucleotide*, or *nucleoside phosphate*, is formed by the attachment of a phosphoric residue to the 5′ position of a nucleoside by an ester linkage. For example, adenine + ribose gives adenosine whereas adenine + deoxyribose gives deoxyadenosine. Such nucleotides can be joined together by the formation of a second ester bond by a reaction between the phosphate of one nucleotide and the 3′ hydroxyl of another, thus generating a 5′ to 3′ *phosphodiester bond* between adjacent sugars; this process can be repeated indefinitely, with the elimination of water, to give long polynucleotide molecules. The order of nucleotides within such chains is known as its *primary structure* and each has a constant polyphosphodiester backbone. The phosphate groups join C-3′ hydroxyl groups to C-5′ hydroxyl groups in all RNAs and DNAs so far identified. The joining of the nucleotides in this regular, defined manner gives *polarity* to the nucleic acid chain. Thus there is a standard convention for writing nucleic acid primary structures with the same polarity so that they can be easily compared. The sequence is written from left to right as 5′-GATC-3′ so that the phosphate group joins the C-3′ hydroxyl group of the left side nucleoside to the C-5′ hydroxyl group of the right side nucleoside. Each polynucleotide will therefore have a free phosphate at one of its ends, and a free 3′ hydroxyl at its other end.

3.2.2 Secondary structure of nucleic acids

In 1953, Watson and Crick, with Wilkins, described the secondary structure of DNA. This molecule exists as a double-stranded molecule in the shape of a *double helix*, in which the bases of the *two* polydeoxyribonucleotide chains lie in the centre of the molecule, with the sugar–phosphate backbones on the outside. The polarities of the two chains (or strands) are *antiparallel*, i.e. they run in opposite directions. A constant distance between the two backbones is achieved by arranging the bases protruding from each backbone perpendicular to the axis so that a purine is always opposite a pyrimidine. Consequently, the sequence of bases in one strand is *complementary* to that in the other. Thymine can form weak hydrogen bonds with adenine, and cytosine with guanine. Guanine and cytosine base pairs are more strongly held together than adenine and thymine because three hydrogen bonds rather than two are involved. X-ray analysis of DNA identified a repeat of 0.34 nm which corresponded to the distance between successive bases in one chain, i.e. the distance between base pairs along the axis. The *pitch* of the helix was found to be 3.4 nm giving ten base pairs per complete turn of the helix. Since the DNA molecule is very long, the number of hydrogen bonds holding the two strands together is very great, thus making a very stable double-stranded molecule. The stable helical structure can be *denatured* by incubation of the molecule in strong acid or alkali conditions or heating to temperatures greater than 65 °C. These properties are frequently used within the laboratory, especially during DNA preparations and double-stranded DNA sequencing.

RNA almost always exists as a single strand, sometimes containing several sequences within the same strand that are complementary to each other, and that can therefore base-pair if brought together by suitable folding of the molecule. This is most obvious in the case of *transfer* RNAs, which have four pairs of complementary sequences within their 70–80 nucleotide lengths; consequently the single strand folds up to give a 'clover leaf' secondary structure. As with DNA, cytosine pairs with guanine, but in RNA adenine pairs with uracil.

Strands of RNA and DNA can associate with each other, if their sequences are complementary, to give double-stranded, *hybrid* molecules (uracil of RNA base-pairs with adenine of DNA). Similarly, strands of radioactively labelled RNA or DNA, when added to a denatured DNA preparation, will act as *probes* for DNA molecules to which they are complementary. This *hybridization* of complementary strands of nucleic acids is very useful for isolating specific fragments of DNA.

3.3 Functions of nucleic acids

DNA is chemically much more stable than RNA and is thus well suited to its genetic role of hereditary continuity. In cells, DNA exists as extremely large molecules and is mostly located in eukaryotic nuclear *chromosomes*. Additionally DNA is also found in mitochondria and chloroplasts. The average molecular weight of DNA in one of the 46 human chromosomes is 7.4×10^{10} daltons. RNA acts as genetic material only in special cases, for example in retroviruses. RNA is usually involved, however, in the transfer of genetic information to the cellular apparatus where this information is converted into proteins. Each block of DNA that codes for a single RNA or protein is called a *gene*, and the entire set of genes in a cell forms its *genome*. The majority of eukaryotic genes contain lengths of non-coding DNA, called *introns*, which interrupt their coding regions (*exons*).

3.4 DNA replication

Chromosomal DNA must be replicated at a rate that will at least keep up with the rate of cell division. Replication begins at a sequence called the *origin of replication*, and involves the separation of the two DNA strands over a short length, and the binding of enzymes, including DNA and RNA *polymerases*. In prokaryotes, RNA polymerase synthesizes a short, complementary RNA chain on each exposed strand, using the DNA as a template. *DNA polymerase III* also uses the DNA as a template for synthesis of a DNA strand, using the short RNA as a primer. Synthesis of the DNA strand occurs in a 5′ to 3′ direction, but, since the two strands of DNA are antiparallel, only one can be synthesized in a continuous fashion. The other is synthesized in relatively short stretches, still in a 5′ to 3′ direction, using an RNA primer for each stretch. These RNA primers are then removed by *DNA polymerase I*, acting as a 5′ to 3′ exonuclease, the gaps are filled in by the same enzyme acting as a polymerase, and the separate fragments are joined together by *DNA ligase* to give a continuous strand of DNA. The replication of eukaryotic DNA is less well characterized; however, both processes involve 5′ to 3′ synthesis of new DNA strands. The net result of the replication is that the original DNA is replaced by two molecules, each containing one old and one new strand; and the process is therefore known as *semi-conservative replication*.

3.5 Classes of RNA

Three classes of RNA are normally distinguished according to function: *messenger* RNA, *ribosomal* RNA, and *transfer* RNA. *Regulatory* RNA may exist in the cell for certain signalling events during gene expression but this is not well defined. RNA is generally assumed to be functionally localized within the cell cytoplasm.

Messenger (m)RNA consists of polymers of ribonucleotides that transfer genetic information for protein synthesis from the nucleus to the ribosomes in the cytosol of the cell. Since three bases are required to determine the incorporation of each amino acid, a typical mRNA molecule might contain 450–1200 bases. Additionally, in most eukaryotes, *post-transcriptional processing* of mRNA is required: at the 3′ end a sequence of up to 200 adenylate residues is added, which may be important in the transport of mRNA out of the nucleus; at the 5′ end a small '*cap*' sequence is added, which may serve to regulate the stability of mRNA.

Ribosomal (r)RNA forms part of the structure of *ribosomes*, which are the sites of protein synthesis. Each ribosome contains only three or four different rRNA molecules, complexed with at least 70 different proteins. All ribosomes are composed of two subunits, which are characterized by their *sedimentation coefficients* in the ultracentrifuge so that, in eukaryotes, the subunits are termed 60 S and 40 S.

Transfer (t)RNAs are relatively small molecules, containing around 70 bases. There is at least one specific tRNA molecule for each amino acid, which is attached covalently to the tRNA. These molecules carry amino acids to the ribosomes, and interact with mRNA in such a way that their amino acids are joined together in the order specified by the mRNA. A characteristic of tRNA molecules is the high proportion of unusual (other than C, G, U, T) bases. The functions of these unusual bases may be to increase the stability of the molecule by their ability to resist degradation by *ribonucleases*. Finally, tRNA is synthesized in the nucleus as a much larger precursor molecular, which is then cleaved by specific ribonucleases in the cytoplasm.

3.6 Transcription (DNA to RNA)

Most prokaryotic genes are made up of three regions. At the centre is the sequence that will be copied in the form of RNA; called the *transcription unit*. To the 5' side (upstream) of the strand that will be copied (the +ve strand) lies the *promoter region*, and downstream of the transcription unit is the *terminator* region. Transcription begins when a *DNA-dependent RNA polymerase* binds to the promoter region and moves along the DNA to the transcription unit. At the start of the transcription unit the polymerase begins to synthesize an RNA molecule complementary to the −ve strand of the DNA, moving along in a 3' to 5' direction, and synthesizing RNA in a 5' to 3' direction, using nucleoside triphosphates. The resultant RNA will have the same sequence as the +ve strand of DNA, apart from the substitution of uracil for thymine. On reaching the terminator region, the RNA strand is released. This process is called *transcription* (since the nucleotide language is transcribed from one form to another) and takes place in the nucleus.

3.7 Translation (RNA to protein)

Following transcription, the mRNA is transported across the nuclear membrane into the cytosol. Here it associates with ribosomes, which are complexes of rRNA and protein. Each ribosome consists of a large (60 S) and small (40 S) subunit, which associate during the process of translation (conversion of the genetic language from nucleotides to amino acids). A sequence of three nucleotides, in a given mRNA molecule, corresponds to a specific amino acid and these *triplet* sequences are known as *codons*. Thus, the amino acids are assembled along the mRNA according to the codon (base) sequence. The successive assembly of amino acids along the mRNA molecule requires the smaller RNA molecules, known as transfer RNAs (tRNA). tRNA molecules are covalently linked to specific amino acids, forming *amino-acyl tRNAs*, and each molecule possesses a triplet of bases that is complementary to the codon for that amino acid. This exposed triplet is known as the *anticodon*, and allows the tRNA to act as an *adaptor molecule*, bringing together a codon and its corresponding amino acid. After binding to a specific *initiation* sequence (AUG) at the 5' end of the mRNA, the ribosome moves towards the 3' end, allowing an amino-acyl tRNA molecule to bind with each successive codon, thereby arranging amino acids in the correct order for protein synthesis. The ribosome forms peptide bonds between the amino acids as it moves along the mRNA, and releases a completed polypeptide chain when it reaches a *termination codon* (UAA, UGA or UAG).

Since the mRNA is read in triplets, an error of one or two nucleotides in positioning of the ribosome will result in the synthesis of an incorrect polypeptide. Thus it is essential for the correct *reading frame* to be used during translation. This is ensured in prokaryotes by base-pairing between the *Shine–Dalgarno sequence* and a complementary sequence of one of the ribosome's rRNAs, thus establishing the correct starting point for the ribosome on the mRNA.

3.8 Proteins

Proteins, like nucleic acids, are informational macromolecules, which form a major constituent (approximately 15% total weight) of cells. However, whereas nucleic acids contain genetic information that is duplicated and separately packaged into daughter cells to ensure inherited characteristics, proteins are the end products of genetic information. Proteins have five main functions: as *enzymes* (proteins with catalytic properties controlling metabolism and releasing energy), *antibodies* (protein subunits concerned with defence against foreign bodies), *structural elements* (maintenance of the cell architecture), *transport devices* (proteins that carry small molecules or ions within the cell or through membranes) and *metabolic regulators* (proteins concerned with biological processes within the cell, acting either directly or indirectly on genetic information).

3.8.1 Structure

The great range of functional activities for proteins is reflected in their structural diversity. However, two structural features are common to all proteins: they contain varying proportions of only 20 *common amino acids*, and they are *polypeptides*; polymers of amino acids joined together by peptide bonds. Occasionally, some *non-standard amino acids* are found in proteins but these are derivatives of the common 20, formed after the protein has been biosynthesized. Proteins usually contain 300–400 amino acids, although they may range in size from about 50 amino acids (insulin) to 2500 amino acids (myosin), and to even larger complexes of molecular weights of several million (haemoglobin).

The common 20 amino acids are α-amino acids with one exception, proline, which is an α-imino acid. It has an -NH- (imino) group in the α-position next to the -COOH (carboxyl) group instead of an -NH$_2$ (amino) group. The general structural formula is $RCH(NH_2)COOH$ for the α-amino acids, where R can be divided into *non-polar*, *uncharged polar*, and *charged polar* groups.

The *non-polar* or *hydrophobic* amino acids include those with aliphatic side chains (alanine, leucine, isoleucine, valine and proline), those with aromatic side chains (phenylalanine and tryptophan) and one containing sulphur (methionine). They are characterized by being hydrophobic, an important property for protein folding and for making lipoprotein structures. The *uncharged polar* amino acids include the hydroxyl (OH)-containing serine, threonine and tyrosine; amide-containing asparagine and glutamine and sulphur-containing cysteine. The group also includes, for convenience, glycine which is the smallest amino acid with no distinctive properties and in which R is a hydrogen atom. This class has an affinity for water which tends to make the protein soluble. The third group consists of *charged polar* amino acids, which are either negatively charged at the physiological pH of 7, aspartic and glutamic acids, or which are positively charged, arginine, lysine and histidine, albeit the last very weakly.

A general property of amino acids is their *optical activity*. Biological molecules are usually asymmetric and have optical activity, i.e. they can rotate the plane of polarized light. In the case of all amino acids (except glycine), they have four different chemical groups attached to the α-carbon, making them asymmetric molecules, and hence optically active. All natural amino acids are L-amino acids.

All proteins are *polypeptides*. Thus they are linear arrangements of amino acids joined through *peptide bonds*:

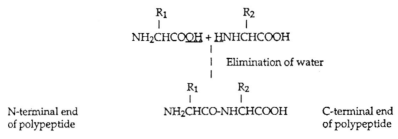

where R_1 and R_2 signify the R groups belonging to different amino acids. This linear arrangement of amino acids is termed the *primary structure*. The two features of the primary structure of proteins relevant to DNA are (1) the peptide bonds linking the different amino acids in such a linear arrangement giving a polypeptide *backbone* to the protein and (2) the diversity of structure, which depends on the different R groups that stick out from the backbone structure. It is the order of the R groups that is genetically determined, the properties of which cause the protein to fold up automatically in a three-dimensional structure and specify the structure and functional activity of the protein.

Secondary structures arise by the ordering of the polypeptide backbone due to short-range weakly electrostatic hydrogen bonds between carboxyl oxygen atoms and amide nitrogen atoms. In proteins two classes of structures result: *α-helical* or *β-pleated sheet* structures. In addition it is found that the polypeptide backbone, even if an α-helix, can be folded into a variety of conformations giving an overall sometimes globular structure which is the *tertiary* or *three-dimensional structure*. Interactions that hold together the tertiary structure can be hydrogen bonds, salt linkages, disulphide bonds, and weak attractive forces called *van der Waals forces*. Finally, a *quaternary structure* is when individual proteins aggregate for a specific function. Aggregate formation, which in haemoglobin, for example, is from four subunits, is brought about by surface interactions between exposed backbones resulting in a geometrical fit between subunits.

Molecular cell tools

4

Tools required for molecular cell analysis

FIONA K. BEDFORD
and JANUSZ A. Z. JANKOWSKI

4.1 Introduction

Perhaps the biggest limitation for the molecular biologist is the availability of equipment (tools). In the vast majority of cases, however, alternative protocols are available that allow most techniques to be performed with a minimum of equipment. As a result it is important for investigators to think in terms of equipment and resources. The outline of this chapter covers four related areas in biological research, in particular the equipment required for experiments, from basic morphological and cellular methods to more advanced molecular analysis or genetic manipulation techniques.

4.2 Tools for tissue analysis and morphological characterization

The most frequent technique employed in biological research today is arguably immunohistochemistry. This technique requires access to a cryotome or refrigerated microtome in addition to fridges (4 °C), freezers (constant -20 °C and -70 °C maintained) and liquid nitrogen tanks for long-term tissue storage. A paraffin wax tissue processor however, greatly simplifies the processing of tissue. If *in situ* hybridization studies are to be performed, facilities must be available for preparation of sterile and RNAase free microscopic glass slides as well as radioactive work with low energy ^{35}S probes.

4.3 Tools required for cellular analysis and cell culture

The equipment required for cell culture comprises a tissue culture laminar flow hood, a CO_2 incubator and an inverted microscope under which to view the progress of cells.

4.4 Tools required for chromosomal and molecular analyses

The tools required for detailed molecular analysis enable proteins, RNA or DNA fragments to be separated by their size. In this regard centrifugation is

used to purify molecules from the surrounding cellular constituents, electrophoretic techniques separate molecules by size and shape and labelling
methods detect and quantitate the molecular species of interest.

4.4.1 Centrifugation

A general formula for conversion of r.p.m. to relative centrifugal field is used.
The radius is calculated from the mid rotor position to the tip of the microcentrifuge bucket in mm; g force $= 110 \times 10^{-7} \times$ radius $\times n^2$ (speed in r.p.m.).

Speed of centrifugation: low 6000 g, high 60 000 g and ultracentrifuge
600 000 g (usually used for density gradients).

4.4.1.1 Types of centrifugation

1. Differential pattern centrifugation (size, shape, and to a lesser extent
   density differentiation).
2. Rate-zonal centrifugation (size differentiation), e.g. monosaccharide gradients such as sucrose, separate cell membranes or cell organelles,
   whereas polysaccharides such as Ficoll separate cells or viruses.
3. Isopycnic centrifugation (density differentiation), e.g. caesium chloride
   gradients – separates protein and nucleic acids.

4.4.2 Utility of nucleotides – high energy labelled probes

4.4.2.1 Radioactive nucleotides and radioactive detectors

(a) γ emitters (high energy electromagnetic waves)
^{125}I high energy emissions travel through 3.3 cm of tissue. Use a lead/
acrylic safety shield.
^{59}Fe (iron) high energy emissions travel through 16 cm tissue, therefore
use a 2 inch lead shield.
^{131}I moderate energy γ and β emitters – use a 1 cm Perspex slab proximal
to the radioactive source and a 2 mm lead shield behind the Perspex.

(b) β emitters (one neutron converted to a proton generates an electron,
β particle)
^{32}P low resolution and high energy particles which travel through 0.8 cm
tissue and 7 m air, therefore use 1 cm Perspex ($T_{1/2}$ 14 days).
^{33}P has moderate resolution and 1/6 the energy of ^{32}P ($T_{1/2}$ 21 days).
^{35}S has weak emissions, stronger resolution and is stopped by 1 mm H_2O
($T_{1/2}$ 90 days).
^{3}Hβ emissions are very weak and are stopped by 1 mm air ($T_{1/2}$ 12.5
years).
^{14}C is the weakest emitter ($T_{1/2}$ 5760 years).

4.4.2.2 Radiation dose

Emission (disintegrations): 1 mCi = 37 MBq.

Absorbance: 1 mSv = 6.25 million β particles each with an energy of 1 MeV are uniformly absorbed in 1 g of tissue (100 mSv may cause sterility, 500 mSv causes depression of haematopoiesis and 500–2000 mSv causes lens cataracts).

The maximum total body dose for non-classified workers is 6 mSv/year; usually the specific organ may be up to five times this amount.

(a) Geiger–Muller detectors

All types detect β and γ rays but caution is required as some detectors may read zero when swamped by radiation while using ^{125}I, ^{14}C, ^{35}S and ^{33}P:

Use 'E' type for high energy β or β/γ emitters ^{32}P and ^{131}I.
Use 'EL' or 'EP15' type for soft β emitters such as ^{35}S.
Use Berthold LB1210B type for all β emitters except tritium.

(b) Scintillation probes

Detect β and γ emitters, but not both at once and not robust.
Can use 542, 542A, 544A, and 544B and can detect ^{125}I, the Bremstrahlung and X-rays.

(c) Ionization chambers

Non portable and only for use on flat surfaces.
Used for low energy β emitters such as tritium.

4.4.3 Gel electrophoresis

4.4.3.1 General guidelines

Always electrophorese samples on gels from the negative to the positive terminal so that the molecules of interest will travel along the electron gradient. At the anode oxygen deposited (large bubbles) while at the cathode hydrogen (small bubbles which rise to the surface readily) or metal is deposited.

4.4.3.2 Agarose gels

Agarose contains D-galactose and L-galactose and can be used for the separation of medium- and large-sized nucleic acids. There are several varieties of agarose, which have individual advantages; NuSieve is best for identifying small oligonucleotides of 20–1,000 bp, SeaPlaque is best for 100–10 000 bp fragments and SeaKem is best for 500–100 000 bp nucleic acids (FMC Bioproducts).

Table 4.1. *Agarose concentrations required to separate nucleic acids of various sizes*

Agarose (%)	Linear nucleic acid size (kb)
0.3	2–70
0.5	0.7–45
0.8	0.4–20
1	0.3–10
1.2	0.4–6
1.5	0.2–3
2	0.1–2

Some of the ultrapure varieties of agarose such as GTG NuSieve (FMC Bioproducts) melt at 70 °C and therefore nucleic acids can be extracted by heating or chemical dissolution of the agarose.

(a) Some notes for pouring and running agarose gels
Beware agarose may boil over during heating in a microwave.
Replace lost volume with water when gel overheats.
Aim to make gels 3–5 mm thick.
Maximum loading of DNA is 0.5 μg/μl of well volume. Overloading will result in trailing and smearing, especially of large DNA fragments.
Bromophenol Blue runs in 0.5 TBE buffer at the same rate as 300 bp linear dsDNA and xylene cyanol runs at the same rate as 4 kb linear dsDNA.
Some high salt buffers such as those for *Bam*HI and *Eco*RI retard migration of DNA and distort the migration of DNA in adjacent wells.

Table 4.1 demonstrates the appropriate agarose percentage in order to best resolve different DNA species.

(b) Electrophoresis required for agarose gels
If the DNA is greater than 100 000 bp in length, pulsed field electrophoresis is usually required. Agarose gels with constant voltage will be able to separate DNA from 200 bp to 50 kb. Different conformations of plasmid DNA have a differential migration on agarose gels; superhelical circular (form I) have the fastest, nicked circular (form II) the slowest and linear DNA (form III) migrates between the two.

(c) Visualization of DNA by using ethidium bromide and UV light
To ensure that bands run as specified above, the ethidium bromide (EtBr) concentration is critical, use 0.1–0.5 μg/ml otherwise closed circular can bind excess EtBr and unwind. EtBr in correct concentration reduces migration of

linear and to a lesser extent nicked circular DNA by 15%. During migration EtBr migrates to the cathode, in the opposite direction to the DNA. Extended electrophoresis can remove much of the EtBr, making detection of small fragments difficult. If this occurs, restain the gel by soaking it in a solution of EtBr (0.5 μg/ml) for 40–45 minutes. UV light at 250–300 nm (purple) detects EtBr bound to DNA, which emits at 590 nm (red–orange). EtBr has, however, low affinity for single stranded nucleotides such as RNA and the fluorescence is poor.

Destaining is not usually required but when assessing very small amounts ($<$ 10 ng), background unbound EtBr is reduced by soaking the stained gel in water of 1 mM $MgSO_4$ for 20 minutes at room temperature.

(d) Applied voltage

Electrophoresis at 5–8 V/cm is usually sufficient but if the DNA is larger than 2 kb no more that 4–5 V/cm should be used, as bands will lose resolution, due to differential increases in mobility of high molecular weight fragments.

Electrophorese gels at room temperature unless they contain less than 0.5% agarose, in this circumstance cool the gel to 4 °C during electrophoresis. Alternatively, make the gel with a sublayer of thicker support below the running gel.

(e) Buffers for non-denaturing (native) gel electrophoresis

Homogeneous buffers (the same buffer in both gel and electrophoresis apparatus) are used as discontinuous buffer systems but do not improve the sharpness of the nucleic acid bands. In addition, the maximum efficiency of electrophoresis is when the gel is run under 1–2 mm of buffer. This overlying buffer helps avoid the drying out of the gel and also provides some degree of cooling.

TBE (Tris–borate EDTA) buffers are better for smaller bases, $<$ 10 000 bp, whereas TAE (Tris–acetate EDTA) buffers are better for bases $>$ 10 000 bp.

(f) Buffers for denaturing agarose gel electrophoresis

Denaturing gel electrophoresis is used to analyse single-stranded nucleic acid molecules, for example, RNA, first and second strand cDNA synthesis and for assessing nicking activity in nick translation. These gels are generally of two types: alkaline gels (usually for DNA), made of 1% (w/v) agarose in 1 mM EDTA and 50 mM NaOH or 5% (v/v) deionised formamide/formaldehyde gels made in 1% agarose (usually for RNA but can also denature DNA poorly).

4.4.3.3 Acrylamide gels – add 8 M urea for denaturing gels

Acrylamide gels have greater resolving power than agarose gels and can be used to assess separation of small DNA fragments (5–500 bp) as well as protein extracts. Preparation of these gels is very different from agarose gels because they are usually cast between two glass plates in order to produce a gel of

0.2–1 mm thickness. In addition, nucleic acids must be visualized by other means as EtBr has very poor resolving ability in acrylamide and in this regard radioactive labels can be substituted. Unfortunately acrylamide gels also have several additional dangers including: the neurotoxic effects of bisacrylamide when making the gel, high voltages when running the gel and disposal of the radioactive waste after completion.

Unlike agarose, which sets spontaneously on cooling, acrylamide requires persulphate to act as a catalyst to aid polymerization. This is relatively unpredictable without another co-substrate, bisacrylamide, which cross-links strands of each polymer accelerating the solidification process dramatically. Different concentrations of acrylamide are required to ensure satisfactory resolution of different sized nucleic acids or proteins: 5% for 200–1000 nucleotides, or 200 kDa proteins; 6% for 100–800 nucleotides and 10% for 50–400 nucleotides, or 100 kDa proteins.

4.4.4 PCR thermocoupler

PCR machines rely principally on two features, the abilities to heat and cool specimens rapidly.

The heating phase usually occurs at 1–2 °C/s but heating can be more uneven in some machines that use a metal heating block. This is particularly unfavourable when rapid cycles are required. The cooling phase usually occurs at 0.5–1 °C/s. In addition, many modern machines that use water or air cooling can go below room temperature and this is useful for keeping the reaction product at 4 °C after thermocycling.

The thermocoupler software must be programmable to different settings such as the following:

> Selection of three different temperatures in the range 25–100 °C.
> Accurate time (range 1 s–5 min).
> Optional 'hot start' and low storage temperatures at the end of the reaction.
> Memory for 40 programmes of up to 99 cycles.

4.4.5 Computer facilities for 'gene banks'

4.4.5.1 Sequence analysis

The technology in this area is changing rapidly but all software packages should allow the genetic sequences under investigation to be compared with known sequences. As a result, it must be possible to analyse several features in the fragments, including: the presence of homo/heteroduplexes, the initiation codon (AUG, methionine), stop codons (UAG, UAA, UGA) and palindromic sequences. The software should also be able to predict the secondary structure and tertiary structure of putative protein sequences extrapolated from known nucleic acid sequences. Finally, the software should be able to compare the DNA sequence with other known genes, so that regions of potentially conserved homology can also be identified.

4.4.5.2 Bioscan software and hardware packages

In addition, other software packages may be necessary to enable automatic sequencing applications, genomic DNA fingerprint analyses and for densitometry of comparative labelled bands on gels.

4.5 Tools required for gene manipulation experiments

4.5.1 Plasmids (covalently closed circular DNA)

Plasmids are self-replicating circular extrachromosomal DNA molecules into which foreign DNA can be inserted. Bacterial plasmids are of two general types: simple bacteria plasmids and phagemids constructed from filamentous phage.

4.5.1.1 Content of plasmids

1. Replicon containing the origin of replication and plasmid RNAs and proteins necessary for replication.
2. Selectable markers, which are usually dominant, include: antibiotic resistance, heavy metal resistance, sugar chromogen analogues of degradation of aromatic compounds.
3. Cloning region – unique restriction sites into which foreign DNA can be inserted, without interfering with the plasmid's ability to replicate or confer the selectable phenotype to its host.

4.5.1.2 Use of plasmids

1. Cloning vector – for cloning insert DNA; bacterial (pBR322 and M13 derived) or eukaryotic.
2. Transcription vector – for *in vitro* transcribing RNA; bacterial (pSP65 as well as most other M13 derived) (Promega).
3. Expression vector – for the expression of proteins produced from the DNA inserts; bacterial (pGEMEX, encodes an inducible promoter) (Promega) in addition to mammalian (pBK-CMV and pBK-RSV) (Stratagene) and insects (baculoviral transfer vector pMbac) (Stratagene).
4. Reporter vectors – for assaying putative promoters and enhancers in extracts from transfected cells; mammalian (pGL2 or pCAT series) (Promega).

4.5.1.3 Types of plasmids

(a) Simple

pBR322 (4.3 kb) was one of the first plasmids to be used and was followed shortly by pBR237 (pAT153, 2 kb), which contained no *bom* gene and therefore could not be transferred from cell to cell. Since the mid-1980s a number of pUC derived plasmids (2.7 kb) have been developed, which exploit the α fragment of the β-galactosidase gene as a selectable marker for cloning success. In this regard, DNA cloned into the β-galactosidase gene inactivates the enzyme's function, which is to cleave its substrate X-gal and produce a blue precipitate, thus allowing the identification of recombinant clones, which remain white. Unfortunately, these plasmids have a maximum DNA insert size of only 6 kb.

(b) Phagemids
Bluescript M13
This is made of the filamentous phages M13 (2.9 kb) and the f1 filosome region and this type of plasmid has a maximum capacity for DNA inserts of up to 15 kb. This plasmid also has powerful RNA polymerase promoters (sp6/T3 or T7) incorporated into either side of a multiple cloning region so that sense and antisense RNA molecules can be transcribed from the DNA insert. These promoters can also be used as primers for sequencing the DNA insert from dsDNA or ssDNA, as the f1 origin of replication can be used for the production of circular ssDNA. Unfortunately other M13-derived vectors may have only 0.1–0.5 the copy number of Bluescript vectors, due to the use of differential origins of replication.

Hybrid plasmids
pEMBL8+ (Stratagene) has a piece of M13 phage placed into a plasmid but offers little advantage over conventional phagemids. The lambda ZAPII system (Stratagene) however, is a phage that has plasmid DNA. This vector comes in the linear form of the lambda phage, which can have insert DNA placed into it. The phage is propagated and when R408 helper phages are co-infected into cells the PBSKII+ (Stratagene) phagemid containing the insert is released. This phagemid can then be purified and infected into similar or indentical *E. coli* and the phagemid turns into dsDNA during replication.

(c) Cosmids
'Cohesive phagemids' contain multiple lambda phage particles linked together and have a cloning capacity of 30–42 kb. They usually carry selectable markers for both prokaryotic and eukaryotic cloning.

(d) Yeast plasmids
The basic 2 μm DNA circle is the native yeast plasmid and gives rise to 60–300 copies per cell. This can be altered artificially to give a series of yeast

expression plasmids such as the YEp351 (or YEp24, Biolabs) which acts as a rapid shuttle plasmid vector between yeast and *E. coli*. Yeasts are a particularly good system to work with when expressing complicated eukaryotic proteins compared with the prokaryotic bacterial expression systems.

4.5.2 Molecular enzymes

4.5.2.1 Cutting DNA

Several mechanisms to cut DNA are possible, using exonucleases, which cut at the ends, or endonucleases, which cut in the middle.

(a) Exonucleases

An example is exonuclease III (*E. coli*), which removes nucleotides from the 3′ termini of dsDNA but not ssDNA and thus 3′-protruding DNA is protected. It can also nick dsDNA to produce single-stranded gaps. This enzyme is usually used to create unidirectional deletions from a linear molecule with one resistant 3′ overhang. Nuclease Bal-31, however, removes 3′ and 5′ termini from both dsDNA and ssDNA, making shorter bilaterally blunt molecules. This enzyme is also highly efficient at cleaving at nicks, gaps and single-stranded regions of duplex DNA and RNA.

(b) Endonucleases

DNase I, for example, is a non-specific endonuclease that degrades both dsDNA and ssDNA. It can be used to degrade selectively DNA in the presence of RNA (RNAase-free) or to study DNA–protein interactions by DNase I footprinting.

Alternatively, S1 nuclease degrades ssDNA or RNA to give 5′ terminated products. It can be used to remove protruding single-stranded termini in double-stranded DNA, for selective cleavage of single-stranded DNA or for mapping RNA transcripts by S1 hybridization mapping.

(c) Restriction enzymes

Type 2 restriction enzymes are the most specific cutters as they cleave at 4–8 base recognition sites, whereas type 1 cut at 1000 bases downstream and type 3 cut 10–20 bases downstream from recognition sites. Restriction enzymes are mainly used to create blunt, 5′ or 3′ overhang DNA ends (Figure 4.1). Each restriction enzyme cleaves at a unique DNA sequence, however there are exceptions: (a) star activity occurs due to contaminated DNA preparation or inappropriate enzyme buffer and leads to unpredictable cleavage sites; (b) some enzymes may be isoschizomers with different recognition sites but cleavage at identical regions, i.e. *Bam*HI G*GATC*C and *Sau*3A1 and *Mbo*1 *GATC*; (c) some enzymes may also act as compatible cohesive cutters, such as *Bam*H1 and *Bgl*II.

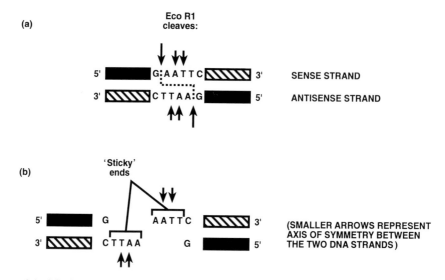

Figure 4.1. Mechanism of action of restriction enzymes.

4.5.2.2 Joining DNA

T4 ligase is the principal enzyme used to join DNA, usually with plasmid vector DNA. The reaction needs a minimum of 0.05–0.1 μg of foreign DNA in a 1:1 ratio with cut cohesive-ended vector DNA (Figure 4.2). The reaction is also ATP dependent. For blunt ligations a 3:1 molar excess of insert DNA compared with the vector DNA is used (usually 1–2.5 μg of insert for 0.5 μg vector DNA) and 15% PEG (polyethylene glycol) 8000 aids the ligation of blunt ended DNA.

4.5.2.3 DNA polymerization

1(a). E. coli DNA polymerase I (5′–3′ polymerization) also has 5′–3′ and 3′–5′ exonuclease activities. The 5′–3′ exonuclease activity removes nucleotides ahead of the growing DNA chain, allowing nick translation.
1(b). Klenow (5′–3′ polymerization) is a proteolytic product of E. coli DNA polymerase I that has retained the polymerization and 3′–5′ exonuclease activity, but has lost the 5′–3′ exonuclease action. The enzyme is used for (a) DNA sequencing by the Sanger sequencing technique; (b) fill-in or labelling of 3′ recessed ends; (c) random primed labelling; and (d) double-strand formation in cDNA cloning.
  2. T4 polymerase (5′–3′ polymerization) is similar to the Klenow fragment of DNA polymerase I but the 3′–5′ exonuclease action is 200-fold stronger. It can be used for: (a) blunt ending protruding 3′ termini and filling in of 5′ termini with or without labelled nucleotides and (b) labelling DNA using replacement synthesis.

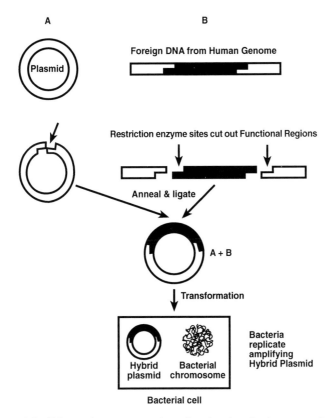

Figure 4.2. Schematic representation of molecular cloning strategies.

3. T7 polymerase (5'–3' polymerization) has 3'–5' exonuclease activity but no 5'–3' activity similar to Klenow and T4 polymerase. It is used for second-strand synthesis in site-directed mutagenesis. A modified form of T7 polymerase that has only 1% of the exonuclease activity is used ideally for sequencing (Sequenase, USB).

4. Thermophilic polymerases such as Taq polymerase work optimally at high temperature (72 °C–80 °C) and have a ten-fold decrease in action at 37 °C. The main advantage of thermophilic polymerases is their ability to read through secondary structures efficiently. They are mainly used in the polymerase chain reaction and for high temperature cycle sequencing. Other thermophilic enzymes such as Pfu (Stratagene) and Vent (Biolabs) have proofreading actions and therefore make fewer mutations during PCR but may require longer extension times because of this action.

5. RNA reverse transcriptase (RT) makes a complementary strand of DNA from an RNA template with no 3'–5' proofreading, therefore there are 1 in 500 errors, but other varieties such as Superscript II (Gibco/BRL) may make fewer errors. RT can be used for first and second strand cDNA synthesis (especially using poly(A) mRNA).

4.5.2.4 Modification of DNA

DNA can also be modified by the addition of linkers, adapters or nucleic acids to the existing ends of DNA strands. In addition, DNA can be phosphorylated or dephosphorylated by specific enzymes.

(a) Linkers

These have blunt ends and a restriction site in the middle and are used to join two pieces of DNA and insert a restriction enzyme site. It may be necessary, however, to methylate the vector DNA in order to inactivate the 'hidden' restriction sites compatible with enzyme linkers.

(b) Adapters

These usually have one blunt end, one sticky end, also a restriction site in the middle and are phosphorylated or dephosphorylated.

(c) Alkaline phosphatase

This enzyme removes the 5' phosphate (i.e. dephosphorylates) from both RNA and DNA. This is used to prevent recirculation of plasmid DNA cut with a single restriction enzyme during ligation. Alkaline phosphate can also replace the 5' phosphate with a radioactive phosphate. 1 unit/100 pmoles of DNA is incubated for 30 min at 37 °C for protruding 5' termini, and 1 unit/2 pmoles of DNA for protruding 3'.

(d) T4 polynucleotide kinase

1. Transfers the gamma phosphate of NTP to the 5'-OH of dsDNA or ssDNA or RNA (i.e. phosphorylates).
2. Transfers phosphate from the 5' end of DNA or RNA to NDP.
3. Removes 3' phosphoryl groups. It can also be used for 5' labelling of DNA or RNA by 'forward reaction'. The 5' phosphate is removed by the phosphatase 'exchange reaction' of DNA/RNA and the unlabelled phosphate is transferred to NDP and replaced with labelled phosphate from labelled NTP.

(e) Terminal deoxy transferase

This can add any d(NTP) onto target DNA but the process is a self-limiting reaction (maximum of 30 bases can be added).

4.5.3 Competent cloning cells and choice of host expression organism

4.5.3.1 Prokaryotic cells

Almost all competent prokaryotic cells are derived from *E. coli* bacteria but each particular strain has been engineered to have certain advantages in each molecular biological application and some examples are given below.

1. Strains for general-purpose cloning – HB101 recA13 (Invitrogen).
2. Strains for single-stranded M13 vectors:
     SURE1 (Suppression of Unwanted Rearrangement Effects) rec B, rec J (Stratagene).
     JM109, rec A1.
3. Strains for lambda vectors – Y1088/90 (Invitrogen/Promega) LE292, LE293 and XL1 blue (Stratagene) – lambda ZAP.
4. Strains for *in vitro* packaging – BhB 2688/2690.

Competent cells should all have:

   absence of recombination genes such as recA–J.
   lac selection lacZ' gene mutations in the plasmids, which require hosts with a partial deletion of the original gene near the lactose operon. The deletion is partially complemented by an engineered F' plasmid which carries proAB+, or better still, lacZdeltaM15 (lac operon minus lacZ' segment).

4.5.3.2 Eukaryotic cells

When expressing complicated mammalian proteins a wide variety of eukaryotic cells are preferable because they can package precursor proteins with their precursor peptides, fold mature proteins correctly, avoid codon bias in the selection of amino acids and secrete proteins appropriately. A wide choice is available but each cell system is invariably more complex to use than *E. coli* prokaryotic systems. Examples, include yeasts S. *pombe* and S. *cerevisiae*, filamentous fungi, insect cells and mammalian cells, usually (monkey) cos cells.

SECTION III

Cellular analysis

5

Initiation and propagation of cultured cell lines

R. IAN FRESHNEY

5.1 Introduction

Cell culture provides a tool whereby cells may be isolated from their normal *in vivo* environment and maintained in defined, or at least reproducible, conditions. This enables the study of the response of the cells to regulatory signals and cytotoxic compounds without complicating systemic effects. It also allows the cell population to be purified, so that specific properties can be attributed to particular cells, and, following purification, different cell types can be recombined to study their interactions. There are also advantages of economy, where several variables can be studied simultaneously and the need for large numbers of replicates reduced by the homogeneity of the experimental sample.

The use of cell cultures has increased dramatically over the past decade, due in large part to their use as substrates, both for viral propagation, already well established at the beginning of the decade, and, more recently, for gene transfection. In addition, advances in molecular and cytological techniques have opened up new areas in somatic cell genetics such that the definition of the whole human genome now becomes a viable proposition. Genetic manipulation, as well as throwing new light on genetics and the regulation of gene transcription, has created a whole new biotechnology industry where biopharmaceuticals can be generated in defined media from validated cell lines more cheaply and with higher purity and lower risk of contamination than from natural sources. There has also been a significant shift towards *in vitro* screening assays for toxicity (see Chapter 7) and biological response. Added to this, the production of antibodies by cloned hybridomas, and the potential for genetic and protein engineering of antibodies, have given cell culture an assured place in the technology of the future for some time to come.

This technology is not without its drawbacks. The limitations of cell culture are apparent in the need for specialized facilities and training to carry out the procedures necessary to maintain cells in a healthy and uncontaminated state. However, with the availability of high grade commercial media, sterile plastics suitable for cell attachment and propagation, the controlled (and not indiscriminate) use of antibiotics, and the provision of a sterile environment using

Table 5.1. *Stages in the generation of cell cultures*

Stage	Time	Manipulation	Outcome	
			Good	Bad
Isolation	min	Dissection or withdrawal by syringe	Isolation of specific tissue from physiological variables of host	Trauma of dissection; removal from specific micro-environment
Disaggregation	min–h	Mechanical or enzymatic	Homogeneity of suspension; high yield of cells	Trauma; loss of cell–cell and cell–matrix interactions
Short-term incubation	h	Assay, e.g. by isotope incorporation	Quick; phenotype close to tissue	Trauma of dissection, heterogeneity, necrosis
Primary culture	d–wk	Incubation and feeding	Retention of major cell types of tissue; expansion of culture	Selection of subpopulation of adherent cells able to survive in culture
Subculture (cell line)	wk	Trypsinization or dilution	Expansion of culture; increased homogeneity of cell type	Selection of stromal cells; loss of specialized cells
Serial passage (cell line)	wk–mth	Trypsinization or dilution	Expansion of culture; characterization and accreditation; storage by freezing	Selection of undifferentiated proliferating cells; possibility for transformation and genetic drift
Senescence	3–6 mth	None		Loss of cell proliferation
or				
Immortalization (continuous cell line or cell strain)	3–6 mth–∞	Spontaneous or induced by transfection or infection	Provision of continuous cell line; increased growth rate and cell yield; creation of permanent resource	Genetic instability; loss of specialized properties; possibility of transformation

filtered laminar air flow cabinets, the need for extensive specialized facilities is greatly diminished. The requirement for careful technique and constant monitoring of the cultures remains, however, to ensure reproducibility and consistency, and avoid the problems of poor cell survival or growth and contamination with microorganisms or other cell lines.

Probably one of the single major problems to overcome is to ensure the survival of a particular cell type of interest. Low stringency conditions will usually permit the propagation of ubiquitous connective tissue cells such as fibroblasts, and special selective conditions are required to maintain most differentiated cells in culture. Even in the culture of tumour cells it is often the stromal compartment that survives, particularly in short-term culture, and only selective conditions, or long-term isolation of a transformed continuous cell line, will ensure the cultivation of the malignant component of the tumour.

Isolation of cells or tissue fragments is the first step in a range of procedures that allow progressively increasing control to be exercised over the extracellular environment. Unfortunately, the longer the culture period the greater the loss of phenotypic control of the cells. Freshly isolated fragments retain the architecture of the parent tissue but are heterogeneous and difficult to maintain uniformly viable for more than a few hours. Propagated cell lines, on the other hand, can be maintained with a high level of homogeneity and viability but are often difficult to relate back to the tissue of origin.

The different stages of culture are described in Table 5.1. With increasing time from isolation, the heterogeneity of the culture decreases and the proportion of cells able to proliferate increases. This is accentuated by subculture until cell proliferation finally ceases or a continuous cell line is formed (see below). The stage after isolation and before the first subculture is referred to as the *primary culture*. Following subculture the culture is referred to as a *cell line*. If a subpopulation is isolated by cloning or physical separation and can have attributed to it certain specific properties, it is referred to as a *cell strain*.

Cultures that are not disaggregated at isolation, other than by mechanical dissection, and are maintained as small fragments of tissue attached to the substrate, at the solid–liquid interface, are referred to as *primary explant* cultures. Such explants can give rise to an *outgrowth* of attached cells, which can be harvested or subcultured, but if explants (or whole embryonic organs) are maintained whole, usually by culture at the gas–liquid interface, outgrowth is minimized, and some tissue architecture is retained. These are referred to as *organ cultures*, may last for several weeks, but cannot be propagated (Figure 5.1).

5.2 Isolation and primary culture

5.2.1 Disaggregation

A primary culture is isolated by mechanically or enzymatically disaggregating the tissue to generate a suspension of single cells and small aggregates. In most

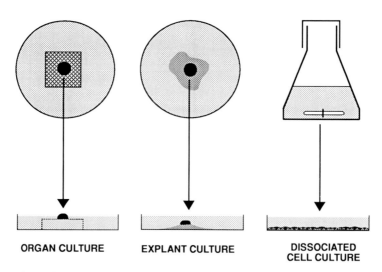

ORGAN CULTURE   **EXPLANT CULTURE**   **DISSOCIATED
                                                         CELL CULTURE**

Figure 5.1. Diagrammatic representation of three main types of culture: organ culture
where an explant is placed on a grid and/or filter at the gas/liquid interface, explant
culture where the explant is attached to a solid substrate and cells migrate outwards
from it, and dissociated cell culture, where enzymatic or mechanical disaggregation
generates a suspension of cells that settle and some of which will attach and grow on a
solid substrate.

cases this need not be a single cell suspension and may even benefit from the
retention of small aggregates (10–1000 cells) as some cells, particularly epithelial cells, survive better if mutual contact is maintained. Usually larger aggregates ($> 100 \ \mu$m diameter) are filtered out or allowed to settle out, as they do
not attach readily and rapidly become necrotic.

Enzymatic disaggregation can be achieved with trypsin, collagenase, Dis-
pase, or Pronase, and DNAase where cell damage is apparent (DNA released
from dead cells can cause reaggregation of dispersed cells). Trypsin is suitable
for many unspecialized cell cultures such as whole mouse or chick embryo but
collagenase is preferred where specific cell types, such as vascular endothelium
or mammary epithelium, are required. Tryptic digestion is carried out in
serum-free medium or Ca^{2+} and Mg^{2+}-free saline, at 37 °C for 1–4 h (harvesting cells at hourly intervals), or at 4 °C for 16–18 h followed by 30 min at 37 °C
after removal of the trypsin (see Protocol 1). Collagenase digestion is carried
out in complete medium, as it is not inhibited by serum, and may require
divalent cations, depending on the source of the collagenase (see Protocol 1).
The incubation period is usually longer (4–24 h) and can be as long as 5 days.
If the incubation period is chosen carefully, connective tissue fibroblasts will
tend to reduce to a single cell suspension, while epithelial (and endothelial)
cells remain as small clusters, which can be separated from the fibroblasts by
allowing them to settle out in a universal container for a few minutes. Removal
of collagenase from the supernatant suspension requires centrifugation, but the

sediment can simply be resuspended in medium and allowed to settle again (see Protocol 1).

5.2.2 *Primary culture*

Disaggregated cells and small clumps are seeded into culture vessels (together or separately, as above) at a concentration of around $1 \times 10^5 - 1 \times 10^6$ cells/ml and a density of $2 \times 10^4 - 2 \times 10^5$ cell/cm^2. This is higher than for normal propagation of a cell line as not all of the cells will survive, and those that do, survive better if cross-feeding and the laying down of extracellular matrix is facilitated by using a higher cell density. The cell number at seeding should be the viable cell count as determined by viability staining.

Conventional tissue culture grade plastic is treated by plasma discharge or γ-irradiation to promote cell adhesion and this is adequate for most cultures. However, certain specialized cell types, such as epithelial cells, vascular endothelial cells and muscle cells, may require pretreatment of the substrate with gelatin, collagen or Matrigel (a proprietory matrix product from EHS sarcoma cells, containing laminin, fibronectin, collagen and proteoglycans). Used as a dried coating they improve cell adhesion, spreading and proliferation, and can be selective, while culturing cells on or in a matrix gel facilitates cell differentiation (see Chapter 6). Some cells, e.g. keratinocytes, will continue to proliferate grown on the surface of a matrix gel but most will show reduced proliferation and often assume three-dimensional structures within the gel.

5.3 Subculture and propagation

Depending on the source of the tissue, primary cultures will grow to form a continuous sheet of cells, a *confluent monolayer*, after 1–3 weeks. At this point the culture may be discarded, and fresh primaries prepared (if the function under observation only appears in the primary culture) or the culture may be *subcultured*. Subculture allows the cell culture to be propagated and expanded, characterized, and, ultimately, stored. The cell population becomes more uniform due to the selection of those cells best adapted to the culture environment.

Subculture of a primary culture, or of an already established cell line, is achieved by dislodging the cells from the substrate, usually with trypsin, generating a single cell suspension for counting, diluting and reseeding. When the culture is seeded, it will generally take about 12–24 h before proliferation starts (Figure 5.2). This is due to the need to attach, spread out and enter the cell cycle. During trypsinization, much of the extracellular matrix is removed and must be resynthesized, the actin cytoskeleton depolymerizes and must repolymerize, enzymes necessary for DNA replication are synthesized, and the cell builds up a microenvironment of autocrine growth factors, and, possibly,

GROWTH CURVE

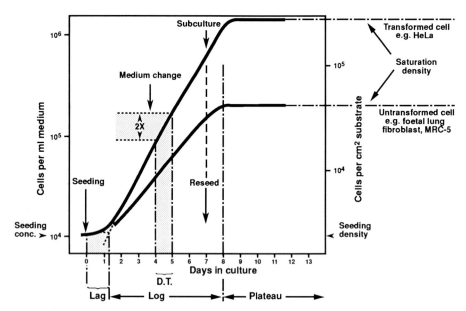

Figure 5.2. The three main phases of the growth curve (lag, log and plateau) with the manipulations carried out at each point (seeding, medium change and subculture) and the proliferation kinetics parameters that can be derived from each phase (lag period, population doubling time (D.T.) and saturation density).

nutrients leaked from the cells. Cells from a primary culture should be seeded at a higher concentration (1.0–2.0×10^5 cells/ml) than for established cell lines (1.0–5.0×10^4 cells/ml).

5.3.1 Growth cycle

Following the lag period, the culture enters the *exponential* or *'log'* phase, during which it will double over a period of time, the *population doubling time*, characteristic of the cell line, under a particular defined set of conditions (see Figure 5.2). When the cells reach *confluence*, i.e. occupy all the available growth surface, they will generally go through one or two more doublings before entering *stationary* or *'plateau'* phase. When in plateau the cells are said to have reached *saturation density*, the maximum density achievable under the defined conditons of medium, serum concentration, growth factors, etc. With a normal, untransformed cell line, plateau represents a stable period of no or minimal cell proliferation, and no or minimal cell loss. This is due to the sensitivity of most normal cells to *contact inhibition* of cell migration and *density limitation* of cell proliferation. With many cells this can be maintained for many weeks, particularly if the medium is replenished with low serum/low growth factor concentrations.

When transformed cells enter plateau, they do so at a higher cell density, due to the piling up of the cells that lack contact inhibition and density limitation. The cells continue to proliferate, though more slowly, and saturation density is nominal and achieved by the rate of cell proliferation equalling the rate of cell loss. Ultimately, given adequate replenishment of the medium, or even perfusion, the cells are limited by diffusion.

It is possible to observe multilayering in normal, untransformed, cell cultures. Normal human fibroblasts, in particular, can secrete extracellular matrix including collagen, which acts as an overlay and permits a new monolayer to be superimposed upon the first. This is distinct from the piling up of transformed cells, as it is a regulated process, and contact inhibition is observed as the second monolayer reaches confluence. Continued proliferation after confluence, beyond one or two population doublings, requires continuous replacement of the medium, particularly the serum and/or growth factors, and is best achieved by perfusion (Kruse, Keen and Whittle, 1970).

A culture should be subcultured when it reaches confluence and not allowed to enter the plateau phase of the growth cycle (see Figure 5.2), as it will take longer to re-enter exponential growth from plateau than from exponential growth. Seeding concentrations vary among cell lines but are usually around $2 \times 10^4 - 1 \times 10^5$ cells/ml and the preferred concentration prior to subculture is usually around 1×10^6 cells/ml.

Some laboratories divide cells at subculture by a set *split ratio*, i.e. the contents of one flask are split between two or more fresh flasks, or only a fixed fraction of the product of one flask is retained and reseeded into a fresh flask. The advantage of using a set split ratio, particularly if a factor of two is used, is that the approximate number of generations can be counted, important if the cell line is finite and will only grow through a set number of generations. Thus a 1:2 split is equivalent to one generation for the cells to regain the same density; 1:4, 2 generations; 1:8, three generations; etc. This assumes, of course, a 100% growth fraction and no cell loss at subculture, neither of which is absolutely true, but the approximation does serve to monitor the age of a culture. Care should be taken to distinguish between *passage number* and *generation number*, when recording the age of a culture in this way. The passage number is the number of times the culture has been subcultured or *passaged* while the generation number is the total number of generations that the cell population has gone through, assuming that all the cells are proliferating and at the same rate (which, of course, they are not). When the method was first employed the two terms were synonymous, as Hayflick (Hayflick and Moorhead, 1961) who first adopted the regime, split his cultures 1:2, but this is often seen as laborious and wasteful of time and materials, and most laboratories use higher split ratios.

It is essential, if using this technique, that a growth curve be performed when the cell line becomes established so that the optimum seeding density at subculture and reseeding can be established, and the culture checked by cell counting, or at least visually, at each subculture to ensure that the cells are subcultured at or just before reaching confluence, and are not allowed to enter plateau, or a prolonged lag period may develop. This can be acute in some transformed cells as viability falls quite rapidly as the cells remain in plateau, and the resulting lag period after subculture can be extended as a result.

Counting cells at subculture also serves as a quality control step, as, under standard conditions, a given seeding concentration should always give the same yield at the next subculture, given that the interval between each subculture remains constant and the culture conditions are kept the same. If the yield is different, then one, or more, aspects of routine culture maintenance are being inadequately controlled.

Viability counting may be performed at subculture and the seeding concentration adjusted accordingly, but this is not usually necessary beyond the tertiary culture stage, as viability should be $> 95\%$.

5.3.2 Selective culture

If no selective conditions are applied, i.e., a simple non-selective medium such as Eagle's MEM with calf serum is used with regular plastic vessels, then the cells most likely to take over the culture will be fibroblast-like cells from connective tissue. If a specific cell type is required then a selective medium may be required (Figure 5.3). Several of these are now available commercially (Clonetics, Tissue Culture Services, LTI/Gibco-BRL, ICN, JRH Biosciences,

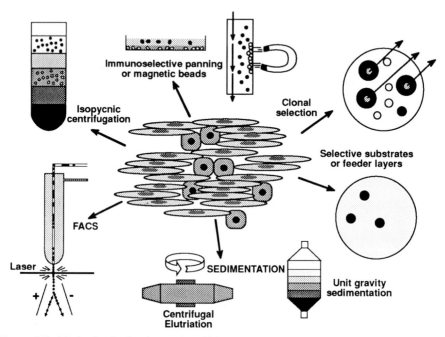

Figure 5.3. Methods of selecting and purifying a heterogeneous population of cells.

Sigma) for the selective culture of epidermal keratinocytes, mammary and bronchial epithelium, vascular endothelium, melanocytes, etc., and others have been reported in the literature for prostatic epithelium, neurones and glia (Maurer, 1992).

The continued use of extracellular matrix (see above) may be required for some specialized cells, such as endothelium, but others may survive adequately on plastic. Selective conditions can also be provided by using a *feeder layer*. This is a preformed layer of cells, usually 3T3 mouse fibroblasts or freshly explanted mouse embryo cells, which are growth arrested by irradiation or treatment with mitomycin C. They provide nutritional support, paracrine growth factors, and extracellular matrix to fresh cells seeded on to them. They may also be selective, by excluding fibroblasts or degrading growth factors such as TGF-β, which inhibit epithelial growth.

5.4 Origin and evolution of cell lines

Following subculture, there is progressive dominance of the culture by a limited range of cell types, continuing with serial propagation. Clearly, if the cell culture is expanded, this must imply that these are proliferating cells. Within each lineage of cells *in vivo*, the proliferating compartment is made up of stem cells or precursor cells, and the differentiated compartment is either incapable of proliferation (e.g. mature keratinocytes, neurones and erythrocytes) or able to dedifferentiate and resume proliferation (e.g. vascular endothelium and fibrocytes). Hence cell lines will represent the proliferating progenitor phenotype of the tissue rather that the fully differentiated cells.

The selection of the wrong lineage (e.g. fibroblasts in an epithelial culture), or expansion of the progenitor cell compartment, accounts for the apparent dedifferentiation of many cell lines following serial subculture. If the wrong lineage has been selected, clearly there is no way of recovering the representative differentiated phenotype, but if the correct lineage is selected, albeit in an undifferentiated state, then recovery of the correct phenotype is possible, given the correct inductive environment (see below).

5.4.1 Senescence, immortalization and transformation

Most normal cells display a fixed lifespan in culture. That is to say, they will go through a set number of cell generations (reflected as population doublings) before they cease to divide. At this point, usually around 50 generations in human fibroblasts, the cells enter *crisis* following which they are said to have *senesced*. They remain static for many months and eventually die out. This appears to be under the control of dominantly acting senescence genes, which, if deleted or mutated, allow the cells to become immortal.

Sometimes a minority population with infinite survival emerges at, or before, the culture enters crisis. Spontaneous immortalization occurs most frequently in mouse cell lines and in tumour-derived cultures from many different species including human, and results in the formation of a *continuous cell line*, with more rapid proliferation, and a higher growth fraction. Tumours may contain cells that are already immortalized, or have acquired sufficient genetic instability that deletion or mutation in one or more senescence genes has a high probability of inducing immortalization. In addition, over-expression of oncogenes, such as *ras* or *myc* favour continued proliferation. However, fusion products of immortal and finite cell lines generally senesce (Pereira-Smith and Smith, 1988), suggesting senescence is under the control of dominant-acting genes.

Immortalization can be induced by infection with SV40 or E1a virus or transfection with the SV40 large T region (Klein *et al.*, 1990; Nachtigal, 1991; Steele *et al.*, 1992). A subpopulation of SV40LT$^+$ cells emerges and continues to proliferate. These then appear to undergo a second crisis, probably involving deletion of the p53 gene, from which a true continuous cell line appears. Immortalization has now been induced in several different types of cells, including human fibroblasts, endothelial cells, keratinocytes and glia (Freshney, 1995) but, at present, the potential to express differentiated properties is often lost. Furthermore, those cells that grow out following infection or transfection are a small minority, and further minority selection is observed when truly immortal cells emerge after crisis.

Immortalization, although often associated with malignant transformation, is not in itself sufficient to induce malignancy. Many immortalized lines, both spontaneously immortalized like the 3T3 mouse fibroblasts or BHK21-C13 hamster fibroblast, or SV40LT transfected, retain contact inhibition of cell motility, density limitation of cell proliferation, anchorage-dependent cell proliferation (will not grow in agar), and low tumorigenicity. A further transformation step is required, such as is seen spontaneously in 3T3 cells kept at a high cell density for several weeks, or induced in BHK21-C13 infected with polyoma virus, which converts the immortalized cell line into a malignantly transformed line. This is accompanied by a loss of contact inhibition, density limitation, and anchorage dependence.

5.5 Continuous cell lines

There are significant advantages in working with continuous cell lines, if the expression of specific differentiated properties is not essential. They are generally easier to handle. Shorter population doubling times combined with higher saturation densities give a much higher yield per flask and their lower serum dependence makes maintenance cheaper and the development of serum-free media formulations easier. The infinite lifespan means that the population can be expanded and disseminated indefinitely, and that cloned lines can be isolated. Continuous cell lines are, however, genetically heterogeneous and unstable. They are both aneuploid (different from their original chromosome complement) and heteroploid (vary within the population). It is advisable to start with a cloned line, expand that to freeze down sufficient ampoules to last the duration of a particular programme, and replace culture stocks regularly. Some examples of continuous cell lines in regular use are given in Table 5.2. Information on others is available from the American Type Culture Collection (ATCC), 12301, Park Lawn Drive, Rockville, MD, USA, or from the European Collection of Animal Cell Cultures (ECACC), CAMR, Porton Down, Salisbury, England.

Because of their rapid and vigorous growth, continuous cell lines are often found as cross contaminants in more slowly growing cell lines and will, consequently, overgrow them. Many cases of 'spontaneous transformation' have been found to be due to cross contamination with a transformed cell line, such as HeLa. It is always important to guard against cross-contamination, e.g. by using separate bottles of media, trypsin, PBS, etc., for each cell line, never sharing media or reagents with anyone else, and not handling more than one cell line at a time. It is particularly important not to handle cultures of a continuous line like HeLa at the same time as more slowly growing cell lines.

5.6 Characterization

Details of methods of characterization will be dealt with in Chapter 6. It is, however, relevant to emphasize the need for characterization. Cell lines can change due to genetic instability, cross contamination, overgrowth by a minority component, or by transformation. Exchanges between laboratories often lead to confusion and accidents can occur when thawing stocks from frozen storage. It is vital that cell lines are characterized to establish their species and tissue of origin and transformation status.

Briefly, species can be established by chromosome analysis, isoenzyme analysis (Hay, 1992) or *in situ* hybridization with species-specific chromosomal paints (Keith, 1994). Lineage is best determined by intermediate filament protein immunostaining with anti-cytokeratin for epithelial cells, anti-vimentin for fibroblasts and other mesenchymal cells, anti-glial fibrillary acidic protein (GFAP) for astrocytes, and anti-desmin for muscle cells. Some surface markers are also lineage specific, e.g. anti-EMA for epithelium, A2B5 for glial cells and a wide range of specific markers for haemopoietic cells.

Transformation status can be determined by an estimation of *saturation density* (cells/cm^2 in plateau under defined conditions) and the *labelling index* at saturation density (continuous labelling at high cell density with ^{3}H-thymidine or BUdR), the ability to clone in agar (*anchorage independence*), production of extracellular proteases such as urokinase-like plasminogen activator (Rifkin *et al.*, 1974), and tumorigenicity in immune-deprived hosts such as the nude mouse.

It cannot be stressed enough that cross contamination is a major recurrent problem. Rapidly growing cell lines can and do infect more slowly growing

Table 5.2. *Examples of commonly used continuous cell lines*

Cell line	Origin	Morphology	Particular Attributes	Reference
3T3-L1	Mouse embryonic fibroblast	Fibroblastoid subconfluent, polygonal at confluence	Contact-inhibited. Used in transformation and growth assays	Green and Kehinde, 1974
A2780	Ovarian carcinoma	Epithelioid	Drug sensitive	Tsuruo *et al.*, 1986
A549	Human lung adenocarcinoma	Epithelioid	Makes surfactant. Lacks CDKN2(p16)	Giard *et al.*, 1972
CHO-K1	Chinese hamster ovary	Fibroblastoid, polygonal at confluence	High plating efficiency, rapid growth	Puck *et al.*, 1958
HeLa	Cervical carcinoma	Epithelioid	Rapidly growing. HeLa-S3 can be made to grow in suspension	Gey *et al.*, 1952 Puck and Marcus, 1955
HT-29	Human colo-rectal adenocarcinoma	Epithelioid	Differentiates in response to planar-polar compounds	Fogh and Trempe, 1975
IMR-32	Human childhood neuroblastoma	Neural	Differentiates readily in response to several inducers	Tumilowicz *et al.*, 1970
L1210	Murine lymphocytic leukaemia	Lymphocytic	Very short cell cycle time	Law *et al.*, 1949
L929	Mouse peritoneal fibroblast	Fibroblastoid	One of the oldest cell lines in existence	Sanford *et al.*, 1948
MCF-7	Human mammary carcinoma, pleural effusion	Epithelioid	Oestrogen receptor positive	Soule *et al.*, 1973
Vero	Monkey kidney	Fibroblastoid	Viral substrate and assay	Hopps *et al.*, 1963
ZR-75-1	Mammary carcinoma, ascites	Epithelioid	Oestrogen receptor negative	Engel *et al.*, 1978

lines and gradually replace them. Regular characterization is essential to confirm the absence of cross contamination, particularly if a change in growth characteristics, morphology, or biochemical properties is observed. Karyotyping and isoenzyme analysis have been traditional and reliable methods but DNA fingerprinting is now accepted as the most reliable method. It does require that the line is fingerprinted at derivation, and that you have fingerprints of other cell lines growing in the laboratory. This is expensive and time-consuming but well worthwhile.

5.7 Monolayer and suspension growth

Most primary cultures, with the exception of cultures from the haemopoietic system, grow attached to a solid substrate. In the case of normal, untransformed, cell lines, attachment and spreading are a prerequisite for cell proliferation. Normal cell lines are said to be *anchorage dependent* when attachment and spreading is required. With transformed cell lines, however, it is possible to propagate the cells in suspension as they are often *anchorage independent*. Anchorage dependence is usually measured by cloning the cells in suspension in soft agar (0.3%) or methocel (0.8%). Transformed cells will plate in suspension with an efficiency usually $> 10\%$ while normal cells either will not plate at all in suspension, or do so with an efficiency of $< 1\%$.

The ability to grow in suspension makes large-scale culture much easier as trypsinization is not necessary for subculture and harvesting and bulk is not limited by the surface area of the culture vessel.

5.8 Scale-up

It is possible to generate $1 \times 10^3 – 5 \times 10^8$ (2 ng–1 g, wet weight) cells by conventional culture methods, by selecting the appropriate size and number of flasks, plates or dishes (Figure 5.4) to suit the numbers of cells, and the number of replicates, required. Increasing the scale above this requires increasing the medium volume and surface area for attached cells and increasing the volume for suspension-growing cells. If the medium in suspension cultures exceeds 10 mm in depth, agitation is required, and if it exceeds 50 mm sparging with a humid mixture of CO_2 (usually 5%) in air is required. In general, scaling up suspension cultures is easier, and a large vessel with a magnetic stirrer (and a gas line if the depth is to exceed 50 mm) is all that is required (Figure 5.5).

The increase in surface area required for monolayers can be achieved by using a multi-surface propagator (e.g., Nunclon Cell Factory), roller bottles that rotate on a special rack and carry the cell monlayer on the inner surface of the walls of the vessel, or by growing the cells attached to small beads called *microcarriers*, which are stirred as for suspension cultures, though slower. Multisurface propagators have a high unit cost (~£50–£200) but low capital investment, useful for one or two experiments. Roller bottles have a lower unit

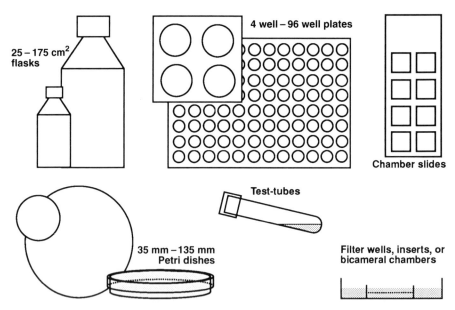

Figure 5.4. Examples of typical tissue culture plasticware. Plates and flasks (on the left) for conventional propagation, as well as experimental sampling and colony-forming assays; multi-well plates and test-tubes (in the centre) for large numbers of replicates; chamber slides (top right) for cytological and immunochemical observations, and filter wells (bottom right) for studies of polarized transport, cell–cell and cell–matrix interaction.

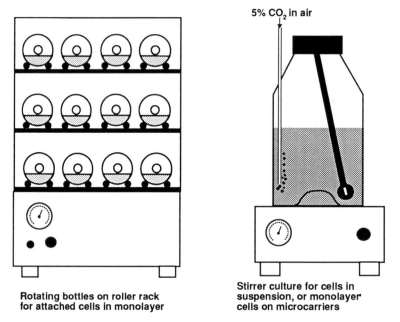

Figure 5.5. Scale-up of cell culture. The roller rack on the left gives an extended surface area for monolayer propagation and the stirrer culture on the right provides large volumes for cells growing in suspension, or attached to microcarriers.

cost, higher capital cost (~£3000 for the roller rack), and are good for an extended series of experiments without much process development. Micro-carriers are, ultimately, the simplest method giving a high yield for a moderate unit cost and low capital cost, but may require a longer development and learning period to adapt your own cell line to the conditions of microcarrier culture (Griffiths, 1992).

5.9 Preservation and banking

Cultures are continuously at risk from contamination (see Chapter 21), mech-anical or electrical failure of incubators, transformation, or genetic drift from prolonged propagation. To protect against these, representative stocks should be frozen in liquid nitrogen. The cell line, fully characterized and shown to be free of contamination, is grown up to around 5×10^7 cells and frozen in liquid nitrogen to form a *seed stock*. An ampoule is thawed, checked for character-istics and contamination, and expanded to around 5×10^8 cells (the actual number depends on how long the project will last, how many people will draw from the stock, and so on) and frozen as a *using stock*. The using stock should be used to replenish cultures every three months. The seed stock should only be used to regenerate the using stock if it becomes depleted, and must not be used for regular replacement of cultures. If the number of ampoules in the seed stock falls to < 5, more seed stock should be generated.

5.9.1 Records

It is vital to record each freezing and log it by cell line and location. The easiest way is to use a computer database that can be searched by cell line designation, freezer space, user name, cell line characteristics, etc. Cell lines in liquid nitrogen last indefinitely, so may be around long after their originator has gone. To be useful, the stock record must therefore be readily understood, and religiously completed.

5.9.2 Cell and databanks

There are repositories of cell lines maintained by the American Type Culture Collection (ATCC, 12301 Parklawn Drive, Rockville, MD 20852-1776, USA), the European Collection of Animal Cell Cultures (ECACC, CAMR, Porton Down, Salisbury, Wilts., SP4 0JG, England). There are also numerous data-banks, information on which may be obtained from either ATCC or ECACC.

5.10 Protocols

(Adapted from Freshney, R. I. (1994) *Culture of Animal Cells*, Wiley Liss, New York)

Protocol 1: Primary culture

Sterile reagents and materials

100 ml dissection BSS (DBSS). Hanks' balanced salt solution (HBSS) without glucose and with 250 U/ml penicillin, 250 μg/ml streptomycin, 50 μg/ml gentamycin, 2.5 μg/ml amphoteracin B (Fungisone).

10 ml 0.25% crude trypsin, *or* 2000 U/ml crude collagenase, in Dulbecco's phosphate buffered saline without Ca^{2+} and Mg^{2+} (PBSA), on ice.

Viability stain (1% (w/v) Naphthalene Black in HBSS).

Complete culture medium with serum.

5% CO_2 supply (for CO_2-buffered media).

Three 9 cm Petri dishes.

Two universal containers.

Culture flasks or dishes.

Pipettes, 1 ml, 2 ml, 10 ml and 25 ml.

Protocol

1. Place tissue in Petri dish with 20 ml DBSS and trim off excess fat and obvious necrotic tissue.
2. Transfer to fresh dish with DBSS.
3. Chop finely with crossed scalpels until 3–4 mm in diameter (for trypsin) *or* ~1 mm in diameter (for collagenase).
4. Transfer pieces to universal container and wash twice by sedimentation with DBSS.
5. *For trypsinization:* remove supernatant DBSS and replace with 10 ml ice-cold trypsin.
6. Place at 4 °C overnight.
7. Remove trypsin next day and incubate for 30 min at 37 °C.
8. Add 10 ml warm medium with serum, pipette up and down until disaggregated.
9. Allow any undisaggregated fragments to settle, pipette off supernatant suspension and count a sample, diluted 50:50 with viability stain, on a haemocytometer.
10. Dilute to 1×10^6 viable cells/ml and 2×10^5 viable cells/ml in medium in two separate flasks.
11. Gas with 5% CO_2 in air and place in incubator.
12. Check after 24 h for attachment and, after 2–3 days, check pH for medium replacement. Non-viable cells are removed at this medium change.
13. *For collagenase:* remove DBSS (at step 5), replace with 4.5 ml complete

culture medium and transfer ~100 pieces to one 25 cm² culture flask and ~20 pieces to a second culture flask. If a pipette is used ensure that it has a wide tip and has been wetted inside first with medium to prevent the pieces sticking to the inside of the pipette.

14. Add 0.5 ml collagenase to each flask.
15. Incubate for 1–4 days.
16. Disaggregation is correct when the tissue fragments seem to 'smear' on the base of the flask, and gentle agitation causes the tissue to break up into single cells and small clusters of cells. Small clusters, usually enriched for epithelium, can be collected at this stage by allowing them to settle under gravity for a few minutes, and removing the supernatant to be cultured separately.
17. Remove the collagenase from the suspension by centrifugation at 100 *g* for 5 min (or by washing the clusters and allowing them to settle).
18. Resuspend the pellet in 6 ml of medium, and transfer 1 ml to a fresh 25 cm² flask and 5 ml to a second.
19. Add 4 ml medium to the first flask.
20. Incubate and observe as in steps 11 and 12.

Note: Agents that will improve disaggregation are 1.0 mM EDTA or 1.0 mM EGTA, which is less toxic to the cells, Pronase, 0.25% (which can be toxic to some cells), or hyaluronidase, 0.1%. DNAase, 0.1%, can also help to disaggregate cells that have aggregated due to cell damage and the release of DNA, subsequently acting as an adhesive.

Protocol 2: Subculture

Sterile reagents and materials

Culture medium with serum (if required)
5% CO_2 supply (for CO_2-buffered media)
100 ml PBSA
0.25% trypsin in PBSA, 3 ml/25cm² of culture surface
Pipettes
Culture flasks

Non-sterile

Haemocytometer or electronic cell counter.

Protocol

1. Check culture by eye and with a microscope to assess status of cells (density, healthy appearance, absence of excessive granularity or vacuolation), freedom from contamination, and need for subculture.
2. Take to a laminar flow hood and remove medium carefully and completely without disrupting the monolayer.

3. Rinse monolayer with PBSA, 5 ml/25 cm^2 and discard the PBSA.
4. Add 0.25% cold trypsin, 3 ml/25 cm^2, and remove after 30 s.
5. Incubate for 15 min, or until cells detach freely and slide across the plastic.
6. Add 5 ml complete culture medium with serum per 25 cm^2 flask or 10 ml per 75 cm^2 flask and wash off the remaining cells.
7. Pipette the cells up and down several times with the tip of the pipette on the base of the flask to generate a single cell suspension.
8. Take a sample and count by haemocytometer or electronic particle counter.
9. Dilute to correct seeding concentration (e.g. 2×10^4 cell/ml for HeLa cells), gas flask with 5% CO_2, seal and place in an incubator. Large flasks (75 cm^2 and above) need to be vented briefly after about 30 min to release the pressure caused by the expanding gas in the flask, otherwise they will not stack evenly.

Protocol 3: Cloning

1. Proceed as for subculture up to step 8.
2. Check that cells are singly suspended. If not, repipette in the flask more vigorously until singly suspended. If still unsuccessful, an alternative disaggregation protocol is required and this should be tested on a new batch of cells (see *Note* following primary culture protocol). EDTA or EGTA can be used as a prewash or incorporated into the trypsin solution.
3. Dilute cells to 10, 50 and 200 cells/ml. (Cells with a high plating efficiency ($> 50\%$) will form colonies at 10 cells/ml, but those with a low plating efficiency ($< 10\%$) will need a higher seeding concentration).
4. Seed Petri dishes, flasks or microtitration plates. If using microtitration plates, seed 100 μl/well at 10 cells/ml for cells with a high plating efficiency, check after 6–8 h and mark wells with only one cell. If using higher seeding concentrations for low plating efficiency cells, check after 3–7 days and mark wells where single colonies are forming.
5. Clones should form in 10–21 days. Pick from Petri dishes by removing medium and placing cloning rings (Bellco), dipped in sterile silicone grease, over colony and trypsinize from within the ring. Harvesting from flasks will require the top of the flask to be removed with a hot wire (flasks are available from Nunc with a removable plastic film). Harvesting from microtitration plates is, by far, the easiest, as trypsinization is straightforward and cells can be reseeded into a 25 cm^2 flask. If the colony is small, or the seeding efficiency at subculture likely to be low, stand the flask on end and culture the cells on the end of the flask in 1–2 ml of medium. They can then be subcultured in the same flask and reseeded on to the usual surface.

Protocol 4: Cell freezing

Reagents and materials

Sterile
As for subculture.
Dimethyl sulphoxide (DMSO) (self sterilizing, but dispense into sterile glass container with glass or polypropylene stopper)
Ampoules

Non-sterile
Canes or racks for storage.
Insulated container for freezing at 1 °C/min, e.g. polyurea-foam pipe insulation for canes or polystyrene foam box for loose ampoules (wall thickness ~12 mm). A freezing box is also marketed by Nalgene.
Gloves
Goggles

Protocol
1. Trypsinize cells as for subculture, above.
2. Resuspend cells and dilute to 5×10^6 cells/ml in complete medium with 10% (v/v) DMSO.
3. Dispense 1 ml into each ampoule and seal ampoule.
4. Clip ampoules on to aluminium freezer canes and place in insulated container.
5. Place insulated cane at −70 °C for 90 min, or until temperature falls to −70 °C.
6. Transfer ampoules rapidly to liquid nitrogen freezer (they must not warm to above −50 °C).
7. Complete records.
8. To thaw, remove ampoule from freezer cane and place immediately at 37 °C in a covered container. *Note: ampoules which have been submerged in liquid nitrogen can inspire liquid nitrogen if not properly sealed. When warmed, they will explode violently. Thawing MUST be carried out in a covered container, and gloves, lab-coat and goggles MUST be worn.*
9. When contents have thawed, swab ampoule with 70% (u/v) ethanol, open ampoule and dispense contents into a 25 cm² culture flask.
10. Add 9 ml medium slowly to flask, so that DMSO is diluted out gradually and osmotic shock is minimized.
11. Residue in ampoule may be stained with one drop of 1% (w/v) naphthalene black to determine viability.
12. Incubate flask for 24–48 h, remove freezing medium and replace with fresh medium. This will remove the preservative and any dead cells.

Note DMSO induces differentiation in some cells (e.g., neuroblastoma, HL60 and Friend erythroleukaemia cells). In these cases, 10% (v/v) glycerol may be substituted for DMSO.

Acknowledgements

I am grateful to Mary Freshney for helpful advice on reading the manuscript and Debi Letham for art work.

References

Engel, L. W., Young, N. A., Tralka, T. S. *et al*. (1978) Establishment and characterization of three new continuous cell lines derived from breast carcinomas. *Cancer Res.* **38**, 3352–64.

Fogh, J. and Trempe, G. (1975) In *Human tumour cells in vitro* (ed. J. Fogh), pp. 115–59. Academic Press, New York.

Freshney, R. I. (ed.) (1995) *Culture of immortalized cells*. Wiley-Liss, New York.

Gey, G. O., Coffman, W. D. and Kubicek, M. T. (1952) Tissue culture studies of the proliferative capacity of cervical carcinoma and normal epithelium. *Cancer Res.* **12**, 364–5.

Giard, D. J., Aaronson, S. A., Todaro, G. J. *et al*. (1972) *In vitro* cultivation of human tumors: Establishment of cell lines derived from a series of solid tumours. *J Natl. Cancer Inst.* **51**, 1417.

Green, H. and Kehinde, O. (1974) Sublines of mouse 3T3 cells that accumulate lipid. *Cell* **1**, 113–16.

Griffiths, J. B. (1992) Scaling up of animal cell cultures. In *Animal cell culture, a practical approach*. (ed. R. I. Freshney), pp. 47–93. IRL Press at Oxford University Press, Oxford.

Hay, R. J. (1992) Cell line preservation and characterization. In *Culture of animal cells, a practical approach*. (ed. R. I. Freshney), pp. 95–148. IRL Press at Oxford University Press, Oxford.

Hayflick, L. and Moorhead, P. S. (1961) The serial cultivation of human diploid cell strains. *Exp. Cell Res.* **25**, 585–621.

Hopps, H., Bernheim, B. C., Nisalak, A. *et al*. (1963) Biologic characteristics of a continuous cell line derived from the African green monkey. *J. Immunol.* **91**, 416–24.

Keith, N. (1994) Fluorescence *in situ* hybridization (FISH) in the analysis of genes and chromosomes. In *Culture of Animal Cells* (ed. R. I. Freshney), pp. 399–402. Wiley-Liss, New York.

Klein, B., Pastink, A., Odijk H. *et al*. (1990) Transformation and immortalization of diploid xeroderma pigmentosum fibroblasts. *Exp. Cell Res.* **191**, 256–62.

Kruse, P. F., Jr., Keen, L. N. and Whittle, W. L. (1970) Some distinctive characteristics of high density perfusion cultures of diverse cell types. *In Vitro* **6**, 75–8.

Law, L. W., Dunn, T. B., Boyle, P. J. and Miller, J. H. (1949) Observations on the effect of a folic acid antagonist on transplantable lymphoid leukemia in mice. *J. Natl. Cancer Inst.* **10**, 179–92.

Maurer, R. (1992) Towards chemically defined serum-free media for mammalian cell culture. In Freshney, R. I. (ed): *Animal Cell Culture, a Practical Approach* 2nd edn. (ed. R. I. Freshney), pp. 15–46. IRL Press at Oxford University Press, Oxford.

Nachtigal, M. (1991) Characterization of human vascular smooth muscle cells transformed by the early genetic region of SV40 virus. *Am. J. Pathol.* **139**, 629–40.

Pereira-Smith, O. and Smith, J. (1988) Genetic analysis of indefinite division in human cells: identification of four complementation groups. *Proc. Natl. Acad. Sci. USA* **85**, 6042–6.

Puck, T. T., Cieciura, S. J. and Robinson, A. (1958) Genetics of somatic mammalian cells: III. Long term cultivation of euploid cells from human and animal subjects. *J. Exp. Med.* **108**, 945–56.

Puck, T. T. and Marcus, P. I. (1955) A rapid method for viable cell titration and clone production with HeLa cells in tissue culture: The use of X-irradiated cells to supply conditioning factors. *Proc. Natl. Acad. Sci. USA* **41**, 432–7.

Rifkin, D. B., Loeb, J. N., Moore, G. and Reich, E. (1974) Properties of plasminogen activators formed by neoplastic human cell cultures. *J. Exp. Med.* **139**, 1317–28.

Sanford, K. K., Earle, W. R. and Likely G. D. (1948) The growth *in vitro* of single isolated tissue cells. *J. Natl. Cancer Inst.* **9**, 229.

Soule, H. D., Vasquez, J., Long, A. *et al*. (1973) A human cell line from a pleural effusion derived from a breast carcinoma. *J. Natl. Cancer Inst.* **51**, 1409–16.

Steele, M. P., Levine, R. A., Joyce-Brady, M. and Brody, J. S. (1992) A rat alveolar type II cell line developed by adenovirus 12SE1A gene, transfer. *Am. J. Resp. Cell. Mol. Biol.* **6**, 50–6.

Tsuro, T., Hamilton, T. C., Louie, K. G. *et al*. (1986) Collateral susceptibility of adriamycin-, melphalan-, and cisplatin resistant human ovarian tumour cells to bleomycin. *Jpn. J. Cancer Res. (Gann)*, **77**, 941–5.

Tumilowicz, J. J., Nichols, W. W., Cholon, J. J. and Greene, A. E. (1970) Definition of a continuous human cell line derived from neuroblastoma. *Cancer Res.* **30**, 2110.

6

Characterization of cultured cell lines

KAREN HENDERSON and SUSAN C. KIRKLAND

It is of the utmost importance when culturing cells, that each cell line is fully characterized using properties such as species of origin, tissue of origin, differentiation features, and transformation status. The reasons for such characterizations are discussed in the appropriate sections below.

6.1 Species of origin

It is not uncommon for tissue culture laboratories to use, and in some cases, co-culture, cell lines from different species and from different members of the same species. In such circumstances, cross-contamination poses a serious threat and has indeed been well documented, particularly with regard to the cervical adenocarcinoma cell line HeLa (Nelson-Rees, Daniels and Flandermeyer, 1981). A cross-contamination event has the potential to nullify an entire research project and it is therefore important that regular assessments of cell lines are carried out. Commonly used techniques for species analysis include anti-species antibody staining, karyology, isoenzyme analysis, and DNA finger-printing. These techniques have been discussed in depth elsewhere (Doyle and Stacey, 1994).

6.2 Tissue of origin

For the most part, whether the source is normal, embryonic or tumour, cell lines are established from tissue consisting of several cell types, such as epithelial, muscle, endothelial and connective tissue elements and so it is necessary to confirm that the cell lines are in fact of the required cell type. In addition, cells of different types, such as epithelial and fibroblastic are often cultured together or present in the same tissue culture room and regular checks guard against cross contamination. Frequently used criteria include morpho-logy, cell surface antigen characteristics and type of cytoskeletal component present. The more extensive the array of techniques used, the more confident the researcher can be with the outcome. In addition, such characterization, when carried out on a regular basis, provides important information on the long-term stability of the cell line.

6.2.1 Morphology

The general appearance of a cell line is usually the first and most fundamental approach. Although it is possible to determine the cell type of many lines to a reasonable degree of accuracy by the experienced cell biologist, it must be borne in mind that morphology may change depending on, for example, culture conditions, i.e. type of media and substrate, cell density, and increasing passage number. Broadly speaking, epithelial cells are very regular, polygonal and display an overall tight pavement pattern (Figure 6.1(a)) with one or two stratified cell layers present in keratinocytes. Mesothelial cells may grow in a dispersed fashion and are not closely adherent as in many epithelial cell types, due to the absence of desmosomes. Human fibroblasts often appear as long spindly cells with a multipolar or bipolar shape. At confluency, swirling patterns (due to cellular orientation) may become obvious when these cells are viewed under low magnifications (Figure 6.1(b)). Neural cells may have numerous fine processes that link and intertwine with adjacent cells. Endothelial cells are frequently described as 'cobblestone' in appearance and a 'hill and valley' growth pattern has occasionally been used to identify muscle cell cultures.

6.2.1.1 Phase contrast microscopy

Cells can be viewed in tissue culture dishes using phase contrast microscopy. The phase contrast microscope permits one to observe considerable detail in untreated material and hence is suited to the study of living cells. Phase contrast microscopy provides the means to study a dynamic cell population on a regular basis. It is sometimes possible to distinguish different cell types using this simple approach.

Procedure: Viewing cell lines
1. Cell lines are directly viewed in the culture vessel in which they are maintained, e.g. flasks, Petri dishes or multiwell plates.
2. Observation should not take more than a few minutes, the cells being returned to the incubator as soon as possible, particularly when using unsealed culture ware, to avoid fluctuations in pH and temperature.
3. Examination should take place on a daily basis to take into account changes in morphology from the time of seeding the vessel to confluency. Morphology, as demonstrated by phase contrast microscopy, of epithelial and fibroblastic cell lines is shown in Figure 6.1.

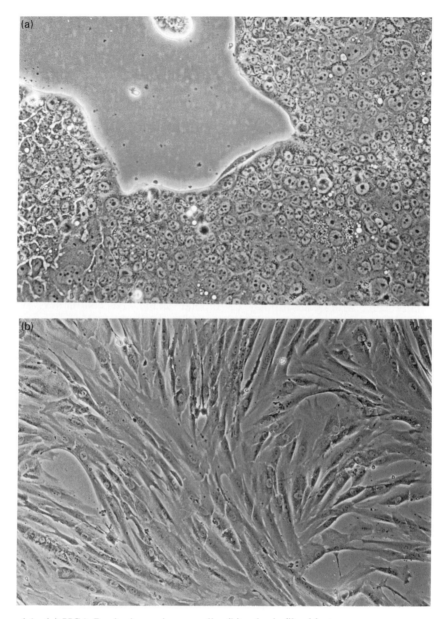

Figure 6.1. (a) HCA-7 colonic carcinoma cells, (b) colonic fibroblasts.

6.2.1.2 Light microscopy

Cells are usually transparent and difficult to view in detail using the light microscope as all parts of the specimen have almost the same optical density. This necessitates the incorporation of colour into the cells and although this usually means that the cells must be fixed, the advantage is that many structures can be visualized that are not apparent under phase contrast. A general staining procedure for cell visualization is outlined below. There are, however, an extensive array of stains that are more specific for different chemical constituents of the cell, for example haematoxylin for cell nuclei, PAS for carbohydrates and Sudan Black for lipids (Bancroft and Stevens, 1982). In addition, specific stains can help to further characterize a cell line with regard to features and products related to differentiation (e.g. mucous, see Table 6.3).

Procedure: General cell staining

Monolayer cultures
1. Following trypsinization (see Chapter 5), seed cells onto convenient substrates, e.g. sterile (autoclaved) glass coverslips or transparent tissue culture coverslips placed into multiwell plates.
2. Culture for required time period (at least 24 h to allow adhesion). Proceed as below.

Suspension cultures
1. Centrifuge cells, resuspend pellet in a small volume of medium.
2. Adhere cells to glass microscope slide using cytospin, drop, filter, or smear method (Freshney, 1994).
3. Rinse cells in PBS.
4. Fix in ethanol (crude or absolute), 10 min.
5. Stain in 1% Toluidine Blue or 1% Crystal Violet in 30% (v/v) alcohol for 1 min.
6. Wash in distilled water.
7. Mount in hydromount, allow to dry and view or alternatively:
8. Immerse in 70% alcohol, 5 min.
9. 100% alcohol 5 min.
10. Clear in TCF (tissue clearing fluid) and mount.

6.2.1.3 Transmission electron microscopy

This enables visualization of desmosomes, tight junctions and microvilli in epithelia (see Figure 6.2), Weibel-Palade bodies in endothelia (Heffelfinger *et al.*, 1992), and other such tissue-specific components. Cells can be processed as intact monolayers, as centrifuged pellets or as suspensions in supports such as agar. Monolayers can be folded up providing many different sectioning angles or uniform transverse and longitudinal sections can be obtained.

Procedure: Handling cell lines for electron microscopy

A detailed description of electron microscopy is beyond the scope of this chapter, in any case many workers have access to electron microscopy units and thus only a general outline aimed specifically at dealing with cell lines is provided below. (For further information see Glauert, 1984, pp. 90–9.)

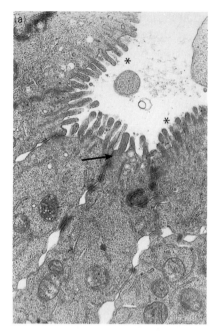 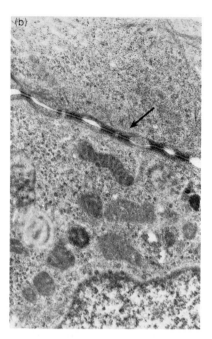

Figure 6.2. HRA-19 rectal carcinoma cells grown as xenografts, (a) demonstrating microvilli (*) and tight junctions (→) in polarized epithelial cells, and (b) desmosomes (→).

1. For suspension cultures pellet approx. 1×10^6 cells by centrifuging at 700 r.p.m. (100 *g*) 10 min. Monolayers can be processed *in situ* in Petri dishes or on glass coverslips or Millipore filters.

2. Remove culture medium and wash in 0.1 M phosphate or cacodylate buffer, pH 7.2, containing 2 mM calcium chloride, at 4 °C.

3. Fix in same buffer containing 1–3% (v/v) glutaraldehyde (dilute from 25% (v/v) stock solution), for 30–45 min at room temp. For suspension cultures, keep the pellet intact by slowly running the fixative down the side of the centrifuge tube.

4. Post-fix in 1% (w/v) osmium tetroxide in buffer.

5. Wash and dehydrate in alcohol. Use a series of solutions containing increasing concentrations of ethanol from 30% (v/v) to 100%.

6. Embed. Inverted capsules full of resin can be positioned over cell monolayers on glass slides. After polymerization, trim away excess resin, remove the coverslip and, if required, re-embed the monolayer for sectioning perpendicular to the plane of culture. Drop pellets into the resin and allow to settle. Following polymerization, trim away excess resin.

7. Cut sections using an ultramicrotome and mount on copper grids, stain with uranyl acetate and lead citrate.

6.2.2 Intermediate filament proteins

The eukaryotic cytoskeleton is broadly comprised of three types of filaments: actin filaments (6 nm diameter), microtubules (23 nm diameter) and the intermediate filaments or IFs (of 7–22 nm diameter). Most IF proteins form a threadlike network around the nucleus, extending out to the cell periphery and originally it seemed likely that their principal function was to provide mechanical support. It is becoming apparent, however, that certain IFs may be involved in specialized functions related to the differentiation state of the cell (Li *et al.*, 1994). IFs are classified on the basis of sequence homologies within their α-helical domains and as many as nine distinct types within five classes have been described (Fuchs and Weber, 1994; Nagle, 1994). Most are tissue specific and as their expression is often conserved *in vitro*, IF analysis is one of the most widely used methods for characterizing cell lines and indeed in some instances has been the only criterion by which cell lines could be distinguished (Möbus *et al.*, 1994). However, there have been an increasing number of reports of aberrant or coexpression of IFs by cell lines and this highlights the need for caution in interpreting results and again demonstrates the importance of using as many means as possible for cell line characterization. The IFs that are most frequently reported *in vitro* are described below.

6.2.2.1 Epithelial cell lines

These can usually be identified by the presence of cytokeratins. There are at least 30 keratins in normal tissue, ranging in size from 40–67 kDa and their distribution is generally restricted to epithelia. There are two subfamilies: the acidic keratins and the neutral or basic keratins. Each type of epithelial tissue expresses a characteristic pattern of keratins and these occur as pairs, an acidic and basic keratin always forming a heteropolymer (Hatzfeld and Franke, 1985). To date, there are 20 keratins in human epithelia (Moll, Franke and Schiller, 1982; Moll, Schiller and Franke, 1990) and eight more in hair and nails. CK8 and CK18 are characteristic of simple epithelia whilst CK5 and CK14 are common to stratified squamous epithelia (Fuchs and Weber, 1994). In addition to epithelial cell lines retaining positivity for their IF class, many have been shown to display a complex pattern of expression resembling that of normal tissue; examples being some cell lines from the ovary (Möbus *et al.*, 1992), breast (Taylor-Papadimitriou *et al.*, 1989), and colon (Rafiee *et al.*, 1992).

6.2.2.2 Cell lines of mesenchymal origin

These express the 52 kDa IF vimentin. Vimentin is conserved *in vitro* by cells including fibroblasts (Franke *et al.*, 1978), lymphoid cells (Zauli *et al.*, 1988), melanoma cells (Trejdosiewicz *et al.*, 1986) and endothelial cells (Heffelfinger *et al.*, 1992). However, many non-mesenchymal cells such as epithelial, muscle and neuronal may also express vimentin in addition to, or in place of, their own tissue-specific IF as soon as they are placed into the culture environment (Lazarides, 1982; Traub, Nelson and Traub, 1983; Pieper *et al.*, 1992).

6.2.2.3 Muscle cells

The 55 kDa protein desmin is found in cardiac, skeletal and smooth muscle cells. It is one of the earliest myogenic markers both *in vivo* and in culture (Kaufman and Foster, 1988) and would appear to play a key role in differentiation (Li *et al.*, 1994).

6.2.2.4 Glial cells and astrocytes

The 51 kDa glial fibrillary acidic protein (GFAP) is a glial-specific IF protein that is expressed in astrocytes in the central nervous system as well as in astrocytoma cell lines (Dickson *et al.*, 1983). However, the majority of these cell lines lose GFAP expression after a few passages *in vitro* although exceptions have been documented (Knott *et al.*, 1990).

6.2.2.5 Neuronal cells

Three neurofilaments are coexpressed in neurones, axons and dendrites: NF-H (heavy), NF-M (medium) and NF-L (light), with respective molecular weights of 200, 150 and 70 kDa. They are well conserved *in vitro* (Lee *et al.*, 1991; Gazzit, He and Graham-Pole, 1992). In general, cell line IFs are analysed using SDS PAGE, Northern or Western blotting, electron microscopy (Spiess, Zimmermann and Lünsdorf, 1987) or by immunocytochemistry. The latter technique is discussed in detail in another chapter, however, a simple protocol specifically for intermediate filaments is given below. Table 6.1 outlines some of the commonly used antibodies to intermediate filament proteins, though there are many more available, particularly for specific keratin polypeptides.

Procedure: Immunofluorescent cytochemistry for intermediate filaments
1. Grow cells on glass (not plastic) coverslips to subconfluent densities (approximately 70% confluence). For suspension cultures cytospin cells onto glass slides.
2. Place coverslips or slides into racks and rinse with PBS.
3. Fix in methanol (5 min, −20 °C) and subsequently 3 min in acetone (−20 °C) (If required, the preparations can now be air dried and stored frozen).
4. Rinse in PBS prior to next step.
5. Block with 10% (v/v) normal serum* (same species as secondary antibody).

Table 6.1. *Commonly used antibodies to intermediate filament proteins*

	Clone	Reference and commercial source
IF		
Keratins (broad range)	AE1/AE3 mix	Woodcock-Mitchell *et al.*, 1982; ICN
Vimentin	V9	Osborn, Debus and Weber, 1984; Dako/ICN/Sigma
Desmin	DE-U-10	Debus, Weber and Osborn, 1983*a*; Sigma/ICN
GFAP	G-A-5	Debus, Weber and Osborn 1983*b*; Sigma/ICN/Dako
NF		
NF-L	NR4	Debus *et al.*, 1983*b*; Sigma
NF-M	NN18	Franke *et al.*, 1991; Sigma
NF-H	NE14	

6. Incubate for 60 min with desired primary antibody* diluted in PBS to recommended concentration (some useful antibodies are listed in Table 6.1). For controls the primary antibody is omitted.
7. Wash with PBS (3 × 5 min).
8. Incubate for 30 min with appropriate fluorescein isothiocyanate (FITC) labelled secondary antibody* in PBS at recommended dilution.
9. Wash with PBS (3 × 5 min).
10. Mount coverslip onto a glass slide using 50% glycerol in PBS.
11. View using fluorescence microscope.

*100 μl volumes are sufficient to cover a 22 × 22 mm coverslip providing that incubations are carried out in a humidity box.

Figure 6.3 demonstrates intermediate filaments visualized using fluorescent immunocytochemistry.

6.2.3 Tissue-specific cell surface antigens

Tissue-specific antigens are characterized by the development of monoclonal antibodies after immunization of mice with cells or cell membrane preparations. Thus a great number of antibodies exist and they provide a further means to characterize cell lines. Some are listed in Table 6.2.

6.3 Differentiation status

Many cells retain the ability *in vitro* to express markers that are specific for their tissue and lineage of origin. These markers can often be used to identify a

Table 6.2. *Tissue-specific cell surface antigens useful for characterizing cell lines*

Tissue	Name of antigen/antibody	References
Epithelial		
	EMA (epithelial membrane antigen)/ PEM (polymorphic epithelial mucin) Recognized by abs HMFG-1 and HMFG-2	Heyderman, Steele and Ormerod, 1979 Taylor-Papadimitriou *et al.*, 1981
	EPG/Epg 34 (epithelial glycoprotein antigen) recognized by ab HEA 125	Moldenhauer *et al.*, 1987
	EPM-1(membrane antigen of epithelial cells)	Dippold *et al.*, 1987
Endothelial		
	Ulex europaeus agglutinin 1 (UEA1) antigens recognized by EN4, and H4-7/33 endothelial cell adhesion molecules CD31 and PAL-E	Hewett and Murray, 1993; Ades *et al.*, 1992
Haemopoietic		
	Leukocyte antigens	Barclay *et al.*, 1993
Muscle		
	Myoblast cell surface antigen 24.1D5 H36	Gower *et al.*, 1989 Kaufman *et al.*, 1985
Neural		
	GE2 and BF7 (glial) A2B5	Schnegg *et al.*, 1981 Eisenbarth *et al.*, 1979

Figure 6.3. Immunofluorescence staining of cytokeratin in human epidermal keratino-cytes. With kind permission of Dr Bill Otto, ICRF Histopathology Unit, Imperial Cancer Research Fund, London, UK.

Table 6.3. *Features and products useful for characterizing cell lines*

Cell lines	Product/feature	Reference
Epithelial		
General	Desmosomes	Freshney, 1992
	Dome formation	
	Mucus	
Colon	Carcinoembryonic antigen (CEA)	Freshney, 1992
	Villin	
	Sucrase isomaltase (SI)	
	Vasointestinal peptide (VIP)	
Epidermis	Involucrin	Freshney, 1992
Prostate	Prostatic acid phosphatase	Freshney, 1992
Liver	Transferrin	Roberts *et al.*, 1994
	α-fetoprotein	
	Glucose-6-phosphatase	
	Serum albumin	
Mesenchymal		
Endothelium	von Willebrand's factor – tube formation	Ades *et al.*, 1992
	on matrigel	Hewett and Murray, 1993
	Angiotensin converting enzyme (ACE)	
	Take up acetylated low-density lipoprotein	
Melanocytes	Melanin for melanocytes	Trejdosiewicz *et al.*, 1986
Others		
Muscle	Myosin	Caviedes *et al.*, 1992
	Myoglobin	
	Formation of myotubes	
Glial	Glutamine synthetase	Knott *et al.*, 1990
Neuronal	Neuron-specific enolase	Ronnett *et al.*, 1990
	Various neurotransmitters	
Endocrine	Hormone peptides, e.g. secretin and	Rindi *et al.*, 1990
	glicentin	

particular cell type (Table 6.3). Sometimes these differentiation markers are expressed by cells growing as monolayers on tissue culture plastic, in culture medium supplemented with 10% foetal calf serum. For example, dome formation, a manifestation of active ion transport, has been observed in monolayers of several cell lines derived from transporting epithelia (Kirkland, 1985) (Figure 6.4). However many cell types require more specialized culture conditions before they are able to express certain differentiated features of a particular tissue or cell lineage. These specialized culture conditions frequently use more physiological substrates, sometimes incorporating one or more basement membrane components with or without modifications to the culture medium.

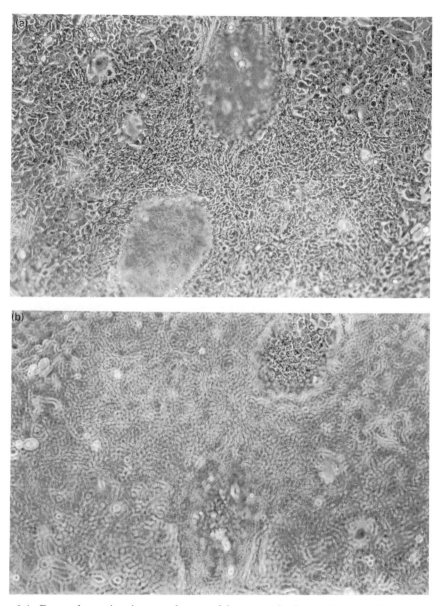

Figure 6.4. Dome formation in monolayers of human colonic carcinoma cells due to vectorial fluid transport, (a) focus on the monolayer, (b) focus on the top of the dome.

6.3.1 Porous cell culture substrates

Many epithelial cells are polarized with two different membrane domains, apical and basolateral. *In vivo* these cells receive nutrients from the basolateral direction and many growth factor receptors are localized exclusively in this membrane domain. When cells are grown on tissue culture plastic the formation of a polarized epithelial monolayer results in the exclusive delivery of nutrients and serum factors to the apical surface. Growth of cells on porous membranes overcomes this problem and allows the delivery of nutrients and other factors to both cell surfaces. Cells grown on porous membranes have altered morphology and protein expression compared to cells grown on tissue culture plastic. Madin–Darby canine kidney (MDCK) renal cells demonstrate a more columnar organization and form a partial basement membrane when transferred to microporous cell culture inserts (Cook *et al.*, 1989). These cell culture inserts are commercially available from many suppliers of tissue culture plastics. These culture inserts or Transwells are composed of a filter mounted in a plastic cylinder (Figure 6.5) and are used in multiwell plates. A range of filters with different pore sizes (0.45 μm, 1 μm and 8 μm) are available for different applications (Collaborative Biomedical Products, Becton Dickinson). Small pore sizes used in experiments will allow access of soluble factors and are useful for co-culture experiments where one cell type is grown on the filter and the other cell type grown in the wall. The 8 μm filters allow cells to pass through and are used in invasion assays as outlined later in this chapter. One great advantage of porous substrates is that the apical and basolateral membrane domains can be accessed independently. This makes such systems powerful tools in the investigation of protein sorting and ion transport in polarized epithelial cells.

6.3.1.1 Ion transport

Epithelial cells grown on coated millipore filters can be mounted in an Ussing chamber for ion transport investigations (Cuthbert *et al.*, 1987). Secretagogues can be applied exclusively to apical or basolateral surfaces to locate the position of functional receptors.

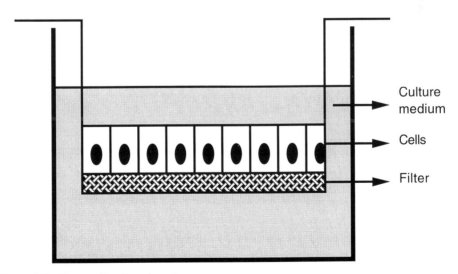

Figure 6.5. Transwell culture insert.

6.3.2 Extracellular matrix

Extracellular matrix (ECM) is the complex network of macromolecules surrounding all cells *in vivo*, and which has been shown to control many aspects of cell behaviour. The components of extracellular matrix include collagens, cell adhesive molecules such as laminin or fibronectin and proteoglycans. Signals from the extracellular matrix are transmitted to the cell largely via integrins, a group of transmembrane receptors composed of α and β subunits (Hynes, 1992). Extracellular matrix is an important regulator of epithelial development and differentiation (Adams and Watt, 1993; Lin and Bissell, 1993). The ability to investigate the effects of extracellular matrix molecules has been greatly improved by the availability of refined cell culture techniques. In cell culture, extracellular matrix molecules can be used to coat tissue culture plastic, to form three-dimensional matrices or to add directly to culture medium.

6.3.2.1 ECM-coated plastic

Tissue culture plasticware can be coated with extracellular matrix molecules by incubating a dilute solution of collagen, laminin or fibronectin with the plastic. The solution is then aspirated and the plastic washed before use. Commercially available matrix molecules will have a recommended coating procedure although this must only be used as a guideline. Preliminary experiments with different coating concentrations and incubation times will be required to arrive at the optimal concentration for a particular application. Cells grown on such coated dishes may differ markedly from those grown on untreated tissue culture plastic. Freshly isolated hepatocytes quickly lose their characteristic morphology on tissue culture plastic. By precoating dishes with Type I collagen, cells assume a polygonal shape (Bissell and Guzelian, 1980).

(a) Adhesion assays

ECM-coated plastic can also be used to determine cell-matrix adhesive properties. Single-cell suspensions are added to coated dishes and incubated for varying periods of time to allow cells to attach, then non-adherent cells are removed by gently washing with PBS (Pignatelli and Bodmer, 1989: Adams and Watt, 1991). The number of adherent cells gives a quantitative assay of the ability of cells to bind to a particular extracellular molecule.

6.3.2.2 Three-dimensional matrices

(a) Type I collagen gels

Three-dimensional gels of Type I collagen have been widely used to investigate the organization and differentiation of a variety of cell types. Cells can be grown on collagen gels or embedded within the gel where organization of cells into three-dimensional structures frequently occurs. To prepare cultures where cells are grown on gels, cell suspensions are added after the collagen has set. Gels can subsequently be left attached to the plastic or detached and floated in the medium. Gene expression in cells on fixed collagen gels has been shown to differ from that on floating gels. Mammary epithelial cells assume a flat morphology on tissue culture plastic or fixed Type I collagen gel and do not express β-casein. However, when the gel is floated, β-casein is expressed (Emerman *et al.*, 1977). This effect was shown to correlate with endogenous basement membrane production by the mammary epithelial cells (Streuli and Bissell, 1990).

Many cell types undergo morphogenesis in collagen gels with the formation of glandular structures by some epithelial cells (Figure 6.6) (Pignatelli and Bodmer, 1988; Kirkland, 1988*a*) and tubular structures by endothelial cells (Montesano, Orci and Vassalli, 1983). Although Type I collagen is commercially available, it is easily prepared from rat tail collagen fibres.

Procedure: Preparation of Type I collagen gel cultures

1. 1 g of rat tail collagen fibres are suspended in 300 ml of 0.01 M acetic acid.
2. The suspended fibres are stirred for 3 days at 4 °C.
3. The solution is centrifuged at 600 *g* for 2 h to remove insoluble material.
4. The resulting clear solution can be stored for several weeks at 4 °C after irradiation in a ^{137}Cs source, or freeze dried and stored at −20 °C until use. Freeze dried collagen is redissolved in 0.01 M acetic acid (0.3 g per 100 ml) before use.

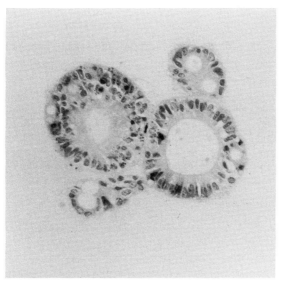

Figure 6.6. Glandular organization by human rectal carcinoma cells grown in Type I collagen gels.

5. To prepare gels, the collagen solution is polymerized by the addition of a mixture of two parts 10 × Eagles medium (Gibco) and one part 0.34 M NaOH, until pH 7 is reached. The colour of the solution will indicate that the correct pH has been reached (Phenol Red indicator present).
6. The neutralized gel is then placed in tissue culture wells and allowed to gel (1 h at 37 °C). Culture medium containing cells is added on top of the gel. If cells are to be embedded inside the gel they are resuspended in the freshly neutralized collagen gel solution and plated into wells. Again the collagen is allowed to gel for 1 h at 37 °C before the addition of culture medium.

(b) Basement membrane-like matrices

Matrigel (Collaborative Biomedical Products, Becton Dickinson) is a commercially available basement membrane extract of the Engelbreth–Holm–Swarm (EHS) tumour (Kleinman *et al.*, 1982). The major components of matrigel are laminin, Type IV collagen, entactin and heparan sulphate proteoglycan. Growth of cells in gels of this basement membrane-like material promotes the differentiation of many cell types including hepatocytes and mammary epithelial cells (Lin and Bissell, 1993). Mammary epithelial cells form alveoli-like structures and express late milk proteins in response to EHS. Similarly, hepatocytes have a more differentiated phenotype on EHS gel than on Type I collagen gel. However, it should be noted that Matrigel contains a variety of growth factors including transforming growth factor-β, epidermal growth factor and fibroblast growth factor. The wide-ranging effects of these factors on cellular behaviour should be taken into account when interpreting results from experiments using Matrigel (Vukicevic *et al.*, 1992).

Procedure: Preparation of matrigel matrices

Matrigel is supplied as a frozen solution and will gel rapidly when warmed. For this reason, Matrigel is thawed at 4 °C and kept on ice during use. Pre-cooled pipettes and plasticware are also required. If cells are to be embedded in Matrigel then they have to be very rapidly resuspended in it in a cooled tube before adding to wells. Gels are incubated for 1 h at 37 °C before addition of culture medium. They should be fed carefully as they are much less robust than collagen gels and can easily be broken.

6.3.2.3 ECM Components as culture medium additives

Extracellular matrix components can be added as dilute solutions directly to culture medium. This can prove a useful way of looking at the effects of ECM molecules without changes in cell shape, which are frequently observed when cells are grown on collagen-coated dishes or within collagen gels. Addition of EHS-derived basement membrane proteins or purified laminin to hepatocytes grown on tissue culture plastic will induce albumin expression with little change in cell shape (Caron, 1990), while addition of Type I collagen to endothelial monolayers promotes tube formation (Jackson and Jenkins, 1991).

6.3.3 Mesenchymal cells

Epithelial–mesenchymal interactions have an important role in both morpho-genesis and differentiation of a variety of epithelial cells (Cunha, 1976; Kedinger *et al.*, 1986). Stromal cells have been shown to play a key role in the control of haematopoiesis (Dexter *et al.*, 1990). One mechanism for this interaction is the production of soluble factors by fibroblasts that could act in a paracrine way modulating cell behaviour in adjacent epithelial cells (Finch *et al.*, 1989).

Feeder layers of mesenchymal cells are useful in the growth of some primary cultures. To prepare feeder layers, cells are treated either with radiation or drugs such as Mitomycin C to prevent cell proliferation then added either before or along with other cell populations. 3T3 cells have been used as feeder cells for the growth of human epidermal keratinocytes (Rheinwald and Green, 1975), human colon carcinoma cell lines (Brattain *et al.*, 1982) and for cell cloning (Kirkland, 1988*b*), where they are thought to secrete 'conditioning' factors that promote the survival of cells in isolation.

6.3.4 Culture medium

The majority of cells are grown *in vitro* in a basic culture medium such as Dulbecco's Eagles supplemented with foetal calf serum (FCS). However, this may not provide the optimal conditions necessary for the proliferation and differentiation of all cell types. For example, growth of epidermal keratinocytes was facilitated by the use of 3T3 feeder cells and addition to the culture medium of 10% FCS, adenine, insulin, hydrocortisone, choleratoxin and EGF (Rheinwald and Green, 1975; Adams and Watt, 1991).

Although most cells require the presence of serum for survival *in vitro*, the complex composition and variability between different batches of serum can complicate the interpretation of experiments. In addition, foetal calf serum has been shown to inhibit cell proliferation in fetal rat intestinal epithelial cells and to promote the growth of mesenchymal cells (Fukamachi, 1992). Great ad-vances have been made in the development of serum-free or defined medium, where serum is substituted by the addition of growth factors, hormones and transport proteins (Hayashi and Sato, 1976). Serum-free conditions have been established for a variety of cell types, however the requirements for long-term cell growth or growth of low density cultures are usually more stringent than short-term or high density cultures. The exact composition of serum free medium will depend upon the individual cell type. Optimal conditions are reached by starting with multiple additions and sequentially removing single additives to determine cell requirements for proliferation and differentiation (Barnes and Sato, 1980).

6.4 Transformation

Cell transformation, *in vitro*, is characterized by permanent phenotypic change that can occur spontaneously or in response to exposure to transforming viruses or chemical carcinogens. A number of phenotypic characteristics are associated with a transformed status although any one cell strain may not express all the characteristics.

6.4.1 Immortalization

Immortalization is the acquisition of an infinite lifespan, whereas normal cells become senescent after a certain number of passages *in vitro* (Hayflick and Moorhead, 1961). Many immortalized cell lines have less stringent growth requirements and frequently display a more autonomous growth control with lower serum dependency. In some cases, cells have been shown to secrete mitogens for which they possess receptors (autocrine growth control) (Sporn and Todaro, 1980). Some immortalized cell lines such as 3T3 cells initially maintain contact inhibition but can subsequently undergo further changes. When 3T3 cells are maintained at confluence for extended periods, then small modified foci with reduced contact inhibition rapidly overgrow the culture. These foci are frequently tumourigenic (Todaro and Green, 1963).

6.4.2 Anchorage independence

Many tumour cell lines demonstrate an ability to grow without attachment to a substrate either in suspension cultures or semi-solid gels of agar or agarose. The ability of a small fraction of primary human tumour cells to grow clonally in agar formed the basis of the widely used clonogenic assay for human tumour stem cells (Hamburger and Salmon, 1977, see Chapter 7, this volume). However the subsequent demonstration that normal endothelial cells (Laug *et al.*, 1980) and fibroblasts (Freshney, 1994) are also able to grow in agar indicates that normal cells are also capable of anchorage independent growth under certain circumstances. Therefore colonies arising from plating cell suspensions from primary tumours should be thoroughly characterized to ensure that they are of tumour origin.

   Although some investigations have shown a correlation between tumourigenicity and anchorage independent growth (Brattain *et al.*, 1981) other studies on a range of human tumours have shown that anchorage independent growth is not a good marker for tumourigenicity (Marshall, Franks and Carbonell, 1977; Paraskeva *et al.*, 1984). In addition, Rheinwald and Beckett (1980) have shown that human squamous carcinoma cell lines failed to grow in suspension and became committed to differentiation, albeit at a slower rate than normal human keratinocytes.

Although anchorage independent growth appears to be a poor predictor of tumourigenicity in carcinoma cell lines, it has been shown to have value in assaying for neoplastic transformation of established cell lines by tumour viruses (Shin *et al*., 1975).

6.4.2.1 *Preparation of suspension cultures*

Cells can also be grown in suspension in standard culture medium by seeding them onto non tissue culture treated polystyrene or dishes coated with a thin layer of agarose or agar (liquid overlay culture) (Yuhas, 1977; Carlsson and Yuhas, 1984), which prohibits attachment of most cell types. For many cell types, particularly those of carcinoma origin, failure to attach to the substrate is followed by aggregation and organization into multicellular spheroids. Spheroids may be propagated on Petri dishes but large spheroids may require growth in spinner flasks to help prevent cell death in the centre of the spheroids. Such spheroids have been used as models of tumour nodules in studies of the effects of radiation and drugs as they more closely resemble poorly vascularized tumour clumps than monolayer cultures (Sutherland *et al*., 1981). Cell polarization within spheroids has been demonstrated for cell types such as thyroid epithelial cells and colorectal carcinoma cells. For these cell types the apical cell surface is at the edge of the spheroid and the basolateral surface is within the spheroid. The reversal of this polarity is observed when these spheroids are embedded in collagen gel (Chambard, Gabrion and Mau-champ, 1981; Kirkland, 1988*a*). Such spheroids are useful models for studying the establishment of such cell polarity and the potential importance of cell polarity on antibody localization during *in vivo* immunotargeting (Pervez *et al*., 1989).

6.4.3 *Tumourigenicity*

Tumourigenicity is measured by the ability of cells to form tumours in either congenitally athymic (nude) mice (Giovanella, Stehlin and Williams, 1974) or mice immune-deprived by thymectomy and irradiation (Selby *et al*., 1980). However it must be stressed that failure to form xenografts does not preclude a transformed phenotype. Cells are usually injected subcutaneously but other sites have been used. The number of cells required to form a tumour varies greatly from one cell type to another but generally between 10^6 and 10^7 cells/site are used. The time taken for tumour formation can also vary greatly between cell types, while most tumourigenic cells form tumours within 3 or 4 months, some cells take considerably longer (Marsh, Stamp and Kirkland, 1993).

Co-injection of cells with basement membrane components (Topley *et al*., 1993) or co-inoculation of fibroblasts with the tumour cells (Picard, Rolland and Poupon, 1986) have been used to improve take rates and tumour growth.

A detailed examination of tumour histology and invasive characteristics can provide a great deal of information on the transformed cell phenotype (Marsh et al., 1993).

6.4.4 Invasion and metastasis

Metastasis of tumour cells to distant sites is a complex biological process involving many individual steps, however the mechanisms involved in this cascade are poorly understood. The clinical importance of metastasis has directed a great deal of research effort into establishing valid in vivo models of metastasis. In addition, the perturbations of cell–matrix interactions that allow tumour cells to invade surrounding normal tissues are being widely investigated using in vitro models of invasion.

6.4.4.1 In vivo models

Experimental models for cancer cell invasion and metastasis have frequently utilized the growth of murine tumours in syngeneic animals, either subcutaneously, following tail vein injection, or by splenic implantation. Using such methods, lung colonies have been described following intramuscular or intravenous injection of Lewis lung carcinoma cells (van Lamsweerde, Henry and Vaes, 1983). Similar results have been obtained with murine carcinoma cells (Tsuruo et al., 1983). Metastasis from xenografts of human carcinomas in nude athymic mice is rarely observed, with only a limited number of carcinomas able to metastasize to distant sites (Kozlowski et al., 1984). Although the inherent characteristics of the carcinoma cells clearly control the ability of cells both to grow and metastasize in nude mice, other factors such as the age of the nude mice and the exact site of tumour implantation play an important part in modulating expression of growth and invasion by the carcinoma cells (Kyriazis and Kyriazis, 1980; Kozlowski et al., 1984). When investigating the metastatic behaviour of cells the possible target organs such as the liver should be investigated histologically, which may reveal small deposits that cannot be observed by macroscopic observation of the whole organ.

An interesting model of metastasis has been developed for colorectal carcinoma, where liver metastases have been demonstrated after the injection of human carcinoma cells into the caecal wall of young athymic mice (Bresalier et al., 1987). In contrast, little or no visceral organ metastasis could be demonstrated by the same cells after subcutaneous injection and introduction of cells intravenously would enable cells to bypass some of the steps in the complex metastatic cascade.

6.4.4.2 In vitro models of invasion

In vitro invasion assays have used organ fragments, cell cultures and membranes such as chorioallantoic membrane of the chick (Easty and Easty, 1974) and human amnion (Liotta, Lee and Morakis, 1980), invasion being measured either histologically or using radiolabelling cells. However, it has proved difficult to generate reproducible data on the invasive capability of cells using these methods and new techniques have been developed using reconstituted basement membrane (RBM). Gels of RBM grown on filters yield a barrier of consistent thickness and formed the basis of quantitative assays for invasion (Figure 6.7) (Hendrix *et al.*, 1987). The most frequently used matrix in these assays is Matrigel, an extract of the Engelbreth–Holm–Swarm tumour containing laminin, entactin, Type IV collagen and heparan sulphate proteoglycans. Transwell chambers coated with Matrigel are available commercially (Collaborative Biomedical Products, Becton Dickinson).

Direct comparison of the human amnion model with the reconstituted basement membrane system showed that fewer cells invaded the human amnion compared to the RBM (Hendrix *et al.*, 1989). In addition, these authors demonstrated substantially more variability in data from the amnion assay, attributed to differences in membrane thickness. One of the most interesting features of the RBM assay is that invading cells can be collected after traversing the matrix, thus enabling isolation of subpopulations with greater invasive features. However, it should be noted that increased invasive behaviour of cells in the RBM system does not always correlate with metastatic potential *in vivo* (Hendrix *et al.*, 1989; Noel *et al.*, 1991), indeed some normal cells are capable of passing through a matrigel barrier (Simon, Noel and Foidart, 1992). The expression of metastatic potential *in vivo* requires cells to complete a complex sequence of events, one component of which is the invasion of basement membrane. Therefore it is not surprising that cells selected solely for their ability to traverse a basement membrane are not always able to complete all the steps in the metastatic cascade. Reconstituted basement membrane assays remain a valid method for studying the complex

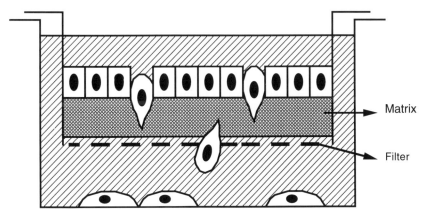

Figure 6.7. Cell invasion chamber. Cell invasion through reconstituted basement membrane on 8 μm filters. Invasive cells frequently attach to the base of the well and can be harvested for further experimentation.

cell–matrix interactions that enable cells to migrate through a basement membrane. Use of these techniques has established a role for motility factors (Stearns and Stearns, 1993) and fibroblasts (Saiki *et al.*, 1994) in the invasion process.

References

Adams, J. C. and Watt, F. M. (1991) Expression of $\beta1$, $\beta3$, $\beta4$ and $\beta5$ integrins by human epidermal keratinocytes and non-differentiating keratinocytes. *J. Cell Biol.* **115**, 829–41.

Adams, J. C. and Watt, F. M. (1993) Regulation of development and differentiation by the extracellular matrix. *Development* **117**, 1183–98.

Ades, E. W., Candal, F. J., Swerlick, R. A. *et al.* (1992) HMEC-1: Establishment of an immortalized human microvascular endothelial cell line. *Invest. Dermatol.* **99**, 683–90.

Bancroft, J. D. and Stevens, A. (1982) *Theory and practice of histological techniques*, 2nd edn. Churchill Livingstone, Edinburgh.

Barclay, A. N., Birkeland, M. L., Brown, M. H. *et al.* (1993) *The leucocyte antigen facts book*. Academic Press, London.

Barnes, D. and Sato, G. (1980) Methods for growth of cultured cells in serum free medium. *Anal. Biochem.* **102**, 255–69.

Bissell, D. M. and Guzelian, P. S. (1980) Phenotypic stability of adult rat hepatocytes in primary monolayer culture. *Ann. New York Acad. Sci.* **349**, 85–98.

Brattain, M. G., Brattain, D. E., Sarrif, A. M. *et al.* (1982) Enhancement of growth of human colon tumor cell lines by feeder layer of murine fibroblasts. *J. Natl Cancer Inst.* **69**, 767–71.

Brattain, M. G., Fine, W. D., Khaled, F. M. *et al.* (1981) Heterogeneity of malignant cells from a human colonic carcinoma. *Cancer Res.* **41**, 1751–6.

Bresalier, R. S., Raper, S. E., Hujanen, E. S. and Kim, Y. S. (1987) A new animal model for human colon cancer metastasis. *Int. J. Cancer* **39**, 625–30.

Carlsson, J. and Yuhas, J. M. (1984) Liquid-overlay culture of cellular spheroids. *Recent Results in Cancer Research* **95**, 1–23.

Caron, J. M. (1990) Induction of albumin gene transcription in hepatocytes by extracellular matrix proteins. *Mol. Cell Biol.* **10**, 1239–43.

Caviedes, R., Liberona, J. L., Hidalgo, J. *et al.* (1992) A human skeletal muscle cell line obtained from an adult donor. *Bioch. Biophys. Acta* **1134**, 247–55.

Chambard, M., Gabrion, J. and Mauchamp, J. (1981) Influence of collagen gel on the orientation of epithelial cell polarity: Follicle formation from isolated thyroid cells and from preformed monolayers. *J. Cell Biol.* **91**, 157–66.

Cook, J. R., Crute, B. E., Patrone, L. M. *et al.* (1989) Microporosity of the substratum regulates differentiation of MDCK cells in vitro. *In Vitro Cell Devel. Biol.* **25**, 914–22.

Cunha, G. R. (1976) Epithelial-stromal interactions in development of the urogenital tract. *Int. Rev. Cytol.* **47**, 137–94.

Cuthbert, A. W., Egleme, C., Greenwood, H. *et al.* (1987) Calcium and cyclic AMP dependent chloride secretion in human colonic epithelia. *Br. J. Pharmacol.* **91**, 503–15.

Debus, E., Weber, K. and Osborn, M. (1983*a*) Monoclonal antibodies to desmin, the muscle-specific intermediate filament protein. *EMBO J.* **2**, 2305–12.

Debus, E., Weber, K. and Osborn, M. (1983*b*) Monoclonal antibodies specific for glial fibrillary acidic (GFA) protein and for each of the neurofilament triplet polypeptides. *Differentiation* **25**, 193–203.

Dexter, T. M., Coutinho, L. H., Spooncer, E. *et al*. (1990) Stromal cells in haemopoiesis. In *Molecular control of haemopoiesis*. Ciba Foundation Symposium 148, pp. 76–86. Wiley, Chichester.

Dickson, J. G., Flanigan, T. P., Kemshead, J. T. and Walsh, F. S. (1983) Monoclonal antibodies reacting specifically with the surface of human astrocytes in culture. *Bioch. Soc. Trans.* **11**, 208.

Dippold, W. G., Bernhard, H., Klingel, R. *et al*. (1987) A common epithelial cell surface antigen (EPM-1) on gastrointestinal tumors and in human sera. *Canc. Res.* **47**, 3873–9.

Doyle, A. and Stacey, G. N. (1994) Authentification. Module 9A. In *Cell and tissue culture: Laboratory procedures* (ed. A. Doyle, J. B. Griffiths and D. G. Newell). John Wiley & Sons, West Sussex.

Easty, D. M. and Easty, G. C. (1974) Measurement of the ability of cells to infiltrate normal tissues in vitro. *Br. J. Cancer* **29**, 36–49.

Eisenbarth, G. S., Walsh, F. S. and Nirenberg, M. (1979) Monoclonal antibody to a plasma membrane antigen of neurons. *Proc. Natl Acad. Sci. USA* **76**, 4913–17.

Emerman, J. T., Enami, J., Pitelka, D. R. and Nandi, S. (1977) Hormonal effects on intracellular and secreted casein in cultures of mouse mammary epithelial cells on floating collagen membranes. *Proc. Natl Acad. Sci.* **74**, 4466–70.

Finch, P. W., Rubin, J. S., Miki, T. *et al*. (1989) Human KGF is FGF-related with properties of a paracrine effector of epithelial cell growth. *Science* **245**, 752–5.

Franke, F. E., Schachenmayr, W., Osborn, M. and Altmannsberger, M. (1991) Unexpected immunoreactivities of intermediate filament antibodies in human brain and brain tumours. *Am. J. Pathol.* **139**, 67–79.

Franke, W. W., Schmid, E., Osborn, M. and Weber, K. (1978) Different intermediate-sized filaments distinguished by immunofluorescence microscopy. *Proc. Natl Acad. Sci. USA* **75**, 5034–8.

Freshney, R. I. (1992) *Culture of epithelial cells*. Wiley-Liss, New York.

Freshney, R. I. (1994) *Culture of animal cells. A manual of basic technique*. Wiley-Liss, New York.

Fuchs, E. and Weber, K. (1994) Intermediate filaments: structure, dynamics, function and disease. *Ann. Rev. Bioch.* **63**, 345–82.

Fukamachi, H. (1992) Proliferation and differentiation of fetal rat intestinal epithelial cells in primary serum-free medium. *J. Cell Sci.* **103**, 511–19.

Gazzit, Y., He, Y. J. and Graham-Pole, J. (1992) A novel methodology for the establishment of neuroblastoma cell lines from metastatic marrow. Expression of surface markers, neurofilaments, MDR-1 and myc proteins. *J. Immunol. Meth.* **148**, 171–8.

Giovanella, B. C., Stehlin, J. S. and Williams, L. J. (1974) Heterotransplantation of human malignant tumors in nude thymusless mice II. Malignant tumors induced by injection of cell cultures derived from human solid tumors. *J. Natl Cancer Inst.* **52**, 921–30.

Glauert, A. M. (1984) *Practical methods in electron microscopy, Vol. 3. Part. 1: Fixation, dehydration and embedding of biological specimens*. North Holland, New York.

Gower, H. J., Moore, S. E., Dickson, G. *et al*. (1989) Cloning and characterization of a myoblast cell surface antigen defined by 24.1 D5 monoclonal antibody. *Development* **105**, 723–31.

Hamburger, A. W. and Salmon, S. E. (1977) Primary bioassay of human tumour stem cells. *Science* **197**, 461–3.

Hatzfeld, M. and Franke, W. W. (1985) Pair formation and promiscuity of cytokeratins: Formation *in vitro* of heterotypic complexes and intermediate-sized filaments by homologous and heterologous recombinations of purified polypeptides. *J. Cell Biol.* **101**, 1826–41.

Hayashi, I. and Sato, G. H. (1976) Replacement of serum by hormones permits growth of cells in a defined medium. *Nature* **259**, 132–4.

Hayflick, L. and Moorhead, P. S. (1961) The serial cultivation of human diploid cell strains. *Exp. Cell Res.* **25**, 585–621.

Heffelfinger, S. C., Hawkins, H. H., Barrish, J. *et al.* (1992) SK HEP-1: a human cell line of endothelial origin. *In vitro Cell. Devel. Biol.* **28A**(2), 136–42.

Hendrix, M. J. C., Seftor, E. A., Seftor, R. E. B. and Fidler, I. J. (1987) A simple quantitative assay for studying the invasive potential of high and low human metastatic variants. *Cancer Letters* **38**, 137–47.

Hendrix, M. J. C., Seftor, E. A., Seftor, R. E. B. *et al.* (1989) Comparison of tumor cell invasion assays: human amnion versus reconstituted basement membrane barriers. *Invasion Metastasis* **9**, 278–97.

Hewett, P. and Murray, J. C. (1993) Human lung microvessel endothelial cells: isolation, culture and characterization. *Microvascular Res.* **46**, 89–102.

Heyderman, E., Steele, K. and Ormerod, M. G. (1979) A new antigen on the epithelial membrane: its immunoperoxidase localisation in normal and neoplastic tissue. *J. Clin. Pathol.* **32**, 35–9.

Hynes, R. O. (1992) Integrins, versatility, modulation and signalling in cell adhesion. *Cell* **69**, 11–25.

Jackson, C. J. and Jenkins, K. L. (1991) Type 1 collagen fibrils promote rapid vascular tube formation upon contact with the apical side of cultured endothelium. *Exp. Cell Res.* **192**, 319–23.

Kaufman, S. J. and Foster, R. F. (1988) Replicating myoblasts express a muscle specific phenotype. *Proc. Natl. Acad. Sci. USA* **85**, 9606–10.

Kaufman, S. J., Foster, R. F., Haye, K. R. and Faiman, L. E. (1985) Expression of a developmentally regulated antigen on the surface of skeletal and cardiac muscle cells. *J. Cell Biol.* **100**, 1977–87.

Kedinger, M., Simon-Assmann, P. M., Lacroix, B. *et al.* (1986) Fetal gut mesenchyme inducer differentiation of cultured intestinal and crypt cells. *Dev. Biol.* **113**, 474–83.

Kirkland, S. C. (1985) Dome formation by a human colonic adenocarcinoma cell line (HCA-7). *Cancer Res.* **45**, 3790–95.

Kirkland, S. C. (1988*a*) Polarity and differentiation of human rectal adenocarcinoma cells in suspension and collagen gel cultures. *J. Cell Sci.* **91**, 615–21.

Kirkland, S. C. (1988*b*) Clonal origin of columnar, mucous and endocrine cell lineages in human colorectal epithelium. *Cancer* **61**, 1359–63.

Kleinman, H. K., McGarvey, M. L., Liotta, L. A. *et al.* (1982) Isolation and characterization of type IV procollagen, laminin and heparan sulfate proteoglycan from EHS sarcoma. *Biochemistry* **21**, 6188–93.

Knott, J. C., Edwards, A. J., Gullan, R. W. *et al.* (1990) A human glioma cell line retaining expression of GFAP and gangliosides recognized by A2B5 and LB1 antibodies after prolonged passage. *Neuropathol. Appl. Neurobiol.* **16**, 489–500.

Kozlowski, J. M., Fidler, I. J., Campbell, D. *et al.* (1984) Metastatic behaviour of human tumor cell lines grown in the nude mouse. *Cancer Res.* **44**, 3522–9.

Kyriazis, A. A. and Kyriazis, A. P. (1980) Preferential sites of growth of human tumors in nude mice following subcutaneous transplantation. *Cancer Res.* **40**, 4509–11.

Laug, W. E., Tokes, Z. A., Benedict, W. F. and Sorgente, N. (1980) Anchorage independent growth and plasminogen activator production by bovine endothelial cells. *J. Cell Biol.* **84**, 281–93.

Lazarides, E. (1982) Intermediate filaments: a chemically heterogeneous, developmentally regulated class of proteins. *Ann. Rev. Bioch.* **51**, 219–50.

Lee, H. J., Elliot, G. J., Hammon, D. N. *et al.* (1991) Constitutive expression of the mature array of neurofilament proteins by a CNS neuronal cell line. *Brain Res.* **558**, 197–208.

Li, H., Choudhary, S. K., Milner, D. J. *et al.* (1994) Inhibition of desmin expression blocks myoblast fusion and interferes with the myogenic regulators MyoD and Myogenin. *J. Cell Biol.* **124**, 827–41.

Lin, C. Q. and Bissell, M. J. (1993) Multi-faceted regulation of cell differentiation by extracellular matrix. *FASEB J.* **7**, 737–43.

Liotta, L. A., Lee, C. W. and Morakis, D. J. (1980) New method for preparing large surfaces of intact human basement membrane for tumour invasion studies. *Cancer Letters* **11**, 141–52.

Marsh, K. A., Stamp, G. W. H. and Kirkland, S. C. (1993) Isolation and characterisation of multiple cell types from a single human colonic carcinoma: tumourigenicity of these cell types in a xenograft system. *J. Path.* **170**, 441–50.

Marshall, C. J., Franks, L. M. and Carbonell, A. W. (1977) Markers of neoplastic transformation in epithelial cell lines derived from human carcinomas. *J. Natl Cancer Inst.* **58**, 1743–51.

Möbus, V., Gerharz, C. D., Press, U. *et al.* (1992) Morphological, immunohistochemical and biochemical characterization of 6 newly established human ovarian carcinoma cell lines. *Int. J. Cancer* **52**, 76–84.

Möbus, V. J., Moll, R., Gerharz, C. D. *et al.* (1994) Establishment of new ovarian and colon carcinoma cell lines: differentiation is only possible by cytokeratin analysis. *Br. J. Cancer* **69**, 422–8.

Moldenhauer, G., Momburg, F., Möller, P. *et al.* (1987) Epithelium-specific surface glycoprotein of Mr 34,000 is a widely distributed human carcinoma marker. *Br. J. Cancer* **56**, 714–21.

Moll, R., Franke, W. W. and Schiller (1982) The catalog of human cytokeratins: patterns of expression in normal epithelia, tumours and cultured cells. *Cell* **31**, 11–24.

Moll, R., Schiller, D. L. and Franke, W. W. (1990) Identification of protein IT of the intestinal cytoskeleton as a novel type 1 cytokeratin with unusual properties and expression patterns. *J. of Cell Biol.* **111**, 567–80.

Montesano, R., Orci, L. and Vassalli, P. (1983) In vitro rapid organisation of endothelial cells into capillary-like networks is promoted by collagen matrices. *J. Cell Biol.* **97**, 1648–52.

Nagle, R. B. (1994) A review of intermediate filament biology and their use in pathogenic diagnosis. *Mol. Biol. Rep.* **19**, 3–21.

Nelson-Rees, W. A., Daniels, D. W. and Flandermeyer, R. R. (1981) Cross contamination of cells in culture. *Science* **212**, 446–52.

Noel, A. C., Calle, A., Emonard, H. P. *et al.* (1991) Invasion of reconstituted basement membrane matrix is not correlated to the malignant metastatic cell phenotype. *Cancer Res.* **51**, 405–14.

Osborn, M., Debus, E. and Weber, K. (1984) Monoclonal antibodies specific for vimentin. *Eur. J. Cell Biol.* **34**, 137–43.

Paraskeva, C., Buckle, B. G., Sheer, D. and Wigley, C. B. (1984) The isolation and characterisation of colorectal epithelial cell lines at different stages in malignant transformation from familial polyposis coli patients. *Int. J. Cancer* **34**, 49–56.

Pervez, S., Kirkland, S. C., Epenetos, A. A. *et al.* (1989) Effect of polarity and differentiation on antibody localization in multicellular tumour spheroid and xenograft models and its potential importance for *in vivo* immunotargetting. *Int. J. Cancer* **44**, 940–7.

Picard, O., Rolland, Y. and Poupon, M. F. (1986) Fibroblast-dependent tumourigenicity of cells in nude mice: implication for implantation of metastases. *Cancer Res.* **46**, 3290–4.

Pieper, F. R., Van De Klundert, F. A., Raats, J. M. *et al.* (1992) Regulation of vimentin expression in cultured epithelial cells. *Eur. J. Biochem.* **210**, 509–19.

Pignatelli, M. and Bodmer, W. F. (1988) Genetics and biochemistry of collagen binding triggered glandular differentiation in a human colon carcinoma cell line. *Proc. Natl Acad. Sci.* **85**, 5561–5.

Pignatelli, M. and Bodmer, W. F. (1989) Integrin receptor mediated differentiation and growth inhibition are enhanced by transforming growth factor-β in colorectal tumour cells grown in collagen gel. *Int. J. Cancer* **44**, 518–23.

Rafiee, P., Ho, S. B., Bresalier, R. S. *et al.* (1992) Characterization of the cytokeratins of human colonic, pancreatic and gastric adenocarcinoma cell lines. *Pancreas* **7**, 123–31.

Rheinwald, J. G. and Beckett, M. A. (1980) Defective terminal differentiation in culture as a consistent and selectable character of malignant human keratinocytes. *Cell* **22**, 629–32.

Rheinwald, J. G. and Green, H. (1975) Serial cultivation of strains of human epidermal keratinocytes: the formation of keratinizing colonies from single cells. *Cell* **6**, 331–44.

Rindi, G., Grant, S. G. N., Yiangou, Y. *et al.* (1990) Development of neuroendocrine tumors in the gastrointestinal tract of transgenic mice. Heterogeneity of hormone expression. *Am. J. Pathol.* **136**, 1349–63.

Roberts, E. A., Letarte, M., Squire, J. and Yang, S. (1994) Characterization of human hepatocyte lines derived from normal liver tissue. *Hepatology* **19**, 1390–9.

Ronnett, G. V., Hester, L. D., Nye, J. S. *et al.* (1990) Human cortical neuronal cell line: establishment from a patient with unilateral megalencephaly. *Science* **248**, 603–4.

Saiki, I., Murata, J., Yoneda, J. *et al.* (1994) Influence of fibroblasts on the invasion and migration of highly or weakly metastatic B16 melanoma cells. *Int. J. Cancer* **56**, 867–73.

Schnegg, J. F., Diserens, A. C., Carrel, S. *et al.* (1981) Human glioma-associated antigens detected by monoclonal antibodies. *Cancer Res.* **41**, 1209–13.

Selby, P. J., Thomas, J. M., Monaghan, P. *et al.* (1980) Human tumour xenografts established and serially transplanted in mice immunologically deprived by thymectomy, cytosine arabinoside and whole body irradiation. *Br. J. Cancer* **41**, 52–61.

Shin, S. I., Freedman, V. H., Risser, R. and Pollack, R. (1975) Tumourigenicity of virus-transformed cells in nude mice is correlated specifically with anchorage independent growth *in vitro*. *Proc. Natl Acad. Sci.* **72**, 4435–9.

Simon, N., Noel, A. and Foidart, J. M. (1992) Evaluation of *in vitro* reconstituted basement membrane assay to assess the invasiveness of tumor cells. *Invasion Metastasis* **12**, 156–67.

Spiess, E., Zimmermann, H.-P. and Lünsdorf, H. (1987) Negative staining of protein molecules and filaments. In *Electron microscopy in molecular biology. A practical approach* (eds J. Sommerville and U. Scheer). IRL Press.

Sporn, M. B. and Todaro, G. J. (1980) Autocrine secretion and malignant transformation of cells. *New Engl. J. Med.* **303**, 878–80.

Stearns, M. E. and Stearns, M. (1993) Autocrine factors, type IV collagenase secretion and prostatic cancer cell invasion. *Cancer Metastasis Reviews* **12**, 39–52.

Streuli, C. H. and Bissell, M. J. (1990) Expression of extracellular matrix components is regulated by substratum. *J. Cell Biol.* **110**, 1405–15.

Sutherland, R. M., Carlsson, J., Durand, R. and Yuhas, J. (1981) Spheroids in cancer research. *Cancer Res.* **41**, 2980–84.

Taylor-Papadimitriou, J., Peterson, J. A., Arklie, J. *et al.* (1981) Monoclonal antibodies to epithelium-specific components of the human milk fat globule membrane: production and reaction with cells in culture. *Int. J. Cancer* **28**, 17–21.

Taylor-Papadimitriou, J., Stampfer, M., Bartek, J. *et al.* (1989) Keratin expression in human mammary epithelial cells cultured from normal and malignant tissue: relation to *in vivo* phenotype and influence of medium. *J. Cell Sci.* **94**, 403–13.

Todaro, G. J. and Green, H. (1963) Quantitative studies of the growth of mouse embryo cells in culture and their development into established lines. *J. Cell Biol.* **17**, 299–313.

Topley, P., Jenkins, D. C., Jessop, E. A. and Stables, J. N. (1993) Effect of reconstituted basement membrane components on the growth of a panel of human tumour cell lines in nude mice. *Br. J. Cancer* **67**, 953–8.

Traub, U. E., Nelson, W. J. and Traub, P. (1983) Polyacrylamide gel electrophoretic screening of mammalian cells cultured *in vitro* for the presence of the intermediate filament protein vimentin. *J. Cell Sci.* **62**, 129–47.

Trejdosiewicz, L. K., Southgate, J., Kemshead, J. T. and Hodges, G. M. (1986) Phenotypic analysis of cultured melanoma cells. Expression of cytokeratin-type intermediate filaments by the M5 human melanoma cell line. *Exp. Cell. Res.* **164**, 388–98.

Tsuruo, T., Yamori, T., Naganuma, K. *et al.* (1983) Characterisation of metastatic clones derived from a metastatic varient of mouse colon adenocarcinoma 26. *Cancer Res.* **43**, 5437–42.

van Lamsweerde, A. L., Henry, N. and Vaes, G. (1983) Metastatic heterogeneity of cells from Lewis lung carcinoma. *Cancer Res.* **43**, 5314–20.

Vukicevic, S., Kleinman, H. K., Luyten, F. P. *et al.* (1992) Identification of multiple active growth factors in basement membrane matrigel suggests caution in interpretation of cellular activity related to extracellular matrix components. *Exp. Cell Res.* **202**, 1–8.

Woodcock-Mitchell, J., Eichner, R., Nelson, W. G. and Sun, T.-T. (1982) Immuno-localization of keratin polypeptides in human epidermis using monoclonal antibodies. *J. Cell Biol.* **95**, 580–8.

Yuhas, J. (1977) A simplified method for production and growth of multicelluar tumour spheroids. *Cancer Res.* **37**, 3639–43.

Zauli, D., Gobbi, M., Crespi, C. *et al.* (1988) Cytoskeleton organization of normal and neoplastic lymphocytes and lymphoid cell lines of T and B origin. *Br. J. Haematol.* **68**, 405–9.

7

Cytotoxicity assays *in vitro*

BRIAN A. BOULLIER and ROGER M. PHILLIPS

7.1 Measurement of cytotoxicity *in vitro*

The widespread use of *in vitro* assays of cytotoxicity is borne out of escalating demand for safety testing of environmental xenobiotics, novel pharmaceuticals, food additives and cosmetics with the general aim of minimizing human risk. In addition, cytotoxicity assays are routinely used in anticancer drug development to identify novel agents that have activity against specific tumour types or to investigate the multitude of factors that influence cellular response to cytotoxic agents. Figure 7.1 presents a generalized overview of the essential steps involved in assessing cytotoxicity *in vitro*. Initial procedures involve the culture of cells and their harvesting, exposure of cells to cytotoxic agents and subsequent assessment of cell damage using either short-term (i.e. immediately after exposure) or long-term (i.e. following a recovery period) survival assays. Numerous endpoints have been developed and this is reflected in the variety of *in vitro* cytotoxicity assays that regularly receive literary review (Fruehauf and Bosanquet, 1993; Bosanquet, 1994; Guzzie, 1994). In general terms the response of cells depends largely upon the conditions of drug exposure, and the class of end-point selected (i.e. short-term or long-term survival). The purpose of this chapter is to provide a broad overview of some of the most commonly used *in vitro* cytotoxicity assays that offer different types of end-points to the investigator. In addition, the advantages and disadvantages of each assay are discussed together with a critical appraisal of some of the factors that influence cellular sensitivity to cytotoxic insult. The emphasis is placed on the use of cytotoxicity assays in the field of anticancer drug development as this reflects the authors' specific research interest. These assays can, however, be applied to other fields with or without minor modifications to the protocols provided.

7.2 Assessment of short-term viability

Many different assays are used to assess experimentally cell viability, immediately following cytotoxic insult, or more routinely, in determining the condition of cultured cells following processes such as enzymic and physical disaggrega-

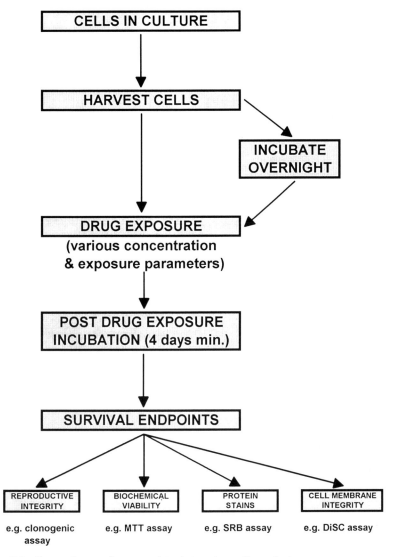

Figure 7.1. General procedures used to determine cell survival.

tion or recovery from cryogenic storage. The most common techniques exploit the capacity of stains such as Trypan Blue, Fast Green and Nigrosin to highlight the cytoplasm of dead or membrane-damaged cells that have lost the capacity to exclude these vital stains. The humble haemocytometer, commonly of the Improved Neubauer design, when used in conjunction with Trypan Blue, provides a convenient method for enumerating the relative proportions of viable cells (typically round and refractile) and inviable cells (typically stained dark blue, larger in size, crenated and non-refractile) (Warburton and James, 1994).

7.2.1 Measurement of viable cell number by dye exclusion using a haemocytometer

1. Thoroughly clean and assemble a haemocytometer slide and coverslip.
2. Prepare a 0.4% Trypan Blue solution in isotonic phosphate buffered saline (note: Trypan Blue is carcinogenic and must be handled with care).
3. Carefully prepare a representative single-cell cell suspension from the stock culture.
4. Gently resuspend the cell suspension before mixing, for example 100 μl, with an equal volume of Trypan Blue solution.
5. Fill both counting chambers on the haemocytometer slide using a Pasteur pipette, allowing capillarity to draw the mixture under the coverslip.
6. Let the cells settle on the counting grids before counting both the viable (unstained) and inviable (stained) cells using a microscope set at $\times 100$ magnification. It is essential to examine the cells within 5 minutes of mixing with Trypan Blue as gradual cell deterioration can otherwise lead to an underestimate of cell viability.
7. Calculate the number of cells/ml and the percentage of viable cells in the sampled cell culture.

In contrast to vital stains such as Trypan Blue, supravital stains can readily diffuse across the plasma membrane of viable cells, where they accumulate in the lysosomes. Dead or membrane-damaged cells exhibit reduced uptake of supravital stains. Cellular uptake of Neutral Red, measured by extraction and spectrophotometric absorption, provides another assay of cell viability and lends itself to the semi-automated evaluation of parallel cultures in multiwell microplates (Cavanagh *et al.*, 1990). Alternatively, viable and inviable cells may be clearly distinguished by exploiting a dual fluorescence technique. Viable cells take up fluorescein diacetate, converting it to fluorescein, which fluoresces green. Inviable cells, in which activity of the prerequisite esterases for such conversion is impaired, can absorb ethidium bromide or propidium iodide (which are excluded by viable cells) and so fluoresce red. Thus viability is expressed as the percentage of cells fluorescing green. Whilst cells may be examined manually using a fluorescence microscope, electronic counting using flow cytometry offers rapid and accurate measurements based on relatively large sample sizes.

7.2.2 Assays of membrane leakage

The leakage of soluble cytosolic enzymes such as lactate dehydrogenase by cells suffering induced membrane damage has been successfully employed in the measurement of cytotoxicity *in vitro*. Indirect spectrophotometric measurement of enzyme activity relies upon conversion of the cofactor NADH to NAD^+ (Guzzie, 1994). Alternatively, membrane leakage can be assessed by measuring the release of radiolabelled compounds, proteins or DNA into the culture medium. Chromium-51 (^{51}Cr) is a common label for this purpose. Only cells in which membrane integrity has been compromised are able to release oxidized $^{51}Cr^{2+}$ into the culture medium (Zawydiwski and Duncan, 1978).

7.2.3 Radioactive precursor incorporation

The intracellular incorporation of radiolabelled metabolites is a quick and sensitive means of estimating cell growth. Most commonly [^{3}H]thymidine or [^{3}H]uridine are used to quantitate DNA and RNA synthesis respectively. It should be noted that these assays are not in themselves direct measures of cell division but rather reflect cell utilization of exogenous nucleotides. Briefly, short-term measurement of incorporated [^{3}H]thymidine is performed as follows:

1. Prepare replicate $1000\ \mu l$ cultures containing 10^4 cells and appropriate concentrations of cytotoxic agent.
2. Incubate cultures at $37\,^\circ C$ for 2 h before adding methyl-[^{3}H]thymidine (5 Ci/mmol) for the third hour of incubation.
3. After 3 h incubation, place cultures on ice, wash by centrifuging and resuspend in ice-cold PBS. Repeat this step three times to minimize non-specific activity.
4. Harvest cells completely onto glass-fibre filter paper.
5. Wash the filters with 5 ml of ice-cold 10% (v/v) TCA (trichloroacetic acid).
6. Remove water from the filters by washing in methanol, methanol/ether and finally ether.
7. Dry the filters and transfer to scintillation vials containing 10 ml scintillant cocktail.
8. Store in the dark for 1 h to minimize the contribution of chemiluminescence to subsequent counts.
9. Count sample activity, typically for 5–10 min.

Equipment permitting, it is increasingly common to perform scintillation counting directly on cultures maintained and treated in 96-well plates. A novel chemosensitivity assay in which [^{3}H]thymidine uptake is measured amongst cells cultured in an agar-based clonogenic assay system has been used to good effect in predicting solid tumour cell drug resistance (Kern *et al.*, 1985; Kern and Weisenthal, 1990). Protein synthesis, as measured by the incorporation of radiolabelled amino acids, most commonly [^{35}S]methionine, is another recognized end-point for cytotoxicity assays (Freshney, 1994).

7.3 Long-term survival tests

The short-term viability assays described above provide a convenient and rapid assessment of cells that are dead at the time of the assay. For many compounds such as anti-cancer drugs, however, the toxic effects of drug exposure take several hours or days to appear and in these cases, long-term survival assays are required. Considerable effort has gone into the design of such assays and there are now a large number of assays available, each one having its own advantages and disadvantages. Most of these assays can be broadly categorized as either biochemical or viability measurements following exposure to toxic insults. Biochemical assays have largely focused upon metabolic properties (e.g. the MTT assay) and protein measurements (e.g. the SRB assay). Viability measurements have emphasized either surface membrane integrity (e.g. the DiSC assay) or proliferative capacity (e.g. the clonogenic assay) with the loss of either parameter taken to represent cell death. The following assays are select examples of survival assays that are routinely used to determine the response of cells to cytotoxic drugs.

7.3.1. The MTT assay

This assay is based upon the ability of viable cells to reduce a water soluble tetrazolium salt (3-[4,5-dimethylthiazol-2-yl]-2,5-diphenyl-tetrazolium bromide) to a water insoluble formazan crystal (Mosmann, 1983). Tetrazolium salts are reduced by various dehydrogenases in active mitochondria (Slater, Sawyer and Strauli, 1963) and so the reaction only occurs in living cells. The formazan crystals can then be dissolved in an organic solvent and the absorbance of the resulting solution measured spectrophotometrically. As the amount of formazan product generated is directly proportional to the number of viable cells, cell survival following treatment can be determined. This assay has been successfully applied to anticancer chemosensitivity and radiosensitivity testing for both adherent and suspension cultures (Carmichael *et al.*, 1987; Pieters *et al.*, 1988; Slavotinek, McMillan and Steel, 1994). The basic protocol is outlined below.

1. Harvest cells from stock cultures in exponential growth and seed 10^3 cells into each well (in 180 μl of medium) of a 96-well plate (leaving the first lane of eight wells free of cells to serve as a blank for the microplate reader).
2. Incubate plate overnight (all incubations are performed at 37 °C in a humidified atmosphere containing 5% CO_2).
3. Add 20 μl of drug to each well (eight wells per drug concentration) to give a range of suitable drug concentrations.
4. For continuous drug exposures no further processing is required. For timed drug exposures, remove the drug at the appropriate time(s), wash the cells with Hanks Balanced Salt Solution and add 200 μl of complete medium.
5. Incubate for 4 to 7 days depending on the growth rate of the cultures involved.
6. Following the recovery period, add 20 μl of MTT (5 mg/ml) to each well and incubate for 4 h.

7. Remove and discard 200 μl of medium from each well.
8. Dissolve formazan crystals in 150 μl of dimethyl sulphoxide (DMSO), mix well and read the absorbance of the resulting solution at 550 nm in a microplate plate reader.
9. Results are expressed in terms of percentage survival taking the absorbance of control wells (drug solvent only) to represent 100% survival.

The basic protocol described above works well on both adherent and suspension cultures, provided that the initial cell density is low (1–2 $\times$ 10^3 cells per well) and the control cultures are not in the plateau phase of the growth curve at the time of the assay. At high cell densities, problems can occur as a result of a reduction in the pH or 'conditioning' of the culture medium. Low pH values cause a reduction in formazan production and a shift in the absorption spectrum of MTT-formazan (from 560 nm at high pH to 510 nm with a smaller peak at 570 nm at low pH). The pH of medium therefore has a significant effect upon the MTT assay and gradients of pH across control and treated wells may lead to a serious underestimation of chemosensitivity (Plumb, Milroy and Kaye, 1989; Jabbar, Twentyman and Watson, 1989). To reduce the risk of error in the MTT assay, conditioned medium may be replaced with fresh medium immediately prior to the addition of MTT (Jabbar *et al.*, 1989) or the pH of the solubilized MTT-formazan product can be raised by the addition of a glycine buffer at pH 10.5 (Plumb *et al.*, 1989). Both of these procedures improve the efficiency of the MTT assay.

7.3.2 The SRB assay

This is a rapid, sensitive and inexpensive method of assessing cytotoxicity, which is currently being employed by the National Cancer Institute (USA) in their disease-orientated anticancer drug discovery programme (Skehan *et al.*, 1990). Sulforhodamine B (SRB) is an anionic protein stain that binds electrostatically to protein basic amino acid sequences under mildly acidic conditions (Skehan *et al.*, 1990). The dye can be extracted from cells and solubilized by weak bases. The intensity of staining is proportional to cell number and a comparison between chemosensitivity data generated using the MTT assay and the SRB assay are similar (Rubinstein *et al.*, 1990). Detailed protocols are described by Skehan *et al.* (1990) and summarized below.

1. Cell cultures are prepared for drug exposure as described previously for the MTT assay (steps 1 to 5).
2. Following the recovery period, cells are fixed with trichloroacetic acid (TCA). For adherent cells, 50 μl of cold 50% TCA (4 °C) is gently layered on top of the growth medium and incubated at 4 °C for 1 h. For suspension cultures, 50 μl of 80% TCA is layered on top of the growth medium, cells are allowed to settle for 5 min followed by incubation at 4 °C for 1 h.

3. Cultures are washed five times with tap water.
4. TCA-fixed cells are stained with 0.4% (w/v) SRB dissolved in 1% acetic acid for 30 min.
5. Cultures are rinsed four times with 1% acetic acid to remove unbound dye and air dried.
6. Bound dye is solubilized with 10 mM unbuffered Tris base (pH 10.5) for 5 min on a shaker.
7. The absorbances of the resultant solutions can be measured within wavelengths of 490–530 nm using a microplate reader.

7.3.3 The DiSC assay

The differential staining cytotoxicity (DiSC) assay is a noteworthy enhancement of traditional dye exclusion assays (Weisenthal *et al.*, 1983; Bird, Bosanquet and Gilby, 1985). DiSC assay end-point is the identification of viable and inviable cells via selective staining and counterstaining of microscope slide preparations of treated and untreated cultures. A four-day assay period facilitates loss of membrane integrity amongst reproductively impaired cells. Hence DiSC assay results compare favourably with other assays of prolonged cell survival. The DiSC assay has generated considerable support amongst investigators of predictive assays of tumour cell chemosensitivity and radiosensitivity (Weisenthal *et al.*, 1989; Hinkley and Bosanquet, 1992; Bosanquet, 1994), particularly in relation to haematological neoplasms (Bosanquet, 1991). Whilst the assay requires careful and time-consuming manual evaluation of cell viability, technical success is not dependent upon pure tumour cell populations, as with the MTT assay, or single cell suspensions, as with the colony-forming assay. The DiSC assay has undergone numerous modifications over the past 10 years, largely to accomodate the experimental evaluation of increasing numbers of anticancer agents but also to improve the commercial effectiveness of the method. A basic protocol is outlined below.

1. Prepare a representative sample of tumour cells, e.g. enzymatically disaggregated solid tumour cells or centrifugally isolated leukaemic lymphocytes.
2. Estimate cell concentration and viability using a haemocytometer chamber and Trypan Blue. Make up an appropriate volume of cell suspension, say 10 ml, at a concentration of 8.89×10^5 cells/ml.
3. Prepare a 96-well multiwell plate with 10 μl PBS in 'control' wells, 10 μl PBS and solvent in 'solvent control' wells, and 10 μl freshly prepared serially diluted drug solutions in 'test' wells.
4. Add 90 μl cell suspension to each well, giving a resultant cell concentration of 8×10^4 cells/well.
5. Incubate at 37 °C in 5% CO_2.
6. After four days incubation add 10 μl of fixed duck red blood cells

(DRBCs) in Fast Green/Nigrosin solution (5×10^6 DRBCs/ml in 2% Fast Green/1% Nigrosin stored at 4 °C in sterile PBS).

7. Cytocentrifuge the well contents onto a microscope slide and counterstain viable cells with a Romanovsky stain such as Wright's stain.
8. DRBCs act as an internal standard and facilitate calculation of relative cytotoxicity, which is expressed as the ratio of viable cells/DRBCs in treated cultures compared to control cultures.

7.3.4 The colony forming assay

The colony forming or clonogenic assay has been extensively evaluated as a method for predicting individual tumour responses to anti-cancer drugs and critical appraisals of this assay have been described elsewhere (Selby, Buick and Tannock, 1983). This assay is designed to measure the effect of cytotoxic drugs upon a subpopulation of tumour cells (stem cells) that are thought to be responsible for tumour growth, the metastatic spread of the disease and tumour repopulation following treatment. Stem cells have high proliferative potential and the clonogenic assay measures those cells that have the ability to multiply and form a colony in culture. Numerous assay techniques have been developed using soft agar to help maintain colony integrity and eliminate the growth of stromal cells from primary tumour cell suspensions (Hamburger and Salmon, 1977; Courtney *et al.*, 1978). For established adherent cell lines, soft agar is not generally required as colony integrity is maintained on plastic culture vessels and details of this assay are described below.

1. Harvest cells from stock cultures in exponential growth by mild trypsinization.
2(a). For continuous drug exposures, seed 10^3 to 10^4 viable cells (depending on the plating efficiency of the cell line in question) into each well of a six-well culture plate (three wells per drug exposure). Add an appropriate volume of drug solution to each well to give the desired range of final drug concentrations. Incubate for 4 days (all incubations are performed at 37 °C in a humidified atmosphere containing 5% CO_2).
2(b). For timed drug exposures, transfer 9.5 ml of cell suspension to a series of sterile universal tubes (six drug exposures and one control are adequate). Add 0.5 ml of drug solution to each of the universal tubes and incubate at 37 °C for the required duration of drug exposure. Following drug exposure, wash the cells twice with Hanks Balanced Salt Solution by repeated centrifugation (1000 rpm for 3 min) and resuspension of the cell pellet. Finally, resuspend the cells in 5 ml of growth medium and count the cell number using a haemocytometer. Seed between 10^3 and 10^4 viable cells (depending upon the plating efficiency of the cell line in question) into each well of a six-well culture plate (three wells per drug exposure) and incubate.

3. Colonies of > 50 cells can either be counted directly using an inverted microscope or colonies can be stained with Crystal Violet (cells are fixed in absolute methanol or 1% (v/v) glutaraldehyde and stained with 1% Crystal Violet for 10 min. Wash stained colonies in water and air dry).
4. Calculate the plating efficiency (the number of colonies formed divided by the number of cells plated) for each drug exposure.
5. Plot percentage survival versus drug concentration taking the control plating efficiency to be 100% survival.

The number of cells plated into each well can be varied so that clonogenic cell survival at high drug concentrations can be determined. However, the assay is labour intensive (particularly when soft agar is required) and suffers from technical problems such as cell clumping (Selby *et al.*, 1983). Despite these limitations, the clonogenic assay is routinely used in many laboratories and provides reliable and reproducible chemosensitivity data.

7.4 Advantages and limitations of cytotoxicity assays – troubleshooting

Each of the assays described has its advantages and disadvantages and the selection of which survival endpoint to use depends largely upon the specific objectives of individual projects. All of the assays are applicable to small-scale laboratory procedures such as testing a limited number of compounds for toxicity or for detailed studies of factors that influence chemosensitivity (i.e. pH, pO_2, growth factors, etc.). For large-scale drug screening programmes, a rapid, inexpensive and semi-automated assay is required and the MTT and SRB assays are suitable for such studies. In the case of the National Cancer Institute's disease-orientated anti-cancer drug discovery programme, over 10 000 compounds are tested annually and the SRB assay has emerged as the most suitable assay for use in this high-volume screen (Skehan *et al.*, 1990). The principle advantages of the SRB assay over the MTT assay are that it is simpler, faster, provides a stable endpoint, better linearity with cell number and it is less sensitive to environmental fluctuations (Skehan *et al.*, 1990). In terms of the dose-response curves obtained using the different survival end-points, estimates of chemosensitivity are comparable provided that the conditions of drug exposure are identical. A close correlation between the results of the MTT and SRB assays and between the MTT and clonogenic assays have been reported (Carmichael *et al.*, 1987; Alley *et al.*, 1988; Rubinstein *et al.*, 1990). Similarly, the results of clonogenic and non-clonogenic assays are comparable in terms of their ability to predict the response of individual patients to chemotherapy (Weisenthal and Lippman, 1985). DiSC assay results compare favourably with those of the clonogenic assay (Bird *et al.*, 1987; Weisenthal *et al.*, 1988), and are unaffected by cell clumping. The DiSC assay also uniquely facilitates the assessment of tumour cell response within a heterogeneous cell culture, as commonly encountered when testing patient-derived samples.

The major limitations of all *in vitro* chemosensitivity assays can be assigned to the fact that cells in culture are growing in a very different environment from the complex structural and biochemical environment from which they were derived. The difficulty lies in the problem of accurately mimicking *in vitro* the specific conditions experienced by cells in tissues (Selby *et al.*, 1983; Phillips, Bibby and Double, 1990). Pharmacokinetic parameters (i.e. rate of change in drug concentrations, duration of drug exposure, metabolism, etc.), for example, are extremely difficult to mimic *in vitro*, particularly when cytotoxic metabolites are generated. The problem of cytotoxic metabolites can to a certain degree be circumvented by the inclusion of a drug metabolizing system into the assay such as S9 liver fractions (Alberts *et al.*, 1984; Benford, Reavy and Hubbard, 1988) although the same pharmacokinetic constraints discussed above still apply. Cultures of primary hepatocytes and several hepatocellular tumour cell lines are metabolically competent with regards to cytochrome P-450 mixed function oxidase activity. Therefore co-cultivation with primary hepatocytes in *in vitro* cytotoxicity assays may also facilitate more accurate evaluation of likely cytotoxic effects *in vivo*.

In the case of solid tumour biology, further complications arise from the fact that tumours are three-dimensional structures that tend to outgrow their vascular supply. This feature of solid tumours introduces several problems that may modify the effectiveness of chemotherapy, such as drug penetration barriers, proliferation gradients and microenvironmental changes (i.e. gradients of pH, pO_2, nutrients and catabolites), all of which vary as a function of distance from a supporting blood vessel. There are therefore major problems that are inherent to the whole concept of chemosensitivity testing *in vitro*, which limit the widespread use of these techniques for predicting toxic effects *in vivo*. In the laboratory, however, the assays described in this text provide a reliable and reproducible method of assessing cellular sensitivity, which are applicable to a diverse range of studies.

7.5 Factors influencing cellular response to toxic insults – troubleshooting

7.5.1 Concentration and duration of drug exposure

Two factors that influence cellular response are the concentration of the drug (C) and the duration of drug exposure (T). The relationship between cell kill and these two parameters depends largely upon the mechanism of action of the drug in question. For cell cycle specific drugs (those compounds that kill cells at various stages of the cell cycle), cell kill is proportional to $C \times T$ (Hill, 1983). For cell cycle phase specific drugs (those compounds that kill cells at specific stages of the cell cycle), cell kill is proportional to T. For compounds with unknown mechanism of action, continuous exposure to a broad range of drug concentrations (4–5 log increments) should be used.

7.5.2 Drug stability

The stability of compounds under both storage and assay conditions is of critical importance in determining cytotoxicity. This is particularly true when continuous drug exposures are employed, as differences in cytotoxicity between a panel of compounds may be due solely to differences in the rate of drug breakdown and not to the inherent toxic properties of the compounds themselves. For anti-neoplastic agents, stability data under a variety of conditions has been extensively reviewed (Bosanquet, 1985, 1986, 1989). For novel compounds where little or no stability data is available, bioassays provide a simple method of obtaining information on drug stability (Hildebrand-Zanki and Kern, 1984). In addition, precautionary measures such as protecting novel compounds from light and avoiding the repetitive freeze–thawing of drug solutions will reduce the risk of drug breakdown during storage.

7.5.3 Cell kinetics

The response of cells to a variety of toxic insults depends upon the position of cells in the growth curve (i.e. exponentially growing or plateau phase) and the type of compound tested. In the case of anticancer drugs, cells in a non-proliferating state may be insensitive to S phase specific antimetabolite drugs whereas the same cells in exponential growth would be sensitive. In addition, the expression and activity of drug metabolizing enzymes such as DT-diaphorase vary with regard to the position of cells in the growth curve (Phillips *et al.*, 1994). Care must therefore be taken to ensure that cells are exposed to drugs at the same stage of the growth curve in order to provide comparative data.

7.5.4 Solvents

Many compounds are insoluble in aqueous media and organic solvents are required. It is must be stressed that organic solvents are in themselves cytotoxic and the use of the minimum concentration of solvent required to keep the drug in solution is essential. Ethanol, DMSO and propylene glycol are commonly used solvents and final concentrations of $< 0.1\%$ are generally non-toxic following continuous exposure (96 hours).

7.5.5 Extracellular matrix

There is increasing evidence in the literature that the extracellular matrix influences both the geometry and biochemistry of epithelial cells (Gospodarowicz, Greenberg and Birdwell, 1978; Bissell, 1988; Lin and Bissell, 1993). To date, little is known about the effect of the extracellular matrix on chemosensitivity although differential toxicity is likely to occur in view of the major biochemical differences that exist between cells grown on plastic and extracellular matrices.

7.5.6 Microenvironmental factors

In the case of many solid tumours, gradients of microenvironmental conditions exist with conditions becoming more stressful as distance from a supporting blood vessel increases. These characteristics can to some degree be imitated *in vitro* by multicellular spheroids (Sutherland, 1988) and numerous studies have been published comparing the response of spheroids and monolayers to cytotoxic drugs. In the case of doxorubicin, for example, major drug penetration barriers have been described and spheroids are significantly more resistant than the same cells grown as monolayers (Sutherland *et al.*, 1979). For other compounds, additional factors such as the 'cell-contact' effect, cell kinetics, hypoxia, low pH and metabolic factors have been suggested as explanations for the relative resistance of spheroids compared with monolayers (Olive and Durand, 1994). It is clear therefore that the properties of the multicellular spheroid have a significant effect on chemosensitivity, suggesting that toxicity data on monolayer cultures should be interpreted with a degree of caution. For compounds that are designed to exploit the tumour microenvironment, such as bioreductive drugs (Workman and Stratford, 1993), the spheroid model offers an ideal experimental tool with which to evaluate this class of drugs.

7.5.7 Source of tissue

The nature of the cultured cells under investigation is also of critical importance. Primary cell cultures, derived directly from animal tissues or organs, may initially represent a better approximation of the complex biochemical functions of the host tissue. However, such cell systems are invariably heterogeneous in nature and thus response to cytotoxic insult is variable and prone to change as the cell population succumbs to artificial selection pressure *in vitro*. In contrast, finite and continuous cell lines can provide cell systems that offer greater reproducibility and consistency of results in this respect. Tissue of origin may also be a significant consideration. In anticancer drug screening studies, for example, attention is focused on cell lines representative of tissues in which likely cytotoxic effects are desirable. Whilst studies of representative tumour cell lines may be informative with regards to tumour cell cytotoxicity, parallel investigation of normal tissue toxicity, using cultured bone marrow haematopoietic cells for example (San Roman *et al.*, 1994), can reveal the true worth of a clinically effective anticancer agent, namely selective tumour activity.

7.6 Conclusions

Cell cultures offer convenient and reproducible manipulation of environmental factors including nutrient supply and pH, divorced from the complex influence of endogenous factors such as hormonal and neural control inherent within *in vivo* systems. However, this absence of systemic response may hamper meaningful interpretation of *in vitro* experimental results. The investigator therefore must give careful consideration to the chosen method(s) of assaying cytotoxicity. The assay end-point must be both specific and well defined, and typically represents quantitative or qualitative measurement of a cytotoxic effect that ultimately reduces cell viability and prolonged survival.

References

Alberts, D. S., Einsphar, J., Lundwig, R. and Salmon, S. E. (1984) Pharmacologic pitfalls in the human tumour clonogenic assay. In *Predictive drug testing on human tumour cells* (ed. V. Hoffman, M. E. Berens and G. Martz), pp. 184–90. Springer Verlag, Berlin.

Alley, M. C., Scudiero, D. A., Monks, A. *et al.* (1988) Feasibility of drug screening with panels of human tumour cell lines using a microculture tetrazolium assay. *Cancer Res.* **48**, 589–601.

Benford, D. J., Reavy, H. J. and Hubbard, S. A. (1988) Metabolising systems in cell culture cytotoxicity tests. *Xenobiotica*, **18**, 649–56.

Bird, M. C., Bosanquet, A. G. and Gilby, E. D. (1985) *In vitro* determination of tumour chemosensitivity in haematological malignancies. *Haematol. Oncol.* **3**, 1–9.

Bird, M. C., Godwin, V. A. J., Antrobus, J. H. and Bosanquet, A. G. (1987) Comparison of in vitro drug sensitivity by the differential staining cytotoxicity (DiSC) assay and colony forming assays. *Br. J. Cancer*, **55**, 429–43.

Bissell, M. J. (1988) Extracellular matrix influence on gene expression: Is structure the message? *Br. J. Cancer* **58**, 223.

Bosanquet, A. (1985) Stability of solutions of antineoplastic agents during preparation and storage for in vitro assays. General considerations, the nitrosoureas and alkylating agents. *Cancer Chemother. Pharmacol.* **14**, 83–95.

Bosanquet, A. (1986) Stability of solutions of antineoplastic agents during preparation and storage for in vitro assays II. Assay methods, adriamycin and other antitumour antibiotics. *Cancer Chemother. Pharmacol.* **17**, 1–10.

Bosanquet, A. (1989) Stability of solutions of antineoplastic agents during preparation and storage for in vitro assays III. Antimetabolites, tubulin binding agents, platinum drugs, amsacrine, L-asparaginase, interferons, steroids and other miscellaneous antitumour agents. *Cancer Chemother. Pharmacol.* **23**, 197–207.

Bosanquet, A. G. (1991) Correlations between therapeutic response of leukaemias and in vitro drug sensitivity assay. *Lancet*, **337**, 711–14.

Bosanquet, A. G. (1994) Short-term in vitro drug sensitivity tests for cancer chemotherapy. A summary of correlations of test result with both patient response and survival. *FORUM Trends Exp. Clin. Med.* **4**, 179–95.

Carmichael, J., DeGraff, W. G., Gazdar, A. F. *et al.* (1987) Evaluation of a tetrazolium based semiautomated colorimetric assay assessment of chemosensitivity testing. *Cancer Res.* **47**, 936–42.

Cavanagh, P. F., Moskwa, P. S., Donish, W. H. *et al.* (1990) A semi-automated neutral red based chemosensitivity assay for drug screening. *Invest. New Drugs*, **8**, 347–54.

Courtney, V. D., Selby, P. J., Smith I. E. *et al.* (1978) Growth of human tumour colonies from biopsies using two soft agar techniques. *Br. J. Cancer* **38**, 77–81.

Freshney, R. I. (1994) *Culture of animal cells*, 3rd edn. Wiley-Liss, New York.

Fruehauf, J. P. and Bosanquet, A. G. (1993) *In vitro* determination of drug response: a discussion of clinical applications. *Princ. Prac. Oncol.* **7**(12), 1–16.

Gospodarowicz, D., Greenberg, G. and Birdwell, C. R. (1978) Determination of cellular shape by the extracellular matrix and its correlation with the control of cellular growth. *Cancer Res.* **38**, 4155–71.

Guzzie, P. J. (1994) Lethality testing. In *In Vitro Toxicology* (ed. S. C. Gad), pp. 57–86. Raven Press, New York.

Hamburger, A. W. and Salmon, S. E. (1977) Primary bioassay of human tumour stem cells. *Science*, **197**, 461–3.

Hildebrand-Zanki, S. U. and Kern, D. H. (1984) A new bioassay for in vitro drug stability. In *Human Tumour Cloning* (ed. S. E. Salmon and J. M. Trent), p. 451. Grune & Stratton, Orlando.

Hill, B. T. (1983) Use of continuous human tumour cell lines to evaluate drugs by clonogenic assays. In *Human tumour drug sensitivity testing in vitro* (ed. P. P. Dendy and B. T. Hill), pp. 129–47. Academic Press, London.

Hinkley, H. J. and Bosanquet, A. G. (1992) The *in vitro* radiosensitivity of lymphocytes from chronic lymphocytic leukaemia using the differential staining cytotoxicity (DiSC) assay: II. Results on 40 patients. *Int. J. Rad. Biol.* **61**, 111–?

Jabbar, S. A. B., Twentyman, P. R. and Watson, J. V. (1989) The MTT assay underestimates the growth inhibitory effects of interferons. *Br. J. Cancer*, **60**, 523–8.

Kern, D. H., Drogemuller, C. R., Kennedy, M. C. *et al.* (1985) Development of a miniaturised, improved nucleic acid precursor incorporation assay for chemosensitivity testing of human solid tumors. *Cancer Res.* **45**, 5436–441.

Kern, D. H. and Weisenthal, L. M. (1990) Highly specific prediction of antineoplastic drug resistance with an *in vitro* assay using suprapharmacologic drug doses. *J. Nat. Cancer Inst.* **82**, 582–8.

Lin, C. Q. and Bissell, M. J. (1993) Multi-faceted regulation of cell differentiation by extracellular matrix. *FASEB J.* **7**, 737–43.

Mosmann, T. (1983) Rapid colorimetric assay for cellular growth and survival: application to proliferation and cytotoxicity assays. *J. Immunol. Meth.* **65**, 55–63.

Olive, P. L. and Durand, R. E. (1994) Drug and radiation resistance in spheroids: cell contact and kinetics. *Cancer Metast. Rev.* **13**, 121–8.

Phillips, R. M., Bibby, M. C. and Double, J. A. (1990) A critical appraisal of the predictive value of in vitro chemosensitivity assays. *J. Nat. Cancer Inst.* **82**, 1457–68.

Phillips, R. M., de la Cruz, A., Traver, R. D. and Gibson, N. W. (1994) Increased activity and expression of NAD(P)H:Quinone acceptor oxidoreductase in confluent cell cultures and within multicellular spheroids. *Cancer Res.* **54**, 3766–71.

Pieters, R., Huismans, D. R., Leyva, A. and Veerman, A. J. P. (1988) Adaptation of the rapid automated tetrazolium dye based (MTT) assay for chemosensitivity testing in childhood leukemia. *Cancer Lett.* **41**, 323–32.

Plumb, J. A., Milroy, R. and Kaye, S. B. (1989) Effects of the pH dependence of 3-(4,5-dimethylthiazol-2-yl)-2,5-diphenyl tetrazolium bromide absorption on chemosensitivity determined by a novel tetrazolium based assay. *Cancer Res.* **49**, 4435–40.

Rubinstein, L. V., Shoemaker, R. H., Paull, K. D., *et al.* (1990) Comparison of *in vitro* anticancer drug screening data generated with a tetrazolium assay versus a protein assay against a diverse panel of human tumour cell lines. *J. Nat. Cancer Inst.* **82**, 1113–18.

San Roman, J. S., Kamali, V., Sibanda, B. *et al.* (1994) Measurement of differential haematotoxicity using long-term bone marrow cultures. *In Vitro Toxicol.* **7**(3), 291–99.

Selby, P. J., Buick, R. N. and Tannock, I. (1983) A critical appraisal of the human tumour stem cell assay. *New Engl. J. Med.* **308**, 129–34.

Skehan, P., Storeng, R., Scudiero, D. *et al.* (1990) New colorimetric assay for anticancer drug screening. *J. Nat. Cancer Inst.* **82**, 1107–12.

Slater, T. F., Sawyer, B. and Strauli, U. (1963) Studies on succinate-tetrazolium reductase system III. Points of coupling of four different tetrazolium salts. *Biochim. Biophys. Acta* **77**, 383–93.

Slavotinek, A., McMillan, T. J. and Steel, C. M. (1994) Measurement of radiation survival using the MTT assay. *Eur. J. Cancer*, **30A**, 1376–82.

Sutherland, R. M. (1988) Cell and environmental interactions in tumor microregions: the multicell spheroid model. *Science* **240**, 177–84.

Sutherland, R. M., Eddy, H. A., Bareham, B. *et al.* (1979) Resistance to adriamycin in multicellular spheroids. *Int. J. Rad. Oncol. Biol. Phys.* **5**, 1255–30.

Warburton, S. and James, R. (1994) Haemocytometer cell counts and viability studies. In *Cell and Tissue Culture Laboratory Procedures* (ed A. Doyle, J. B. Grifiths and D. G. Newell), pp. 4B:1.1–4B:1.5. John Wiley, Chichester.

Weisenthal, L. M. and Lippman, M. E. (1985) Clonogenic and non-clonogenic *in vitro* chemosensitivity assays. *Cancer Treat. Rep.* **69**, 615–32.

Weisenthal, L. M., Marsden, J. A., Dill, P. L. and Macaluso, C. K. (1983) A novel dye exclusion method for testing in vitro chemosensitivity of human tumors. *Cancer Res.* **43**, 258–64.

Weisenthal, L. M., Nagourney, R. A., Kern, D. H. *et al.* (1989) Approach to the clinical circumvention of drug resistance utilizing a non-clonogenic in vitro assay measuring the effects of drugs, radiation, and interleukin II on largely non-dividing cells. *Adv. Clin. Oncol.* **1**, 91–7.

Weisenthal, L. M., Su, Y., Duarte, T. E. and Nagourney, R. A. (1988) Non-clonogenic, *in vitro* assays for predicting sensitivity to cancer chemotherapy. In *Prediction of Responses to Cancer Therapy* (ed. T. C. Hall), pp. 75–92. Alan R. Liss, New York.

Wilson, A. P. (1992) Cytotoxicity and viability assays. In *Animal cell culture. A practical approach*, 2nd edn (ed. R. I. Freshney), pp. 183–216.

Workman, P. and Stratford, I. J. (1993) The experimental development of bioreductive drugs and their role in cancer therapy. *Cancer Meth. Rev.* **12**, 73–82.

Zawydiwski, R. and Duncan, G. R. (1978) Spontaneous ^{51}Cr release by isolated rat hepatocytes: An indicator of membrane damage. *In Vitro* **124**, 707–14.

8

The production of antibodies using synthetic peptides as immunogens

T. RAJKUMAR and WILLIAM J. GULLICK

8.1 Introduction

Antibodies are useful reagents that have found diverse applications in science. Depending on the method of production, the antibodies can be classified into polyclonal and monoclonal. Polyclonal antibodies have usually been raised in rabbits, goats, sheep and swine. Monoclonal antibodies can be raised by using hybridoma technology or by phage display. The latter is gaining popularity due to its versatility but will not be discussed in this chapter. Monoclonal antibodies have been raised in mice and rats or by using peripheral blood or lymph node lymphocytes from humans exposed to the antigen of interest. The main advantage of a monoclonal antibody is its specificity, unlimited supply and absence of batch-to-batch variation, which is often seen with polyclonal antibodies. However, in some instances monoclonal antibodies may be too specific, for instance where the epitope recognized is polymorphic in the population. In addition, the cost of producing monoclonal antibodies is greater than that for raising polyclonal antibodies. In this chapter we will focus on the production of antibodies to growth factor receptors using synthetic peptides as immunogens. The principles and techniques described, however, also apply more widely.

8.2 Monoclonal antibody production

8.2.1 Immunization

Broadly, antigens used for immunization can be classified into natural or synthetic. The most commonly used synthetic antigens are peptides, the sequence of which is derived from the protein of interest. The major advantages with peptide antigens are that they are pure, available in unlimited quantities, are simple to use in the screening procedure, are generally capable of evoking antibodies recognizing both native and denatured proteins and can be prepared with the knowledge of the DNA sequence of the putative protein alone. Their disadvantages are, conversely, that the sequence of the antigen has to be known and in general it is difficult to raise monoclonal antibodies (using peptides as antigens) against certain sequences such as the extracellular domains of receptor proteins, since they are folded and post-translationally modified by glycosylation, as a result of which the antibody may not be able to access the epitope in the protein in the native state.

Natural antigens can also be used for immunization. However, it is important that the antigen is as pure as possible, since this simplifies screening, but this may not be important if a good screening assay is available. Monoclonal antibodies against the extracellular domain of epidermal growth factor receptor (EGFR) have, for example, been raised by using whole cells (expressing high levels of EGFR) as immunogen and then employing a screening assay based on blocking radiolabelled EGF binding. Another approach in raising monoclonal antibodies to the extracellular domain of growth factor receptors is to transfect NIH3T3 cells with the cDNA of the antigen and select clones that express high levels of the antigen and use them to immunize syngeneic mice. In this instance, the syngeneic mice mount an immune response only to the transfected antigen but not to the NIH3T3 cells. Hence, the immune response is specific for the antigen and screening can also be simplified by using an ELISA with transfected and untransfected NIH3T3 cells.

This chapter will describe only the peptide-based immunization approach.

8.2.1.1 Coupling of peptides to keyhole limpet haemocyanin (klh)

(a) Materials and reagents required
1. Keyhole limpet haemocyanin (klh), as a slurry from Calbiochem.
2. Phosphate buffered saline (PBS): 140 mM sodium chloride, 20 mM sodium phosphate buffer, pH 7.2.
3. 50 mM sodium borate/HCl buffer, pH 9.0.
4. 100 mM sodium phosphate buffer, pH 8.0.
5. 25% (v/v) glutaraldehyde solution from Sigma, EM grade.
6. 1 M glycine/HCl, pH 6.0.
7. Quartz cuvette.
8. Spectrophotometer.
9. Dialysis tubing (Spectrum Medical Industries).

(b) Procedure
1. 0.5 ml of klh (Calbiochem) is dialysed against 2 l of PBS, overnight at 4 °C. (The large volume of PBS is required because the klh is provided in a high concentration of ammonium sulphate.) Once dialysed, the klh can be stored at 4 °C for several days.
2. To a quartz cuvette, 1 ml of sodium borate buffer, pH 9.0, is added and used to calibrate a spectrophotometer at 280 nm. 10 μl of the dialysed klh is added to the cuvette and mixed and the absorbance measured. The amount of klh present is then calculated as follows.

If the absorbance value was 0.1, then 10 μl gives an absorbance of 0.1.
1 ml will therefore have an absorbance of $0.1 \times 100 = 10$.
Normally, 1 mg/ml of klh gives an absorbance of 1.4, therefore the amount of klh present $= 10/1.4 = 7.14$ mg/ml.

3. A stock solution containing 10 mg/ml of peptide is prepared in 100 mM sodium phosphate buffer, pH 8.0. A 1:1 mixture of klh and peptide is added to an Eppendorf tube, e.g. 2.5 mg of peptide (250 μl) and 2.5 mg of klh (350 μl), and mixed.

4. Glutaraldehyde is used to couple the peptide with klh. 2.5 μl of 25% (v/v) glutaraldehyde is added to the Eppendorf tube, mixed well and incubated for 15 min at room temperature. A further 1.25 μl of 25% glutaraldehyde is added, mixed and incubated for the same period of time.

5. In order to neutralize the excess glutaraldehyde, 100 μl of 1 M glycine is added and incubated for 10 min

6. The final volume of the klh-peptide is made up to 1 ml with PBS and stored at $-20\,°C$.

8.2.1.2 Immunization schedule

(a) Mice

We have preferred to use female Balb/c mice, four per peptide antigen, which are injected subcutaneously at multiple sites, with 250 μg of peptide coupled to klh (100 μl) at two weekly intervals (Days 1 and 14). The first injection is given with 100 μl of complete Freund's adjuvant and the second with 100 μl of incomplete Freund's adjuvant. It is essential that preimmune sera are obtained (day 0 or 1) from the mice before they are immunized and thereafter immune sera is obtained on days 14, 28 and 56. An ELISA is then done to assess the immune response. If the absorbance value obtained at 1/1000 dilution is above 1.000 in the presence of a low background, then that would generally indicate that a good immune response has been achieved. If it is unsatisfactory, it may be better to wait for an additional month, rebleed the mice and test again. Additionally, in some instances a third dose of the immunogen can be given in incomplete Freund's adjuvant. If the mice have developed a good immune response, additional assays, for which the monoclonal antibody would finally be used, are performed. If these results are satisfactory, the mice can then be boosted for fusion.

(b) Rabbits

After drawing the preimmune bleed (10–15 ml of serum), two rabbits are injected with 2.5 mg of peptide-klh (1 ml) mixed with 1 ml of complete Freund's adjuvant for the first injection or with 1 ml of incomplete Freund's adjuvant for the second and third injection, subcutaneusly at multiple sites, on days 1, 14 and 28 respectively. Test bleeds are obtained on days 35 and 56.

(c) Precautions

Gloves and eye protection should be worn when handling Freund's adjuvant.

8.2.2 Screening assays

Assuming the mice have developed a good serum titre, the success of the fusion depends heavily on the screening assay. A simple and specific assay almost certainly ensures that the fusion would result in more than one monoclonal antibody being detected. An ambiguous assay is rarely likely to succeed and in most instances would result in wasted time and resources. The major advantage of the peptide antigen is that a simple and highly specific ELISA-based screening assay can be used to screen the hybridomas. Before one attempts to do the fusion, this ELISA assay should be optimized. An example of an indirect ELISA assay is described below.

8.2.2.1 ELISA

50 μl of positive or negative antigen (such as an irrelevant peptide) in either 200 mM sodium carbonate buffer, pH 9.6, or PBS is added to alternate columns of wells in a 96-well Immulon plate (Dynatech, Virginia, USA) and incubated at 37 °C for 2 h or at 4 °C overnight. The plates are then washed with PBS containing 0.05% (v/v) Tween 20 (polyoxyethylene sorbitan monolaurate). 200 μl of 5 mg/ml BSA in PBS is added to all the wells and incubated for 1 h at room temperature to block non-specific binding. The plates are then washed as before and 50 μl of varying dilutions (1/50, 1/200, 1/800, 1/3200 and 1/12800) of preimmune and immune sera in PBS is added to each pair of antigen positive and negative wells, in quadruplicate. PBS only is added to the wells of the first column (negative control). The first antibody is incubated overnight at 4 °C or for 2 h at room temperature. The plates are washed as described above and then 50 μl of horseradish peroxidase (HRP)-conjugated rabbit anti-mouse antibody (Dako, High Wycombe, UK) diluted in PBS containing 1% BSA is added at 1:500 dilution and incubated for 2 h at room temperature or 30 min at 37 °C. The plates are washed as before and 100 μl of the substrate solution (2 mg/ml *o*-phenylenediamine in 100 mM sodium phosphate buffer, pH 6.0, containing 1 μl/ml 30% hydrogen peroxide) is added. The reaction is stopped with 50 μl/well 1 M sulphuric acid and the absorbance measured using a Titertek Multiscan MCC/340 plate reader at 492 nM.

The absorbance values obtained (*y*-axis) with each sample are plotted against dilutions (*x*-axis). Modifications can be made to this assay to optimize it, by varying either the antigen concentration or the blocking agent used and its concentration, or the concentration of the peroxidase-conjugated second antibody.

Additional assays such as immunoprecipitation, immunofluorescence, immunohistochemistry and Western blotting can be done at this point.

8.2.3 Boosting the animal

The mouse that shows the best immune response is boosted intraperitoneally by adminstering 125 μg of peptide coupled to klh without any adjuvant, 3–4 days prior to the fusion.

8.2.4 Myeloma cells

We prefer to use NSO myeloma cells for fusion. These cells are HPRT negative and the hybridomas are selected by growing them in hypoxanthine–aminopterin–thymidine (HAT)-containing medium. A frozen stock of NSO cells should be grown at least two weeks prior to the fusion, in RPMI medium containing 10% (v/v) fetal calf serum (FCS). The cells must be maintained in an exponential growth phase by subculturing every third or fourth day, depending on the density of the cells (never allow the cells to become dense and overgrown). A rough indicator would be the colour of the medium, which would be orange initially but becomes yellow when cells are overgrown. The ideal time to split the cells would be when the medium changes to an orange-yellow colour. On the day prior to the fusion, the cells are split and resuspended at a concentration of 3×10^5 cells/ml in RPMI–15% FCS. At least 1×10^8 cells should be available for fusion (although only one-tenth of these will be used during the fusion).

8.2.5 Reagents and materials required for the fusion

1. PBS-J. This reagent prevents the splenocytes from clumping during teasing of the spleen. One gram of potassium dihydrogen phosphate (KH_2PO_4), 4.28 g of disodium hydrogen phosphate (Na_2HPO_4) and 2.25 g of sodium chloride (NaCl) are dissolved in 500 ml of distilled water and filter sterilized. The solution can be stored at room temperature.
2. Polyethylene glycol 1500 (PEG). This can be purchased from Boeringer–Mannheim as a sterile solution. It should be stored at 4 °C.
3. HAT and HT medium. These can be purchased as sterile 50 × concentrates from ICN Flow. Solutions are stored as 2 to 3 ml samples at −20 °C.
4. RPMI medium, five 2000 ml bottles.
5. 50 ml of FCS.
6. A non-immunized Balb/c mouse for laying down feeder cells.
7. At least 25 96-well Falcon tissue culture plates, to seed the hybridomas.
8. Four 5 ml sterile syringes; four 19 gauge sterile needles; 15 and 50 ml Falcon centrifuge tubes; 9 cm Petri dishes; 1 ml, 2 ml, 10 ml and 25 ml sterile disposable pipettes.

All the media and reagents are warmed to 37 °C in a water bath, 2 h prior to the fusion procedure.

8.2.6 Fusion

The NSO myeloma cells are harvested and washed twice in serum free RPMI medium. The cells are resuspended in 10 ml of serum free RPMI medium and a cell count done. At least 90% of the cells should be viable. The immunized mouse is killed by cervical dislocation. The mouse is placed on its back and the abdominal skin swabbed with 70% (v/v) alcohol. Using sterile forceps and scissors the skin is cut open to reveal the peritoneal cavity. 70% alcohol is squirted over the peritoneal layer prior to making an incision with a fresh pair of forceps and scissors. The spleen is identified and dissected free of stomach and pancreas and transferred to a sterile Petri dish containing 10 ml of PBS-J. The entire dissection is done in a room other than the one having the sterile hoods. The Petri dish with the spleen is transferred to the sterile hood after swabbing the bottom of the Petri dish with a tissue soaked in 70% alcohol.

Two of the 19 gauge needles are mounted on two 5 ml syringes and then used to tease apart the spleen, using one needle to fix the organ and the other to tease and break up the spleen. This is continued until the entire spleen has been broken down completely. The contents of the Petri dish are then transferred to a sterile 15 ml Falcon tube. An additional 5 ml of PBS-J is added to the Petri dish and rinsed by pipetting up and down repeatedly and then transferred to the same Falcon tube.

The tube is centrifuged at 1500 r.p.m. (3000 g) using a swing-out rotor for 5 min to sediment the cells. The cells are washed twice in serum free RPMI medium and then resuspended in 10 ml of serum free RPMI medium. 0.2 ml of the resuspended splenocytes is transferred to a plastic Biijou bottle using a sterile 1 ml pipette. 50 μl of the cell suspension is transferred to another Biijou containing 450 μl of 1% (v/v) acetic acid in PBS and mixed well. 500 μl of 0.2% (w/v) Trypan Blue is added to this cell suspension, mixed well and 10 μl is transferred to a Neubauer haemocytometer. The number of cells in the four large squares (which are subdivided into 16 smaller squares) is then counted. The value is divided by four to obtain the mean (a). The number of cells obtained = $a \times 10^4 \times 20 \times 10/\text{ml}$, where 20 represents the dilution done and 10 represents the total volume in which the cells have been resuspended.

The splenocytes and the myeloma cells are mixed in a ratio of 10 viable spleen cells to one viable myeloma cell (10:1) and spun down. The supernatant medium is removed as completely as possible and the cell pellet resuspended by flicking the tube.

The tube is placed in a beaker of water at 37 °C and 1 ml of PEG warmed to 37 °C is added dropwise into the tube, gently over a period of 1 min while continually stirring the mixture. Immediately, 1 ml of serum free RPMI is added over 1 min as above. Next, 4 ml of serum free RPMI is added over 3 min, down the side of the tube without agitation. The contents of the tube are gently tipped into a 50 ml Falcon tube. 20 ml of serum free RPMI and 20 ml of RPMI–15% FCS are added gently down the side of the tube without agitation over 10 min. The cells are then incubated at 37 °C for 1–2 h.

Feeder cells are meanwhile obtained from an unimmunized Balb/c mouse. Feeder cells can be obtained either from the spleen or from peritoneal macrophages. The latter is a more demanding procedure since the risk of puncture of the intestine is higher. Briefly, the mouse is killed as before and the skin incised to expose the peritoneal cavity. 5 ml of ice-cold RPMI is injected into the peritoneal cavity using a 25 gauge needle. The medium is then aspirated gently, keeping the bevelled edge of the needle pointing outwards to reduce the risk of perforation of the gut. Alternatively, one can obtain feeder cells from the spleen by dissecting out and teasing apart the spleen as before.

The feeder cells (peritoneal macrophages or splenic feeder cells) are washed twice in serum free RPMI and counted as before.

The hybrids and feeder cells are spun down and resuspended in 220 ml of RPMI–10% FCS medium containing $1 \times$ HAT. Using a multichannel pipette with sterile tips, 100 μl of the cell suspension is added to the wells in the first 11 columns of 22 to 25, 96-well Falcon tissue culture plates.

100 μl of NSO myeloma cells at 1×10^4 cells/ml in either RPMI–10% FCS or RPMI–10% FCS containing $1 \times$ HAT is added in quadruplicate to the last plate. These serve to assess the efficiency of HAT medium in killing the NSO myeloma cells.

On day 8, 100 μl of RPMI–10% FCS medium containing $1 \times$ HAT is added to the wells.

Generally, between days 10 and 15, the wells can be screened for clones secreting the required monoclonal antibody.

8.2.7 Screening of hybridomas

8.2.7.1 Reagents and materials required for screening

1. Falcon FAST ELISA system containing 15 troughs with beaded lids; 15 troughs with plain lids and 15 96-well plates for developing the reaction. Two units of this system are required (30 troughs each and 30 96-well plates).
2. PBS.
3. 200 mM sodium carbonate buffer, pH 9.6.
4. 3% (w/v) bovine serum albumin (BSA) in PBS, prepared and filter sterilized one hour before use.
5. 1:200 dilution of immune sera in PBS, filter sterilized.
6. PBS–0.5% Tween 20.

7. Rabbit antimouse IgG-horseradish peroxidase-conjugated antibody (Pierce, Chester, UK) at 1:6000 to 1:10 000 dilution in PBS–1% BSA.
8. Substrate solution: 2 mg/ml o-phenylenediamine in 100 mM sodium phosphate buffer, pH 6.0, containing 1 μl/ml of 30% hydrogen peroxide. This is to be prepared just before use.

8.2.7.2 Procedure

A 10 mg/ml stock solution of the peptide antigen is prepared in sodium phosphate buffer, pH 8.0. A final dilution of 5–10 μg/ml of the peptide antigen is prepared in sodium carbonate buffer, pH 9.6, and filter sterilized.

20 ml of the antigen solution is added to each beaded lid trough.

The beaded lids are incubated first in a trough containing the antigen for 2 h at 37 °C.

During this incubation, 50 μl of a sterile 1:200 dilution of immune sera or sterile PBS alone (or 1:200 dilution of preimmune sera) are added to two of the wells in the last column of each 96-well plate containing the hybridomas. These serve as positive and negative controls for the screening assay.

The beaded lids are then washed and blocked to prevent non-specific binding by incubating in sterile PBS–3% BSA in another trough for 45 min at 37 °C.

The beaded lids are then transferred to the plates with the growing hybridomas and incubated for 2 h at 37 °C in an incubator.

The lids are then washed in PBS–0.05% Tween 20 and incubated in troughs (the ones that were used to block non-specific binding are washed and reused) containing rabbit anti-mouse IgG–HRP conjugated antibody (Pierce, Chester, UK) at 1:6000 dilution in PBS–1% BSA for 30 min at 37 °C.

The beaded lids are washed as before and then placed in a 96-well microtitre plate containing 100 μl of the substrate solution (prepared as described above).

8.2.7.3 Precautions

1. It may be better to do the screening in batches over two days (e.g. 12 plates on the first day and the remainder on the second day). This ensures that if for some reason contamination occurs on one day, only half the hybridomas would be lost.
2. It is absolutely necessary to carry out all the procedures in a sterile environment, until the beaded lids are removed from the hybridoma plates.
3. The 96-well plates containing the hybridomas should be numbered both

on the sides and on the lids. The beaded lids should be numbered appropriately when placed into the respective hybridoma plate. Similarly the numbering on the developing plate should match that of its beaded lid.

4. The lids of the hybridoma containing 96-well plates should be used to cover the trough containing sterile BSA. These lids should then be used to cover the hybridoma plates once the incubation of the beaded lids in the hybridoma plates is completed.

The wells showing an unambiguous positive reaction are identified and their corresponding hybridoma wells are visualized to see if they contain cells. Hybridoma cells in strongly positive wells are transferred gently into a 24-well plate containing per well 1 ml RPMI–10% FCS containing 1 × HT. These cells are allowed to grow to sufficient numbers before the supernatant is retested, if necessary by a different assay, preferably the assay for which it would be used. The cells are then transferred into a 6-well plate and then into a 75 cm flask. The cells are maintained in RPMI–10% FCS containing 1 × HT for at least one month.

At each point of expansion it may be worthwhile to freeze some of the cells. However, the cells frozen prior to single cell cloning do not survive as well as those frozen after cloning.

It is also necessary to decide how many positive clones are to be taken forward. Up to two to three clones can be handled by a single person relatively easily. If one has initially more than three, then based on the assays done, about three should be chosen, preferably of different iso-types and the remainder frozen down.

Hybridomas secreting the required monoclonal antibody are cloned by limiting dilution at least twice so as to make sure that the clones obtained are really derived from a single cell.

8.2.8 Cloning by limiting dilution

Although it is not mandatory to have feeder cells during growth after fusion, it is essential during cloning by limiting dilution. Feeder cells are obtained as described earlier.

The hybridoma cells are counted and resuspended at a density of 80 cells/ml. 200 μl of the cell suspension is then plated into each well (16 cells/well) of the first two rows of a 96-well plate. 100 μl of feeder cell suspension at a density of 2–5 × 10^5 cells in 2 × HT medium is seeded into each of the other wells of the plate.

100 μl of the hybridoma cell suspension from the first row is transferred to the wells of the third row and mixed by pipetting. 100 μl of the cell suspension from the wells in the third row is then transferred and mixed with the cells in the wells of the fifth row. The process is repeated until

the eleventh row and the final 100 μl is discarded. The same procedure is repeated sequentially with the hybridoma cell suspension in the second row.

The first two rows will have 8 cells/well, the next two rows of wells will have 4 cells/well (rows 3 and 4). Rows 5 and 6 will have 2 cells/well; rows 7 and 8, 1 cell/well; rows 9 and 10, 0.5 cell/well and rows 11 and 12, 0.25 cell/well. Therefore, the rows 7–12 have a greater chance of having a clone derived from a single cell.

The plates are then incubated at 37 °C in a 5% carbon dioxide incubator. On day 8, 100 μl of 1 × HT medium is added to all the wells. Between days 10 and 14, screening for positive clones is done.

125 μl of supernatant from the wells of rows 7–12 are transferred to the wells of a 96-well ELISA plate which has been blocked with 1–3% BSA (w/v) in PBS. The blocking is done to prevent loss of any antibody binding to the ELISA plate. The plates are marked so as to facilitate identification of the parental clone. The screening can be done by an indirect ELISA as described earlier. Alternatively, the screening can be done with the FAST ELISA system.

Once the single cell cloning is done, the cells from a clone are expanded and the supernatant from the cells used for additional assays. The remaining positive single cell clones are frozen and stored under liquid nitrogen. A second cloning is done by repeating the procedure described above.

Once a large volume of hybridoma supernatant is obtained (200–1000 ml), purification of the monoclonal antibody can be done by protein A Sepharose affinity chromotography. The purified antibody can then be reassesed by different assays, including immunoprecipitation, Western blotting, FACS analysis and immunohistochemistry.

8.3 Polyclonal antibodies against peptides

Rabbits are commonly used to raise polyclonal antibodies. The immunization schedule and assessment of immune response have been described earlier. It is often useful to purify the antibodies from whole serum, particularly for immunostaining.

8.3.1 Affinity purification of polyclonal antibody

8.3.1.1 Reagents and materials required

1. Reactigel Hw-65F (Pierce)
2. Peptide antigen
3. 100 mM sodium borate, pH 9.0
4. 100 mM sodium phosphate, pH 8.0

5. 50 mM glycine, pH 2.2 (pH brought to 2.2 using hydrochloric acid)
6. Serum from the animals
7. Econo columns from Biorad
8. Fraction collector (Pharmacia)
9. Flat bottomed 2 ml screw-capped Eppendorf tubes or any other tubes to suit the fraction collector
10. Spectrophotometer
11. Quartz cuvette
12. Pasteur glass pipettes
13. Rubber teats
14. PBS
15. Universal tubes
16. Spectrum dialysis membrane

8.3.1.2 Coupling the peptide to Reactigel

1. 4 ml of Reactigel is washed twice with sodium borate (10 ml per wash). After removing the supernate from the second wash, 1 mg/ml of peptide antigen in sodium borate buffer is added (1 mg of peptide to 1 ml of Reactigel) to the Reactigel and mixed well. The mixture is tumbled overnight at 4 °C. The gel is washed several times (4–5) in sodium phosphate, pH 8.0, and twice with 50 mM glycine, pH 2.2. A final wash with sodium phosphate buffer, pH 8.0, is done.
2. 4 ml of thawed serum is added to the Universal tube containing the Reactigel coupled to the antigen and tumbled at 4 °C for 2–4 h. A small volume of the serum (15 μl) is set aside for ELISA later ('Before' sample).
3. The tube is spun briefly (one minute at 2000 r.p.m.) and the supernatant removed carefully without disturbing the Reactigel and stored separately ('After' sample).
4. The Reactigel is washed twice with sodium phosphate buffer, pH 8.0, and then transferred to a column after resuspending in about 4–5 ml of the same buffer.
5. After the Reactigel is packed into the column, it is washed with two volumes of sodium phosphate buffer, pH 8.0. A fraction collector is programmed for a 2 minute/fraction run. To the tubes loaded in the fraction collector, 50–100 μl of 1 M disodium hydrogen phosphate, pH 9.0, is added to neutralize the highly acidic eluate.
6. A few fractions are collected representing the last few ml of the sodium phosphate wash. Without allowing the column to run dry, 50 mM glycine buffer, pH 2.2, is added dropwise gently without disturbing the column and then filled up and connected to a reservoir source of glycine buffer. Fractions are collected and the absorbance at 280-nm measured in a spectrophotometer and the values plotted against the sample numbers.

7. The contents of the tube showing peak activity and the sample tubes before and after this tube are pooled and their pH checked. If the pH is less than 6.0, additional buffering is done to bring the pH to above 6.5.
8. The antibody solution is dialysed against PBS at 4 °C, overnight. The antibody can then be stored in measured samples at −20 °C.

The specific activity of the antibody can be determined using an ELISA assay similar to the one described earlier, using varying dilutions of before, after and antibody samples. The protein content is determined using Bradford assay.

Suggested reading

Campbell, A. M. (1985) *Monoclonal antibody technology. The production and characterisation of rodent and human hybridomas*, 1st edn. Elsevier Science Publishers B. V, Amsterdam.

Harlow, E. and Lane, D. P. (1988) *Antibodies, a laboratory manual*. Cold Spring Harbour, New York.

Tissue analysis

9

Modern morphological techniques in molecular cell analysis

LEE GORDON and ANNE E. BISHOP

9.1 General introduction

Major advances in molecular cell analysis over the past two decades have caused a revolution in various fields including the study of the diffuse neuroendocrine system. These advances were made with the introduction of reliable immunocytochemical methods and nucleic acid technology. Molecular cell analysis thus encompasses not only genes and mRNA but also the peptides/proteins encoded by them. In the following chapter, a practical guide is offered to the researcher or diagnostician wanting to phenotype or genotype cells. Although neuroendocrine cells are used as illustrative material in this chapter, the methods can be applied to a range of other cell types.

9.2 Immunocytochemistry

9.2.1 Introduction

Immunocytochemistry is a useful, easily reproducible tool in morphological analysis which is based on simple principles; an antibody, labelled in some way to allow its visualization, is applied to some biological material (cells, tissues, etc). The distribution of the resultant antibody–antigen reaction is then observed using microscopy. The specific localization of a fixed tissue antigen by an immunological method was first demonstrated in 1955 by Coons, Leduc and Connolly who used direct labelling of an antibody with fluorescein. Since then, the basic method has been modified, improved and adopted in various fields of morphological investigation. Most modifications have aimed at increasing the sensitivity of the method and the resolution of the final immunostain. Current methods allow sensitive qualitative and quantitative assessment of cells and their components but carry certain caveats, mainly that interpretation of immunostains can only be made satisfactorily if the antibody used is both specific and well characterized and that a negative or poor result may not mean that a cell does not contain a certain antigen but, at the time of staining, it contains too little for adequate immunoreaction. Thus, combination of immunocytochemistry with *in situ* localization of endogenous nucleic acid species is providing not only more accurate identification of cell types but also insight into the biology and activity of the cells.

The following describes the application of immunocytochemistry to the study of the diffuse neuroendocrine system.

9.2.2 Tissue fixation/processing

As immunocytochemistry involves the immersion of cells/tissues in a series of aqueous solutions, it is necessary to fix tissues prior to immunostaining. This must be done with optimal preservation of both the morphology of the tissue and the antigenicity of the molecule under study. Originally, fixation of tissues for immunocytochemistry of neuroendocrine products, mainly small peptides, was a complex, specialized procedure involving such methods as snap freezing, freeze-drying and vapour fixation. This was necessary largely because of the relatively poor quality of the antibodies. Peptide antigenicity was often maximized at the expense of tissue morphology. However, as the immunocytochemical techniques increased in specificity and the available antibodies in quality, it became clear that more straightforward, routine tissue fixation methods can often be used. Normal endocrine cells, or their tumours, can usually be detected in formalin-fixed, wax-embedded tissues using currently available antibodies (see Appendix for recommended sources) to specific peptides, amines or general neuroendocrine cell markers (e.g. neuron-specific enolase, protein gene product 9.5, chromogranins, etc.). The main proviso for use of routinely fixed tissues is that sections should be picked up on slides coated with an adherent (such as poly-L-lysine; Protocol 1) and allowed to air dry at room temperature. Where satisfactory immunoreactivity cannot be obtained, some kind of 'unmasking' procedure may be required. In the past, this has usually taken the form of some form of proteolytic digestion employing such substances as trypsin (Towle, Lauder and Joh, 1984), Pronase (Denk, Radaszkiewicz and Weirich, 1977) or pepsin (Reading, 1977), all of which can have obvious deleterious effects on tissue morphology. A more recent innovation has been the application of microwave technology to the process of antigen retrieval (Gown, de Wever and Battifora, 1993; Leong and Milios, 1993). Routinely processed sections are immersed in citrate buffer and exposed to microwaves in a standard oven. Although this procedure is reported to increase antigenicity for a range of antibodies, it has yet to be adopted widely in routine practice.

Specialized methods of fixation and tissue processing are still required, however, in certain circumstances. Firstly, certain antibodies to neuroendocrine cell products are known to favour a particular fixation procedure. For example, optimum immunoreactivity for the inducible form of the free radical generator nitric oxide synthase is obtained in tissue fixed in weak (1% v/v), freshly prepared formaldehyde solution (Buttery *et al.*, 1993). Thus, tissue fixation/processing should always be considered as a possible variable when immunostaining with newly acquired antibodies is being optimized. Secondly, immunostaining of particular types of tissue/cell preparations may require specialized processing. Antigenicity in separated cells, for example, is often best preserved by post-fixation of frozen sections of a spun pellet.

The other main area where specialized fixation methods are needed is the immunostaining of nerves. With good antibodies, some immunoreactivity for small antigens (e.g. peptides, GABA (gamma amino butyric acid), etc.) and adequate immunoreactivity for larger ones (e.g. neurofilaments, etc.) can be obtained in routinely fixed material. However, for maximum antigenicity coupled with resolution of the immunostaining that allows examination of the fine morphology of the nerves, a different approach is needed. A number of fixatives have been recommended for the immunocytochemistry of nerves, one of the best and most widely used being Zamboni's solution (Protocol 2). Once fixed, sections can be cut in a cryostat or Vibratome. To allow better study of neural projections, whole mount preparations are preferred (Davis, 1993) (Figure 9.1). To aid antibody penetration into thick sections or whole mount preparations, the addition of a surfactant (e.g. 0.2% (v/v) Triton X-100) to antibody dilutions and washing buffers is recommended. Also, initial dehydration of the tissues through graded alcohols to inhibisol and back to an aqueous

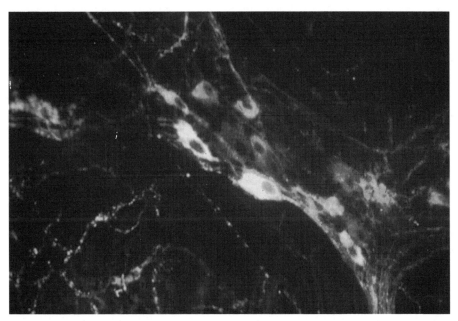

Figure 9.1. Nitric oxide synthase localized to ganglion cells and fibres of the myenteric plexus. Indirect immunofluorescence was performed using antibodies to the constitutive neuronal form of nitric oxide synthase on whole mount preparations of the myenteric plexus of the guinea pig stomach. Each focal plane of the thick preparation was scanned by a confocal laser microscope and the figure shows the reconstructed image.

solution can aid penetrance. Thick preparations are best immunostained free-floating.

Satisfactory tissue and antigen preservation obviously depend on the freshness of the original tissue/cell sample. Thus, samples for immunostaining of neuroendocrine cell products must always be taken and fixed as soon as possible after death of or removal from the individual. For experimental animals, it is best to perfuse the tissue with fixative wherever possible and then complete the fixation procedure by immersion of the dissected tissues for the required length of time. When large surgical samples of human tissues are available for study it is also preferable to fix these by perfusion (Wakefield *et al.*, 1989; Domoto *et al.*, 1990).

9.2.3 *Immunocytochemistry procedures*

9.2.3.1 *Antibodies*

The vast majority of the antibodies used in current immunocytochemical procedures are of the IgG class. Antibodies to peptides and proteins recognize an epitope of only three to eight amino acids, so thorough characterization of antibodies is a basic prerequisite of reliable immunocytochemistry of neuro-endocrine cells. Despite the advent of hybrid myeloma technology for the production of monoclonal antibodies, polyclonal antisera are still the most widely used form. This is despite the fact that polyclonal antisera carry a series of complications regarding their specificity. For example, the peptide/protein immunogen may be purified or synthesized. If 'purified', or extracted, it may still contain impurities that give rise to contaminating antibodies. Furthermore, peptides are not very immunogenic and, in order to obtain high titre antisera, they must be coupled to large immunogenic carrier proteins. These, in particular serum albumin, are notoriously impure and it is advisable to preabsorb possible contaminants by incubating sera with the carrier. However, affinity purification of antibodies, with Sepharose or polyacrylamide beads, is a preferred means of cleaning up an antiserum.

In view of the small size of the epitope, a polyclonal antiserum to an entire peptide will contain a mixture of antibodies recognizing different parts of the molecule. The disadvantage of this is that sequences shared by different peptides/proteins can give rise to cross-reacting antibodies. This can be avoided by raising antibodies to specific, small sequences of the peptide/protein. However, in view of possible antigen damage or masking during tissue processing, it is best to use an antiserum that contains antibodies against different, specific regions of the molecule under study. Thus, in theory, a pool of monoclonal antibodies to synthetic, specific peptide sequences would appear to be the ideal solution but, unfortunately, such reagents are not yet widely available.

9.2.3.2 Testing antibody specificity

As mentioned above, detailed knowledge of the characterization of an antibody is required if any sound conclusions are to be drawn from the results of an immunostain. This begins with information on the original immunogen, the means by which antibodies were raised and the form of antibody purification, if any. There are several ways in which an antibody can be tested subsequently for its specificity. Prior to immunostaining, the antiserum can be incubated with purified antigen. At optimal dilutions, the immunostaining should be quenched by formation of maximum antibody–antigen complexes. However, the results are only as reliable as the antigen; an impure antigen may absorb contaminating as well as specific antibodies. In theory, radioimmunoassay could be used but the conditions under which the antibodies are used and the requirements for each technique are not parallel. Immunodiffusion and immunoelectrophoresis can be used for precipitating antigens. ELISA provides a simple rapid means for testing antibody specificity but the results can be affected by the conformational presentation of the antigen. Immunoblotting, either dot blot assay or Western blotting, can also provide quantitative data. Western blotting is a comparatively complex and lengthy procedure but it gives additional information on antibody specificity and is particularly useful for polyclonal antibodies where there is a mixture of epitopes. The method also allows for assay of tissue extracts, providing information on the possible cross-reaction of a particular antiserum with related or unrelated substances.

9.2.3.3 Immunostaining methods

(a) Immunofluorescence

The main drawback to directly labelled antibodies, as initially used by Coons *et al.* (1955), is that each antibody has to be labelled separately. This is hardly practical in today's immunocytochemical laboratories where the number of antibodies in use can run into hundreds. Direct labelling was soon superseded by the introduction of a double layer method where the first antibody layer is raised against the molecule of interest and the second consists of a labelled antibody recognizing the IgG of the species that donated the first antibody. This indirect approach is exemplified by the indirect immunofluorescence technique (Protocol 3), a relatively simple procedure. The immunostains are not permanent, although addition of various substances, including polyvinyl alcohol, Entellan and sodium dithionite, to the mounting medium has been reported to extend their 'shelf-life'. In addition, a fluorescence microscope is required for viewing the preparations. However, the technique gives strongly contrasted immunostains and, when counterstains such as Evans Blue or Pontamine Sky Blue are used, can provide very striking images for photography (Figure 9.1). Multiple immunostains can be carried out using antibody complexes from different species labelled with fluorochromes that are visible in light of different wavelengths. Another advantage of fluorescence immunocytochemistry is that immunostained preparations can be analysed quantitatively using a confocal laser microscope (see later) (Figure 9.1).

(b) Immunoenzyme methods

The first immunostains visible in transmitted light microscopy were based on the use of an immunoenzyme system. An indirect, two-step method was developed using horseradish peroxidase as a label on the second antibody (Nakane 1967; Nakane and Pierce, 1967; Sternberger, 1979) (Protocol 4). The classic technique for visualizing this enzyme is the diaminobenzidine (DAB) method of Graham and Karnovsky (1966) (Protocol 4); peroxidase catalyses the conversion of hydrogen peroxide to water and oxygen and, when this reaction is carried out in the presence of a light-visible coupler (e.g. DAB), a coloured product is deposited at the sites of immunoreaction. As well as being light-visible and far less prone to fading, the other advantage of this immuno-enzyme approach over immunofluorescence was the increased efficiency of antigen detection. In other words, because the enzyme reaction can be ampli-fied and enhanced, an adequately visible reaction product can be obtained with less primary antibody than is needed for immunofluorescence.

DAB is a suspected carcinogen and needs to be handled with appropriate safety precautions. Chromogens other than DAB can be used for peroxidase histochemistry, e.g. 3-amino-9-ethylcarbazole, 4-chloro-1-napthol. As well as being possible alternatives to DAB, these other chromogens provide the opportunity for carrying out multiple immunostains (see later).

An obvious means to intensify the labelling of immunoreaction sites using peroxidase labelling was to increase the number of molecules of enzyme attached to IgG. Chemical conjugation of peroxidase to antibody was often associated with a loss of catalytic or antibody activity. This was avoided by the introduction of a method that used immunological binding of peroxidase to antibody; the so-called unlabelled antibody or peroxidase anti-peroxidase method (Sternberger, 1979) (Protocol 5) (Figure 9.2). In this procedure, the first layer antibody received a second layer of unlabelled IgG raised against the donor species of the primary antibody. The final layer consisted of two anti-peroxidase IgG molecules linked by their attachment to a peroxidase molecule and with additional molecules on their free Fab fragments. Thus, three peroxidase molecules were present at each site of attachment of one primary antibody molecule. Further amplification of the immunostain can be achieved by repeating the second and third layers, although in practice this increases background staining. As with the two-step peroxidase method, differ-ent chromogens can be used, allowing multiple immunostaining.

Although the peroxidase methods have been the most popular, other im-munoenzymatic techniques exist. Alkaline phosphatase can be used as an antibody label and developed with substances such as Fast Red, Violet or Blue and, as has recently been reported, Salmon and Magenta Phosphate (Avivi,

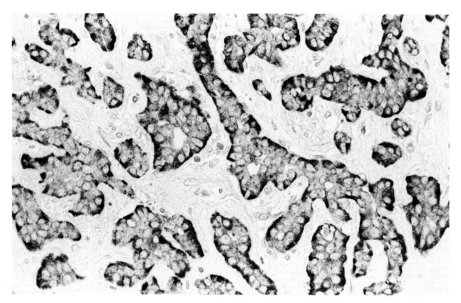

Figure 9.2. Chromogranin immunoreactivity in cells of a gastric neuroendocrine tumour. The general neuroendocrine marker protein chromogranin was immunostained using the peroxidase anti-peroxidase technique in a neuroendocrine (enterochromaffin-like cell) tumour infiltrating the oxyntic mucosa of the human stomach. The preparation was counterstained with weak Haematoxylin.

Rosen and Goldstein, 1994). Glucose oxidase can also be used and detected by formazan precipitation. Using primary antibodies from different species, the various immunoenzyme methods can be obtained to give multiple immunostains.

(c) Avidin–biotin complexes

Avidin, a protein derived from egg white, shows high affinity for biotin and, thus, any substance labelled with biotin can be detected using an avidin-conjugated, visible substance. This is the basis of two immunocytochemical methods: the labelled avidin–biotin technique (Guesdon, Ternynck and Avrameas, 1979) and the more widely applied avidin–biotin (peroxidase) complex (ABC) method (Hsu, Raine and Fanger, 1981*a*, *b*) (Protocol 6) (Figure 9.3). In the latter, a first layer antibody to the molecule under study is applied to a tissue. The second layer consists of IgG molecules raised against the first donor species and each labelled on the Fc fragment with four biotin molecules. The third layer is formed by avidin, which has four binding sites for biotin conjugates. Each avidin molecule has three sites occupied by biotin labelled with peroxidase. The fourth binding site is free and ready to bind with a biotin molecule on the IgG of the second layer. Thus, twelve peroxidase molecules are present at each antibody binding site.

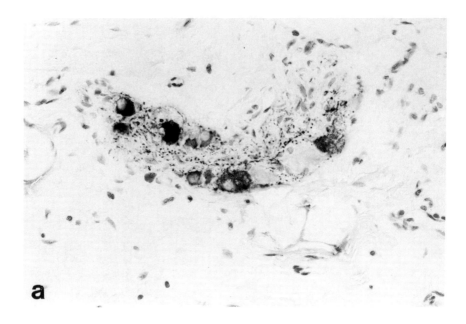

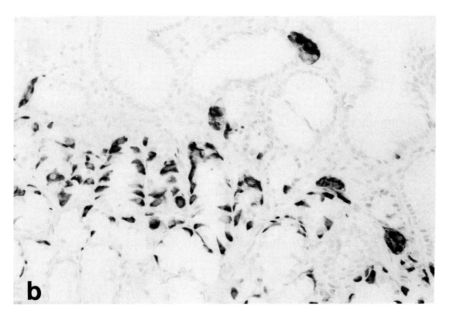

Figure 9.3. Vasoactive intestinal peptide (VIP) (a) and gastrin (b) localized to components of the neuroendocrine system of the human stomach. Using the avidin–biotin–
peroxidase complex (ABC) technique, the neuropeptide VIP was immunostained in
nerve fibres and ganglion cells of the submucous plexus of the human stomach (a).
Gastrin was immunostained in hyperplastic G cells in the antral mucosa using the same
technique (b).

9.2.3.4 Controls

Specific control procedures for each of the immunostaining techniques mentioned are given in the respective appendices. However, certain general points exist and are worth emphasizing. Firstly, even the most accomplished technician may occasionally make an error, e.g. in diluting an antibody, forgetting a layer, so it is important to include a known positive control tissue and relevant antibody, wherever possible. The use of non-immune serum as a primary layer or omission of each of the antibody/label layers in turn will provide preparations showing sites of unwanted labelling of tissue components, so that these can be eliminated when evaluating the final immunostaining results. This will also show whether there are any background reactivities (e.g. endogenous enzyme activity, autofluorescence) that escape the blocking steps in the immunostaining technique.

9.2.3.5 Multiple immunostains

These have already been mentioned in the preceding passages. Reasons for using multiple immunostains include examination of the relationship between two or more cell/tissue types within a single section and investigation of possible dual localization of antigens. Probably the easiest method of making multiple immunostains on the same section is to use two or more antibody systems, each recognizing a separate antigen and raised in different species. The final labels can be fluorochromes activated at different wavelengths of light or chromogens of different colours. For establishing dual localization of antigens, double immunofluorescence can be used and then the same tissue component can be examined under different UV filter systems (Pryzwansky, 1982). The use of light-visible immunostains is more complicated. Although double staining is used for co-localization of antigens the results are not always clear-cut, as a mixture of two chromogens can give indistinct results. Serial sectioning, where possible, is often preferable. There are methods that rely on antibody elution. In these, a section is immunostained, the positive components are photographed and then the antibody complex is eluted (Nakane, 1967; Tramu, Pillez and Leonardelli, 1978). The immunostain for the second antigen is carried out, photographs are taken and the results of the two stains compared. In theory, this technique should be useful for use when the only antisera available are from the same species. However, in practice removal of antibody complexes often reduces antigenicity and compromises tissue morphology. A recently proposed method for overcoming the problem of using antibodies from the same species in multiple immunostains involves using a secondary polyclonal monovalent Fab antibody in the first layer (Negoescu et al., 1994). This blocks the epitopes of the primary antibody and prevents attachment of the secondary antibody in the second layer. As the fragment is monovalent, the primary antibody in the second step will not adhere.

Immunoenzyme and immunofluorescence techniques can be combined. For example, combination of an alkaline phosphatase method with immunofluorescence has been reported to provide good morphological identification of doubly stained cells in wax sections (Tao *et al.*, 1994).

9.2.3.6 Quantification

There has been a continuing need in the field of immunocytochemistry for a means to provide accurate, precise quantification of results. Early attempts at image analysis were based on point or component (e.g. cell) counts. These involved an observer or, preferably, two or more independent observers, viewing preparations under a microscope and carrying out counts that were related to parameters such as field area, tissue area or length or intersection with an overlying grid. Some of the data derived were unreliable. For example, counts were compared from dissimilar tissues (e.g. non-inflamed and inflamed) and differences in results were more likely to be due to general variations in the tissues rather than in the population being counted. Counts were generally made on tissue sections and the results were often of questionable significance for assessment of the tissue as a whole, despite the application of mathematical formulae. However, used on a comparative basis, with strict controls, simple counts retain a useful place in quantitative immunocytochemistry where individual tissue components are under study.

Where more sophisticated and detailed quantification is needed, for example in the investigation of ramifying neural networks, computer-assisted image analysis is a necessity (Figure 9.1). The state-of-the-art integration of computers and optics is confocal laser microscopy, which allows quantitative, three-dimensional analysis of immunofluorescent tissue preparations. In the microscope, illumination is by a single spot of light, which scans across the planes within the preparation and retains an image of each. The nature of the light makes scatter minimal and the computer digitally enhances the data collected. The images are reconstructed in three dimensions. Recent developments in the application of confocal laser microscopy have moved immunocytochemical quantification further towards its final goal of fully automated image analysis. Algorithms have been presented that provide the means for fully automated tracing of nerve pathways (Cohen, Roysam and Turner, 1994) and quantitation of cell populations (Roysam *et al.*, 1994).

9.3 *In situ* Hybridization

9.3.1 Introduction

Since its conception in 1969, *in situ* hybridization (ISH) has been extensively adapted and applied to the localization of nucleic acid within a wide range of tissues. ISH offers a major advantage over other molecular cell biology techniques in that it provides precise information of the site of gene expression to cells within a heterogeneous population, thereby complementing the immunohistochemical demonstration of antigen. By combining these approaches it is possible to localize mRNA species to a particular cell population and determine whether its translated protein is rapidly degraded or transported, or is stored but not synthesized by the cell.

Numerous methodological approaches have been adopted to address a range of inquiries in the field of pathology and in fundamental research, which may require the localization of DNA sequences on chromosomes, or the detection of mRNA and viral nucleic acid. However, while many aspects of the technique may have been adapted for the specific requirements of scientists or clinicians, the various protocols all demand attention to three main factors: First, maximum sensitivity must be attained and this is dependent on the accessibility of the target, the method of probe labelling and detection and the type and length of probe. A critical requirement of the technique is to achieve the correct balance between target accessibility and the preservation of tissue morphology. Second, the unequivocal determination of the specificity of the hybridization signal is dependent on the stringency conditions imposed on the technique and the adoption of rigorous controls. Finally, signal resolution requires consideration and this is dependent on the method of probe labelling and detection.

This aspect of the chapter will focus mainly on the protocols that we have adopted for the localization of mRNA to tissue sections using complementary single-stranded RNA probes, labelled either by isotopic or non-isotopic methods. However, alternative strategies for the morphological detection of nucleic acid will be outlined and further recommended reading is cited. The protocols detailed at the end of the chapter are in no way definitive and may be adapted according to the probe used and the type of tissue to be investigated.

9.3.2 Generation of single-stranded RNA probes

The cDNA of interest is cloned in the polylinker region of an appropriate vector which is flanked by two different polymerase initiation sites. The plasmid is linearized with selected restriction enzymes so that the vector sequences are not transcribed, which would otherwise result in high background as a consequence of inappropriate hybrid formation. Single-stranded RNA probes are then synthesized (Protocol 9) using the appropriate polymerase to transcribe sequences downstream of the initiation sites, thereby allowing the generation of either 'antisense' or 'sense' RNA. The 'sense' form of RNA is useful as a negative control in subsequent hybridization experiments (see below). DNA generated by the polymerase chain reaction (PCR) may also be subcloned into an appropriate vector system (for example, using Stratagene's PCR-Script SK(+) Cloning Kit) prior to linearization and RNA probe generation. Alternatively, primers for PCR may be designed to incorporate a promoter region (Sitzmann and LeMotte, 1993), which eliminates the possibility of transcribing 'contaminating' vector sequences and the need for time-consuming cloning experiments. The main advantage of employing RNA probes is the strength of the hybrids formed compared with those that include DNA. However, their use must be accompanied by stringent precautions to prevent RNAase contamination, including the use of baked glassware, and autoclaved microcentrifuge tubes and DEPC (diethylpyrocarbonate)-treated solutions.

Alternative strategies for probe production include the enzymatic synthesis of DNA probes, which may be double or single stranded. The use of single-stranded DNA in ISH generally provides greater sensitivity as there is no competition between target and the complementary sequence, so the optimum concentration of probe is available for hybridization. Double-stranded probes can be prepared by nick-translation, random priming or PCR in the presence of labelled nucleotide and are denatured thoroughly before use. Single-stranded DNA probes can be enzymatically generated by primer extension on single-stranded templates or by PCR (Hannon et al., 1993), or by chemical synthesis of oligonucleotides. Oligonucleotides are attractive for use in ISH in that they are commercially available and may be produced in large quantities to high purity. Due to their small size they can penetrate cells and tissue easily, but the main disadvantage of their use is that relatively few labelled nucleotides may be attached to their 3' or 5' termini, which consequently reduces the sensitivity of the technique. This limitation may be overcome by employing a cocktail of oligonucleotides which are complementary to different regions along the target molecule.

9.3.3 Probe labelling

Probe labelling may be achieved using radioactive isotopes, which employ an autoradiographic detection system, or with non-isotopic labels that require immunohistochemical detection. The choice of isotope will depend on the requirements of resolution, speed of detection and stability of probes. While ^{3}H-labelled probes offer optimal subcellular resolution, prolonged autoradiographic exposure is a limitation of its use. The ultrastructural localization of nucleic acid requires the selection of a label with maximum resolution: Trembleau, Calas and Fevre-Montagne (1990) have used synthetic oligonucleotides that were 3' end labelled with tritium to localize oxytocin mRNA within the cells of rat hypothalamus. The speed of detection is optimized when using ^{32}P-labelled probes, but the relatively poor resolution achieved with this isotope far outweighs its advantages in the morphological detection of nucleic acid. However, a good compromise is found in the use of ^{35}S-labelled nucleotides in probe labelling as they provide resolution to single cells and offer detection times of about one week. A drawback of using this isotope is in the requirement for reducing agents such as dithiothreitol in the hybridization medium and in subsequent washing solutions, to avoid background problems that are associated with sulphur oxidation. Nucleotides labelled with the isotope ^{33}P are now available (Amersham International and DuPont) for use in probe generation, which exhibit characteristics similar to ^{35}S-nucleotides, without the inherent background problems experienced with the latter, although cost may preclude their use.

While the use of isotopes in ISH offers a number of advantages, especially in the area of sensitivity, the main disadvantages in their use are their inherent instability, as well as being expensive and requiring defined safety protocols for their handling and the disposal of waste. Non-isotopic labelling has therefore been developed and the three main systems currently in common usage employ the haptens digoxigenin, biotin and fluorescein (see Boehringer Mannheim Biochemica: Application Manual, 1992). Hybrid visualization is then achieved by an immunohistochemical detection system using enzymes (alkaline phosphatase, peroxidase), fluorescent dyes (fluorescein, rhodamine) or colloidal gold, coupled with high affinity antibodies (see below).

The incorporation of isotopic label during probe generation may be assessed by scintillation counting of a small volume of the reaction following precipitation, compared with the amount of isotope originally added. For non-isotopically labelled probe the amount synthesized may be assessed by spotting a sample onto a 1% (w/v) agarose gel with ethidium bromide and comparing its optical density with that of standards.

9.3.4 Probe size

Probe length influences the sensitivity of the technique and level of background obtained: While longer probes are less able to penetrate the tissue the specific hybrids formed will be of greater strength due to the higher number of hydrogen bonds. By contrast, short probes will penetrate the tissue or cell preparation easily but will form less stable hybrids and may result in background problems due to non-specific binding. We find that the optimal length of RNA probe is about 200–500 bases, however, optimal probe length may vary according to the tissue system being investigated (Wilkinson, 1992). Shorter RNA probes can be produced by limited alkaline hydrolysis (Protocol 10) or by the cloning of shorter specific fragments of cDNA for generating riboprobes in the first instance, using the appropriate restriction enzymes. Alternatively, penetration difficulties experienced with longer probes can be overcome by more extensive permeabilization of the tissue with detergents and proteolytic enzymes (see below), however this may result in inferior tissue morphology.

9.3.5 Tissue preparation

The approach adopted for the preparation of tissue is dependent on whether the technique is to be performed on tissue sections, whole mounts or cell cultures. We generally preserve excised tissue by immersion-fixation in 4%

(w/v) paraformaldehyde for about 6 h at 4 °C, depending on the sample size, and subsequently process for cryostat sectioning at a thickness of 10 μm (Protocol 11). However, many groups adopt alternative strategies with respect to the choice of fixative and embedding material (for further reading on the choice of fixative see Kiernan, 1990). If the ISH technique is to be combined with immunocytochemistry, some consideration must also be given to the preservation of antigen (see below).

Paraformaldehyde is a cross-linking fixative, and these are generally used in preference to precipitating fixatives for the optimum retention of nucleic acid. Often a compromise must be attained between accessibility of target and the preservation of tissue morphology, which in turn will influence the retention of nucleic acid. Alternative cross-linking fixatives are formaldehyde and glutaraldehyde, but while these allow for better preservation they may be less useful as they limit probe accessibility, although penetration may be subsequently enhanced by treating the tissue with proteases and detergents. Frozen tissues can be fixed after sectioning but there may be a greater loss of target nucleic acid by ubiquitous RNAases than if the tissue is fixed prior to freezing and subsequently sectioned. While the technique using wax-embedded tissue is apparently less sensitive than with frozen tissue (Saeger et al., 1991), there are obvious advantages to its use, not the least in that archival tissue is available for such studies. This approach also allows for superior morphological preservation, thinner sectioning and storage at room temperature. Thomas and Rose (1991) demonstrated that archival material fixed in 10% buffered formal saline gave superior preservation of human papilloma virus DNA when compared with fixation by formalin in acid alcohol.

The glass slides onto which tissue sections are lifted may be subbed using a number of compounds including poly-L-lysine and gelatin, but we are currently using Vectabond-treated slides (Vector Laboratories) as this treatment proves effective in retaining tissue throughout the technique. Acetylation of the slides may also reduce non-specific adherence of probe to glass.

9.3.6 Tissue pretreatment

Prior to hybridization the tissue is treated to optimize the efficiency of hybridization while limiting non-specific background to a minimum (Protocol 12). For wax-embedded tissue sections, organic solvents are employed to remove all traces of wax. Next, target accessibility is maximized by detergent and protease digestion. Enzymatic digestion must be carefully monitored to ensure that the morphology of the tissue is not destroyed, and the optimum incubation time should be initially determined for each tissue investigated. For cell culture preparations this treatment is unnecessary and, moreover, may be detrimental to the localization of nucleic acid. Following protease treatment the enzyme is inactivated and a brief post-fixation step included in the protocol to prevent excessive leakage of target RNA. Non-specific, electrostatic interactions between probe and positively charged amino groups within the tissue section yield background problems that may be prevented by treatment with acetic anhydride. There is considerable divergence of opinion whether in fact this step is actually necessary, but it is probably more useful in applications with probes of longer length and at higher than optimal concentration to distinguish differential mRNA expression.

To establish the appropriate conditions for the preparation and pretreatment of a given tissue it may prove useful to employ a labelled synthetic oligonucleotide, composed of consecutive thymidine residues, to assess the level of retention of poly(A)$^+$ RNA within the section (Pringle *et al.*, 1989; Montone and Tomaszewski, 1993). Similarly, Montone, Budgeon and Brigati (1990) have determined the availability of DNA for ISH using a consensus sequence for the human genome repeat alu.

9.3.7 *Hybridization conditions*

The optimum conditions for hybrid formation in solution may be predicted by calculating its T_m, or melting temperature, which is defined as the temperature at which half of the hybrids dissociate. T_m is influenced by a number of factors including the nature of the probe and target, probe length and base composition and the components of the hybridization medium (Wilkinson, 1992). The presence of a tissue matrix inevitably influences the conditions required for probe hybridization, reducing the stability of the hybrids, possibly due to steric hindrance, and modifying the reaction kinetics. For this reason the appropriate hybridization temperature must largely be determined by carrying out empirical tests. Hybridizations using RNA probes are generally carried out under conditions of moderate–low stringency, about 25 °C below the estimated T_m, and the reaction is usually complete after about 6 h. By contrast, higher stringency conditions require prolonged incubation of probe with section, for about 16–24 h. Formamide has a destabilizing influence on the formation of hybrids and when included in the hybridization medium allows a reduction in the temperature required for specific annealing of probe with target: an increase in formamide concentration of 1% reduces the T_m by 0.61 °C for DNA:DNA hybrids and 0.35 °C for RNA:RNA hybrids. This has advantages for maintaining tissue morphology although in some instances the presence of formamide may reduce the sensitivity of detection when using non-isotopically labelled oligoprobes (Thomas, Davis and Williams, 1993).

Probe concentration is generally in excess of target to allow saturation, which occurs at 0.3 ng/μl/kb for RNA probes. Probe concentration that is greatly in excess of target should be avoided as high levels of background or 'noise' may ensue. The effective probe concentration can be increased by using dextran sulphate, which is an exclusion polymer.

9.3.8 Post-hybridization washes

The hybridization conditions selected are usually of moderate stringency, which can result in higher levels of background. However, this is offset by a more thorough washing protocol, which is intended to optimize signal-to-noise ratio. Sections are washed in solutions of increasing stringency, that is increasing temperature with decreasing salt concentration, to remove non-specifically bound probe or probe that is bound to sequences that are similar to but distinct from the target. We generally carry out a protocol (Protocol 14) that includes washes at room temperature as well as at the temperature used for hybridization. Detergents such as sodium dodecyl sulphate (SDS) may be included in the washing solutions to further reduce background. In ISH with RNA probes that have a tendency to be 'sticky', RNAase may be used to digest single-stranded molecules and further reduce background. Prior to this treatment all RNAase inhibitors, which include SDS, dithiothreitol and formamide, must be eliminated from the washing solutions. (Take precautions when using RNAase to avoid contamination of future ISH experiments!)

9.3.9 Hybrid detection

The method of hybrid detection depends on the type of probe label originally selected. For radioactive probes autoradiographic techniques are required, including access to darkroom facilities. Low-level image resolution may be obtained by exposure to autoradiographic film, such as Amersham's β-max Hyperfilm, for a time interval of 6–16 h for ^{35}S-labelled preparations, depending on the specific activity of the probe and density of signal (Figure 9.4). Greater resolution is obtained by dipping the slide in liquid nuclear track emulsion (Figure 9.5). The slide is air-dried, exposed for the required time interval at 4 °C and developed (Protocol 15). This is followed by some form of counterstaining of the tissue to precisely localize the site of gene expression. The dipping procedure is critical as artefacts may result from mechanical stresses (see below).

This visualization of non-isotopically labelled hybrids may be achieved by using enzyme-conjugated hapten binding protein and a substrate that yields a coloured precipitate (Figure 9.6). While non-isotopic ISH is generally considered to offer lower sensitivity (Xerri *et al.*, 1992) it is finding an increasing number of applications with the development of various enzyme systems and is used particularly in the detection of viral nucleic acid (Fleming, 1992; McQuaid and Allan, 1992). Increasing sensitivity has been obtained by various methodological adaptations including high concentrations of probe, thorough denaturation of probe and target (for DNA) and prolonged incubation of sections with substrate. Of the enzymes used in the detection of hybrids, alkaline phosphatase coupled with Nitroblue Tetrazolium and bromochloroindolyl phosphate is currently considered to provide greatest sensitivity (Protocol 16). Endogenous alkaline phosphatase is usually abolished by levamisole pretreatment (Lloyd,

Figure 9.4. Atrial natriuretic peptide (ANP) mRNA localized within the human fetal heart. ISH was performed using a complementary (antisense) RNA probe labelled with ^{35}S-UTP. Hybridization was carried out at 42 °C and the section was exposed to autoradiographic film for 6 h at room temperature to obtain a low power perspective of the distribution of nucleic acid. 1.5 M dithiothreitol was included in the hybridization buffer and stringency washing solutions to minimize the background that is associated with the use of this isotope.

Jin and Fields, 1990). The peroxidase system requires a preliminary blocking step with hydrogen peroxide and employs the diaminobenzidine/imidazole reaction. This produces a brown product, which is insoluble in alcohol, so the preparations may be dehydrated before counterstaining. Various adaptations of the visualization protocols will produce precipitates of different colours, which allow for co-localization studies of different nucleic acids (Herrington *et al.*, 1989; Dirks *et al.*, 1990).

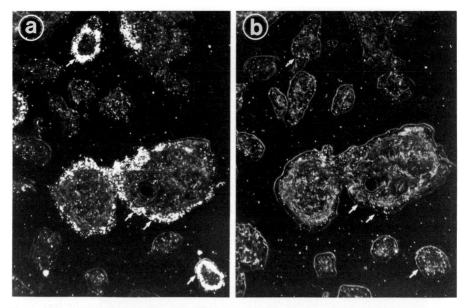

Figure 9.5. ANP receptor type C mRNA localized to the syncytiotrophoblast (arrows) of human placenta (a). ISH was carried out using an ^{35}S-labelled probe, which was incubated with the section at 50 °C. The preparation was dipped in nuclear emulsion to provide good microscopical resolution. Relatively little signal was observed after 7 days exposure so the slide was viewed using dark-field microscopy. An adjacent section (b) was incubated with sense probe to provide a negative control.

Figure 9.6. Insulin mRNA in a pancreatic islet. RNA probe was labelled with digoxigenin and visualized using the alkaline phosphatase system. Maximum resolution is achieved with this approach and coloured precipitate is found within the cytoplasm of single cells, whereas the nuclei do not exhibit hybridization. (Photograph reproduced with the permission of Mr P Facer, Department of Histochemistry, RPMS.)

Fluorescent labels such as fluorescein and rhodamine give high resolution and low background but are limited in their use as they offer less sensitivity than enzyme labelling systems. Despite this, fluorescence ISH has found widespread application in the analysis of chromosomes and interphase nuclei as well as in primed *in situ* extension (PRINS; Koch *et al.*, 1989) used in gene mapping, diagnostics and fundamental genetics.

9.3.10 Quantification of ISH signal

It may be necessary to quantify the ISH signal as an indication of relative levels of gene expression for comparative studies or to estimate the amount of nucleic acid. The approach will depend on the type of visualization protocol adopted and there are now a number of software packages available, for computerized image analysis of silver and immunogold grain counting and the distribution and density of immunocytochemical product (Nunez *et al.*, 1989; Baskin and Stahl, 1993; Bolender, 1993; McCabe and Bolender, 1993). In addition the quantification of specific DNA sequences and ribosomal RNA using fluorescent ISH and flow cytometry has been accomplished (Baumann and Bentvelzen, 1988; Nederlof *et al.*, 1992; Ballard and Ward, 1993).

9.3.11 Controls and Troubleshooting

To demonstrate the specificity of the experimental results of ISH unequivocally it is essential to incorporate a rigorous battery of controls. When commencing a study it is necessary to ensure that linearization of the cDNA template is complete and the size of the generated probe is correct and this may be determined by running samples out on a 1% agarose gel. Preliminary dot blot or Northern analysis should also be performed to validate the identity of the probe.

Establishing a model system to determine the optimal conditions for ISH using the same probe type/labelling/detection system may prove fruitful (Martinez-Montero *et al.*, 1991). As mentioned earlier, the use of a Poly(dT) oligodeoxyribonucleotide probe may also be a useful means of standardizing tissue processing to optimize nucleic acid accessibility while limiting deterioration of morphology. Similarly, probing for a nucleic acid that is known to be present in the tissue system of interest may be a useful positive control.

Non-specific hybridization can be determined in a number of ways, for example by including unlabelled or inappropriate/unrelated probe. For ISH using RNA probes, the latter may be conveniently provided by generating the 'sense' strand, although non-specific signal may occasionally arise with *in vitro* transcription of 'contaminating' vector sequences, which is particularly encountered with Bluescript (Witkiewicz, Bolander and Edwards, 1993). Excessive probe concentrations should be avoided as they result in high background. To determine specificity when background problems are encountered the stringency of hybridization and subsequent washing procedures should be modified.

Strategies for visualizing specific hybrids also require a number of controls to eliminate spurious results. Artefacts may arise when using nuclear track emulsion, resulting in physical stresses produced when dipping or drying the slide. The emulsion must be at the correct temperature (45 °C) prior to dipping and the slide is warmed to a similar temperature on a hot plate prior to dipping to ensure that the emulsion thickness is minimal, as this will permit optimum resolution of the signal to the site of gene expression. Background problems that arise through the immunocytochemical detection of signal may be the consequence of endogenous enzyme such as alkaline phosphatase or hapten such as biotin. This has proved a limitation in the use of the latter in tissue such as the kidney, which has high levels of biotin. Endogenous enzyme may be blocked and to ensure that this procedure is effective the probe should be omitted from a control slide. Standard immunocytochemical controls (see earlier in this chapter) may be adopted for establishing non-specific antibody interactions.

To establish that an RNA probe is in fact identifying RNA species, a preliminary RNAase treatment of a slide may be performed although all necessary precautions must be adopted to ensure that other slides, incubation buffers and equipment do not become contaminated. Low signal may be the result of RNAase contamination and inhibitors and sterilization procedures must be included in the technique.

9.3.12 *Ease of application*

While the technique is undoubtedly of great value, it is time consuming and often unsuccessful. It is imperative that the reasons for adopting the technique should be sound (Figure 9.7) and that the alternative techniques (immunocyto-chemistry, Northern analysis, receptor autoradiography) are insufficient to address the question(s) in hand (Strickler *et al.*, 1990). Having said that, the technique is becoming increasingly simplified by the use of probe labelling and hybrid detection kits, which are available from a number of sources including Boehringer Mannheim Biochemica and AMS Biotechnology Ltd. These will allow the standardization of the technique and the possibility of applying the technique in routine pathology. In addition, the technique has been success-fully automated (Hybaid) for the rapid processing of a large number of slides, although for most applications automation is probably unnecessary.

9.3.13 *Future developments*

It is widely acknowledged that the main limitation of the ISH technique is its lack of sensitivity, although this is to a large extent being overcome by various immunocytochemical amplification procedures (McQuaid and Allan, 1992).

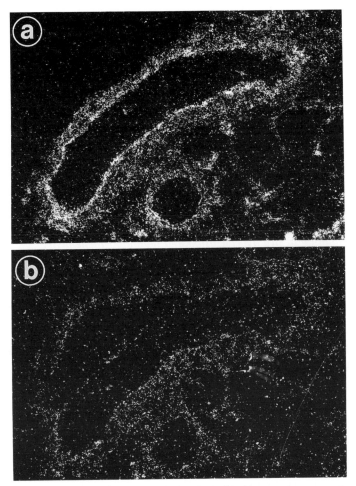

Figure 9.7. Neuropeptide tyrosine (NPY) receptor type 1 mRNA localized to the intima and media of a submucosal blood vessel. Although NPY receptors have been demonstrated in peripheral mammalian tissues, difficulties, which may be associated with differential peptide metabolism, have been encountered in localizing specific binding sites to sections of human tissue (Wharton *et al.*, 1993). ISH was therefore employed, using a complementary RNA probe labelled with ^{35}S to identify sites of expression of the NPY Y1 receptor subtype. Low levels of signal were visualized using dark-field illumination. ISH with sense probe produced a little non-specific signal (b).

There has been much debate on the subject of the development of the ISH technique in combination with the polymerase chain reaction (*'in situ*-PCR'), and there are relatively few publications that present convincing data in tissue sections. Indirect *in situ*-PCR involves the amplification of target sequences and visualization of the PCR products by ISH (Long *et al.*, 1993) To date, *in situ*-PCR seems to be most effective in the detection of DNA in single cell preparations although the quantification of results is presenting some difficulties. Artefacts that arise through diffusion and the extracellular generation of PCR products significantly limit the application of the technique.

Primed *in situ* extension (PRINS) uses an oligonucleotide primer that is annealed to chromosomal DNA and extension occurs *in situ* with the incorporation of labelled nucleotides (Koch *et al.*, 1989). Spurious results may be obtained with PRINS in tissue sections where labelled nucleotides are incorporated into fragmented genomic DNA or cDNA, possibly due to DNA repair and internal priming. The sensitivity of the procedure may be increased by cycling the PRINS reaction (Gosden and Hanratty, 1993) although there is again the limitation of diffusion artefacts unless the amplified fragments are sufficiently large to become immobilized.

9.3.14 Combining in situ hybridization and immunocytochemistry

Immunocytochemistry and ISH may be performed on alternate, serial sections, but the precise localization of antigen and nucleic acid to the same cell may prove difficult using this approach. To perform both techniques on the same section requires consideration of factors such as sterility (RNAase-free conditions with RNA probes) and antigen denaturation. If ISH is carried out first the protocol requires little adaptation. Precautions must be taken to avoid denaturation of antigen and for this reason the hybridization and stringency washing temperatures should not exceed 45 °C, even if the optimum for the probe is greater. With the washing stages of the immunocytochemistry the technique overall will be of increased stringency, which compensates for the reduced incubation temperatures. Protease treatment must also be limited to avoid degradation of antigen. After visualization of the antigen the slide is dipped in nuclear emulsion, exposed and developed as usual (Figure 9.8). The combination of the two techniques may provide important information regarding the production, storage and transport of a given peptide (DeBelleroche *et al.*, 1990).

9.4 Protocols

Protocol 1: use of poly-L-lysine (PLL)

1. Dissolve PLL in distilled water (1.0 mg/ml). Solution may be stored frozen in 1 ml batches. Thawing and refreezing is not harmful. Mix thawed solution well before use.
2. Ensure glass slides are precleaned.
3. Apply *small* drop (10 μl or less) of PLL solution to clean surface.
4. Spread thin film over whole of surface so that it dries instantly.
5. Tissue sections can be picked up onto coated surface immediately afterwards.

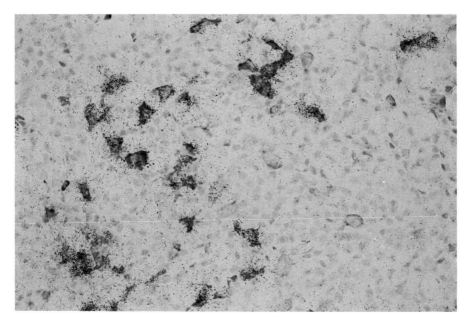

Figure 9.8. Thyroid stimulating hormone (TSH) mRNA and peptide localized to the same cells within rat pituitary. A combined ISH and immunocytochemical approach was adopted to localize TSH mRNA and peptide. While many cells exhibit colocalization there are some where only peptide is detected. This may be a result of the greater sensitivity of the immunocytochemical technique. (Photograph reproduced with the permission of Dr JH Steel, Department of Histochemistry, RPMS.)

6. Dry paraffin sections by placing slides overnight in 37 °C incubator. Cryostat sections should be dried at room temperature for up to 1 h.

N.B Optimum molecular weight of PLL, at least 300 000. (Supplier: Sigma Chemical Co. Order no. P1524).

PLL-coated slides are best used immediately, but batches may be prepared and stored, in boxes to avoid dust, at room temperature for up to 1 week (in this case the coated side should be marked).

Reference: Huang *et al.*, 1983.

Protocol 2: Zamboni's fixative and fixation
85 ml of 2% (w/v) paraformaldehyde in 0.1 M phosphate buffer (pH 7.4)
15 ml of saturated picric acid

Dissolve the paraformaldehyde in 0.1 M phosphate buffer with heat (do not boil) and stir until the solution is clear. Make up to volume and allow to cool. Add appropriate amount of saturated picric acid and store at 4 °C until used.

Fixation:

Immerse tissue in fixative overnight (9–16 h) at 4 °C.

Samples should not be more than 0.5–1 cm thick.

Transfer to phosphate-buffered saline containing 15% (w/v) sucrose and 0.1% (w/v) sodium azide.

Change the sucrose solution several times, over 1–2 days, until it remains clear.

Store at 4 °C.

May be transported at room temperature.

For long-term storage (weeks/months) it is better to keep the tissue in the form of cryostat blocks and store in liquid nitrogen.

Protocol 3: indirect immunofluorescence

1. The 'sandwich' or indirect method is used. Pre-fixed paraffin sections are cut at 5 μm and floated on a warm (not hot) water-bath. They are picked up on poly-L-lysine-coated slides and dried first on a warm surface (surround of water-bath) and then in an oven at 37 °C overnight. (If more than one section is picked up on one slide, maximize the distance between them.)

2. Cryostat sections from pre-fixed material must be picked up on poly-L-lysine coated slides. The sections are allowed to dry for at least 15 min at room temperature. Drying for longer than one hour is not recommended as some antigenicity may be lost. Pre-fixed cryostat sections may be soaked in a dilute detergent solution, e.g. 0.2% (v/v) Triton X-100 in PBS, for 30 min to aid antibody penetration.

3. Paraffin sections may be de-waxed in three changes of inhibisol and dried without passing through alcohols. They need not be rinsed. Ensure that the sections are completely free of wax by allowing them to dry. If they have no wax they will have a chalky, white appearance. Alternatively, they may be brought to water in the normal way.

4. Sections are placed on racks in covered Petri dishes containing water to create a damp chamber and avoid evaporation of the antiserum. A drop of antiserum, appropriately diluted in 0.01 M phosphate-buffered normal saline, pH 7.1–7.4 (PBS) containing 0.1% bovine serum albumin and (for storage of diluted antibody) 0.01% sodium azide, is applied to the section with a Pasteur pipette and is left for 45 min to 1 h at room temperature, or overnight at 4 °C, according to the dilution of the antibody. Put only the smallest amount of antiserum on the section, i.e. enough to cover the section without spreading when the slide is moved. Place the dishes where they will not be disturbed.

5. The section is rinsed in three changes of PBS of 5 min each. The slide is then wiped dry except for the area of the section and replaced in the Petri dish. Do not let the section dry as PBS will precipitate on it.

6. A drop of the appropriate fluorescein-labelled anti-gamma-globulin, optimally diluted in PBS, is placed on the section, which is left for 30 min to 1 h at room temperature, and washed as before.
7. The section is mounted in buffered glycerine (9 parts of glycerine to 1 part of PBS) and viewed under UV light. 2.5% (v/v) 1,4-diazobicyclo-(2-2-2)-octane (DABCO) may be incorporated in the mountant to preserve the fluorescence.

Protocol 4: indirect immunoperoxidase staining

1. De-wax sections in inhibisol or xylene and rehydrate to water through graded alcohols. If fixative contained mercury, remove pigment with iodine and bleach in sodium thiosulphate.
2. Block endogenous peroxidase reaction (e.g. of red blood cells) by immersing sections for 30 min in H_2O_2, 0.3% (v/v) in 0.01 M phosphate-buffered saline, pH 7.1–7.4 (PBS, paraffin sections) or in methanol (frozen sections).
3. Wash in PBS.
4. Block possible background reaction by layer of diluted serum of donor animal of second layer, e.g. if second layer is to be peroxidase-conjugated goat anti-rabbit globulin, the blocking should be done with normal goat serum at a dilution of 1:30 in PBS for 10–15 min. Drain off serum *but do not rinse slide after this step*.
5. Wipe slides dry except for area of section. Apply primary antibody at suitable dilution in PBS containing 0.05% BSA and 0.01% sodium azide. Incubate overnight at 4 °C.
6. Wash several times in PBS (three changes of 5 min each).
7. Dry slides except for area of section.
8. Stain with peroxidase-conjugated anti-rabbit (or anti-guinea pig) immunoglobulin. For this step use diluent that does *not* contain azide. Optimal dilution must be tested for each batch of conjugate.
9. Rinse in PBS (three changes of 5 min each).
10. Carry out the reaction for peroxidase by immersing the slides in incubating solution, freshly made up just before use.

Incubation solution (care–DAB may be carcinogenic!)

Make a solution up in a dish and check for contamination with bleach. Dissolve 25 mg of DAB (diaminobenzidine) in 100 ml of PBS (DAB is kept frozen in aqueous solution in aliquots of 25 mg/ml). Then add 50–100 μl of 100 volume (30%) H_2O_2 to each 100 ml of DAB solution and incubate slides. Be careful to add maximum of 100 μl of H_2O_2 per 100 ml of solution. More H_2O_2 may block the peroxidase label. The development time is rather short – 1–5 min. Therefore, control very carefully by examining sections under microscope. Rinse sections in buffer before examining with microscope to prevent contamination of microscope by DAB. If enhancement of DAB reaction is required, e.g. for immunostained nerves, make up the DAB solution without H_2O_2. Place the sections in the solution for 10 min, then add H_2O_2 and continue as usual.

11. Wash slides in PBS or distilled water.
12. Wash sections in water, counterstain lightly with haematoxylin if required.
13. Dehydrate through graded alcohols, clear in xylene, mount in DPX.

For alternative methods for development of peroxidase see Protocol 7. Enhancement of immunostaining is described in Protocol 8.

Protocol 5: peroxidase-antiperoxidase (PAP) method

1. *Wax sections* – De-wax slides in several changes of inhibisol or xylene, rehydrate through absolute alcohol, 90% alcohol, 70% alcohol to distilled or deionized water (1–2 min in each solvent). Remove mercury pigment with iodine if necessary.
    Pre-fixed cryostat section – Allow to dry at room temperature for 30 min to 1 h, soak in detergent if necessary, rinse in PBS, then continue with step 2.
2. To block endogenous peroxidase, immerse slides for 30 min in a solution of 0.3% H_2O_2 (1 ml of 100 volume (30%) H_2O_2 per 100 ml) in PBS, or methanol is preferable for cryostat sections.
3. Rinse well in PBS.
4. To block background staining, apply normal goat serum 1:30, 30 min. *Do not rinse*, drain off serum onto a paper tissue.
5. Apply primary antibody (rabbit or mouse) highly diluted in PBS containing 0.05% bovine serum albumin and (for storage of diluted antibody) 0.01% sodium azide, 4 °C overnight, or a lower dilution for $\frac{1}{2}$–1 h at room temperature may be used.
6. Rinse three times for 5 min each in PBS.
7. Apply **unconjugated** goat anti-rabbit immunoglobulin – 30 min at the dilution recommended, at room temperature.
8. Rinse three times for 5 min each in PBS.
9. Apply PAP layer for 30 min at the dilution recommended. Use diluent which does **not** contain sodium azide.
10. Rinse three times for 5 min each in PBS.
11. For incubation solution see Protocol 4.
12. Wash well with PBS and then with tap water.
14. Counterstain lightly, if desired, in Haematoxylin.
15. Dehydrate, clear and mount in DPX, Pertex or similar resinous medium.

Protocol 6: ABC immunoperoxidase staining procedure

Follow Protocol 5 until:

7. Incubate sections for 30 min with diluted biotinylated antibody solution (anti-rabbit or anti-mouse IgG, according to primary antiserum).
8. Rinse three times for 5 min each in PBS.
9. Incubate sections for 30 min with Vectastain ABC reagent (make this up 30 min before use; add equal volumes from vials A and B to PBS containing 0.05% BSA but **no** sodium azide).
10. Rinse three times for 5 min each in PBS.
11. For incubation solution see Protocol 4.
12. Wash well with PBS and then with tap water.
13. Counterstain lightly, if desired, in Haematoxylin.
14. Dehydrate, clear and mount in DPX, Pertex or similar resinous medium.

Protocol 7: alternative methods for development of peroxidase

None of the methods given below is as satisfactory as the diaminobenzidine method but the reagents used are probably less likely to be carcinogenic and some give coloured end-products different from that of DAB, so that they may usefully be used in double immunoperoxidase staining. All incubating solutions must be prepared just before use.

1. Hanker's method (Hanker et al., 1977)

Hankers–Yates reagent: *p*-phenylene diamine HC1-1 part pyrocatechol-2 parts.

Dissolve 75 mg of the mixture in 100 ml 0.1 M Tris buffer, pH 7.6 (or PBS) and add 100 μl of 30% H_2O_2.
Incubation time is longer than for DAB.
Colour of end-product is blue–black.
Background is minimal (the reagent is said to be specific for plant peroxidase so endogenous animal peroxidase does not affect the result). Sections may be osmicated if desired and can be dehydrated, cleared and mounted in DPX.
A red nuclear counterstain is effective (Neutral Red or Carmalum).

2. 4-chloro-1-naphthol method (Nakane, 1967)

Dissolve 30–40 mg of 4-chloro-1-naphthol in 0.2–0.5 ml absolute alcohol (not more – alcohol reduces peroxidase activity).
Add with stirring 100 ml 0.5 M Tris buffer, pH 7.6 (or PBS) containing 50–100 μl of 30% H_2O_2. A white precipitate forms. Filter the solution before use.

Incubate at room temperature, checking microscopically from time to time. Counterstain with Carmalum.

In a modified procedure resulting in a stronger stain, the incubating medium is heated to 50 °C (not more) before use. Filter (coarse filter paper) and use the filtrate while hot.

The reaction-product is dark blue–purple. It is soluble in alcohol and xylene, so cannot be mounted in DPX. A water-soluble mountant such as glycerine jelly is suitable. The reaction-product is said to be more stable if mounted in acid-buffered glycerine, pH 2.2.

Protocol 8: enhancement of immunostains

Nickel enhancement of DAB

1. Incubate with primary antiserum and follow protocols for the PAP or ABC methods until the end of the third layer incubation.
2. Wash sections in PBS, 2 × 5 min.
3. Rinse in 0.1 M acetate buffer (pH 6.0) 2 × 5 mins (see below).
4. Incubate in glucose oxidase–nickel–DAB solution for 5–20 min (see below).
5. Rinse in 0.1 M acetate buffer 2 × 5 mins.
6. Dehydrate and mount.

0.1 M acetate buffer, pH 6.0
13.61 g sodium acetate trihydrate per litre distilled water. pH to 6.0 using glacial acetic acid.

Glucose oxidase–nickel–DAB solution
Ammonium nickel sulphate 10 g in 100 ml 0.1 M acetate buffer. Immediately before use add the following:

B-D-Glucose 200 mg
Ammonium chloride 40 mg
DAB 25 mg (or 1 ml of concentrated aliquot stored in freezer).
Glucose oxidase (Sigma, Type III) 0.5–1 mg.

Enhancement of DAB reaction by cobalt ions

1. Carry out immunostain as usual.
2. DAB–PBS solution (standard concentration) but omit H_2O_2: 2–3 min.
3. Add to the above solution 1% cobalt chloride ($CoCl_2$) in deionized water, 2 ml/100 ml of mixture, and mix. Incubate 5 min continue mixing.
4. Add (30%) H_2O_2 50–100 μl/100 ml mixture. Continue incubation

3–4 min or until the sections turn pale blue/grey. Check sections micro-scopically.

5. Wash in tap water for 10 min.
6. Counterstain as desired (e.g. Mayer Carmalum, 1 min).
7. Rinse briefly in tap water.
8. Dehydrate, and mount in DPX.

Results
PAP reactive granules – dark blue
Cytoplasm, CT elements – dark purple
Nuclei – light red
The counterstain provides a relatively light background as compared with traditional DAB staining

Protocol 9: generation of complementary RNA probe

1. cDNA, cloned in an appropriate vector, is purified by phenol:chloroform extraction or by using one of the DNA preparation kits that are commer-cially available (Maniatis, Fritsch and Sambrook, 1989).
2. The vector is linearized at a point in the cloning site of the polylinker, using a convenient restriction enzyme, to allow for the generation of probe of the required orientation (sense or antisense) and is finally purified of enzyme and resuspended in Tris/EDTA buffer (Tris, 10 mM; EDTA, 0.1 mM, pH 8) to a concentration of 1 μg/μl.
3. The following components are added to a sterile microcentrifuge tube at room temperature (spermidine causes precipitation of DNA at low temp-eratures):

5× transcription buffer (see below)	2 μl
Dithiothreitol (DTT)	1 μl
Linearized DNA probe template (1 μg/μl)	1 μl
Isotopic-labelled nucleotide, e.g. ^{32}P-CTP (20 μCi/μl)	5 μl
Non-isotopic-labelled nucleotide, e.g. Digoxigenin-UTP, 10 mM	1 μl
RNA polymerase	1 μl

   Add water to a final volume of 10 μl, mix gently and incubate at 37 °C for 1 h
4. Transcription is terminated by removing the template: add 1 μl DNAase (RNAase-free) and incubate for 10 min at 37 °C.
5. The total volume is made up to 200 μl with RNAase-free water and unincorporated nucleotides and enzymes are removed by phenol:chloro-form (1:1, saturated with Tris) and chloroform:isoamyl alcohol (24:1) extractions.

6. After separating off the supernatant the RNA probe is precipitated by adding:

Ammonium acetate	final concentration of 2.5 M
Ethanol	2.5 volumes (ice cold)
Glycogen (20 mg/ml)	2 μl

Store at $-20\,^{\circ}$C for 2 h and spin for 15 min, wash pellet in 70% alcohol and spin again for 5 min, dry pellet and resuspend in sterile water or Tris–EDTA

Transcription buffer: 200 mM Tris/HCl (pH 7.5), 30 mM $MgCl_2$, 10 mM spermidine trihydrochloride, 50 mM DTT, 0.5 mg/ml bovine serum albumin (BSA), 500 U/ml ribonuclease inhibitor, 2.5 mM of each unlabelled nucleotide, e.g. ATP, GTP, UTP (or CTP).

Protocol 10: probe hydrolysis

1. Dissolve probe in sterile water and add the following:

Water	160 μl
$NaHCO_3$ (0.4 M)	20 μl
Na_2CO_3 (0.6 M)	20 μl

Mix well and incubate at 60 $^{\circ}$C for the appropriate time interval: see below.
2. The incubation time for hydrolysis is calculated as:

$$t = L_0 - L_f/k.L_0.L_f$$

where t = time in minutes; L_0 = original length of probe (kb); L_f = required length of probe (kb); k = rate constant for hydrolysis (0.11 scissions/kb/min)
3. Hydrolysis is terminated by adding sodium acetate to a final concentration of 0.1 M and glacial acetic acid to 0.5% (v/v).
4. Probe is precipitated (as in Protocol 9) and resuspended to a final concentration of 5 ng/μl for isotopic-labelled probe or 25 ng/μl for non-isotopic probe after estimating probe recovery.

Protocol 11: tissue preparation

1. For immersion fixation the tissue of interest is dissected and placed in 4% (w/v) paraformaldehyde in 0.1 M phosphate buffer (pH 7.2) for 6 h at 4 $^{\circ}$C. The ratio of tissue to fixative should be no greater than 1:10.
2. The tissue is then washed thoroughly in 15% sucrose in PBS containing 0.1% sodium azide over the course of 2–3 days.
3. Cryostat blocks are prepared by orientating the tissue onto cork mats, surrounding it in a mounting medium, such as TissueTek (Miles Inc, Elkhart), and rapidly freezing it in melting dichlorodifluoromethane

(Arcton, ICI, UK) suspended in liquid nitrogen. Storage of the tissue blocks is at −70 °C.

4. The tissue is sectioned to a thickness of 10 μm in a cryostat at −25 °C, the sections are lifted onto Vectabond-treated glass slides and are dried thoroughly before processing for ISH. If they are to be stored at this point they are placed in boxes with silica gel and frozen at −70 °C.

Protocol 12: tissue pretreatment

1. Cryostat sections are rehydrated by immersion in detergent: 0.2% (v/v) Triton X-100 in phosphate buffered saline (PBS) for 15 min.
2. Wash in PBS for 2 × 5 min.
3. Protease digestion is carried out by incubating sections in 0.1 M Tris/0.05 M EDTA (pH 8) containing Proteinase K at a final concentration of 1 μg/ml for up to 20 min (The appropriate incubation time should be determined in a preliminary experiment to ensure that tissue morphology is not destroyed. Some tissue, e.g. pancreas, and cells do not require protease treatment).
4. Inactivate enzyme by immersing slides in 0.1 M glycine in PBS for 5 min.
5. Briefly (3 min) post-fix with 4% paraformaldehyde in PBS, to prevent possible diffusion of target nucleic acid.
6. Wash in PBS, 2 × 5 min.
7. Immerse slides in 0.25% (w/v) acetic anhydride in 0.1 M triethanolamine (pH 8) for 10 min with shaking (acetylates tissue and prevents non-specific binding of probe).
8. Rinse briefly in water and dry sections.

A RNAase treatment control section may be included to confirm the nature of the material to which probe is bound: Before protease treatment a section (which has been isolated from the others) is incubated with a solution of RNAase (100 μg/ml, DNAase-free) for 30 min at 37 °C. After this the slide is processed as normal. Ensure that the remaining slides do not become contaminated with RNAase!

Protocol 13: hybridization

1. Probe is diluted 1:10 in hybridization buffer to a final concentration of 0.5 ng/μl (isotopic) or 2.5 ng/μl (non-isotopic):
   Hybridization buffer (may be stored in pre-measured volumes, at 20 °C. Do not repeatedly freeze–thaw): 50% (v/v) deionized formamide; 5× standard saline citrate (SSC, see below); 10% (w/v) dextran sulphate (added at 50 °C); 5 × Denhardt's (see below); 2% (w/v) sodium dodecyl sulphate (SDS); 100 μg/ml herring sperm DNA (denatured and sheared). Denhardt's (100×): 10 g Ficoll, 10 g polyvinylpyrrolidone (PVP), 10 g BSA, 500 ml sterile water.

Standard saline citrate (20×):175.30 g sodium chloride and 88.2 g sodium citrate to 1 litre with double distilled water (pH 7.0).

2. 10 μl of probe diluted in hybridization buffer is applied to each section and a siliconized coverslip is placed on top before incubating the slide at the desired temperature in a humid chamber for 16–24 h. The temperature selected may be from 37 °C (for oligoprobes) up to 60 °C. We usually carry out hybridization at 42 °C (moderate stringency) but conditions can be adapted in retrospect.

Protocol 14: stringency washes

1. The coverslip is gently removed by immersing the slide in 2× SSC/0.1% (w/v) SDS.
2. Wash in 2× SSC/0.1% SDS for 4 × 5 min at room temperature with gentle shaking.
3. Wash in 0.1× SSC/0.1% SDS for 2 × 10 min at the temperature adopted for hybridization.
4. Rinse thoroughly in 2× SSC to remove all traces of SDS.
5. Incubate sections in RNAase (10 μg/μl) in 2× SSC at 37 °C for 15 min.
6. Rinse in 2× SSC.
7. If processing for autoradiography, dehydrate in graded alcohol (70%, 90% and 100% ethanol) containing 0.3 M ammonium acetate and proceed to Protocol 7.

If processing slides for immunocytochemical development proceed to Protocol 8.

Protocol 15: autoradiography

Appropriate safe-light filters for use with autoradiographic emulsion/film are given in the manufacturers' product specifications.

1. Sections may be apposed to autoradiographic film (Amersham β-max hyperfilm) for the desired time (depending on isotope). The film is developed in Kodak D19 for 5 min at 20 °C, washed in a 0.2% (v/v) solution of glacial acetic acid and fixed for 10 min in a 20% (w/v) solution of Amfix. The film is then washed in running tap water followed by a rinse in distilled water (to avoid water marks) prior to drying. The film may then be used directly in photography (see Figure 9.1).
2. Nuclear track emulsion (Amersham LM1, Ilford K5) is melted at 45 °C and poured into a miser cup/vessel where it is allowed to settle to allow air bubbles to disperse. The slides are warmed on a hot plate to the same temperature to ensure a thin coating of emulsion and are gently immersed and maintained in a vertical position when dipping and drying. Excess emulsion is blotted from the lower edge of the slide. A careful and

standardized dipping procedure will minimize the occurrence of artefacts. After dipping the slides are dried at room temperature for 1–2 h and placed in the light-tight boxes with silica gel for storage at 4 °C for the required exposure interval. The slides are developed for 3 min in Kodak D19 at 18 °C, followed by a rinse in water and fixation for 8 min. The slides are thoroughly washed before counterstaining in Haematoxylin and mounting for microscopy. (Processing must be gentle to avoid damaging the coating of emulsion). Determining of the appropriate exposure intervals for both film and emulsion is by trial and error and by the use of replicate slides.

Protocol 16: visualization of digoxigenin-labelled probes

1. Slides are rinsed in buffer containing 0.1 M Tris (pH 7.5), 0.1 M NaCl, 2 mM $MgCl_2$ (Buffer 1) for 2 × 3 min.
2. Non-specific antibody binding is blocked by immersion in Buffer 1 containing 3% BSA for 10 min.
3. Alkaline phosphatase-labelled digoxigenin antisera diluted 1:500 in Buffer 1 is applied to the section, which is incubated for 2 h at room temperature.
4. Sections are washed in Buffer 1 for 3 × 3 min.
5. Sections are then placed in Buffer 2 which contains 0.1 M Tris (pH 9.5), 0.1 M NaCl and 0.05 M $MgCl_2$ prior to incubation in substrate buffer.
6. The slides are then incubated in substrate buffer (Buffer 2 containing 1 mM levamisole (to block endogenous enzyme), 330 μl NitroBlue Tetrazolium solution (NBT: 10 mg/ml) and 33 μl 5-bromo-3-chloro-3-indolyl phosphate stock solution (BCIP: 25 mg in 1 ml of dimethylformamide) for the required interval, in the absence of light. A number of groups have found that incubation at 4 °C overnight is required to visualize digoxigenin-labelled hybrids.

Alternative strategies for non-isotopic hybrid visualization are detailed in Boehringer Mannheim's Application Manual (see Reference list).

Appendix. Recommended sources of antibodies for neuroendocrine cell immunocytochemistry

Antibodies to: Source:
Chromogranin   Boehringer Mannheim (UK)
                                  Bell Lane
                                  Lewes,
                                  Sussex BN7 1IG, UK.

Neuron-specific enolase   DAKO Ltd.
Synaptophysin   16 Manor Courtyard
Serotonin   Hughenden Avenue
                                  High Wycombe,
                                  Bucks HP13 5RE, UK.

Protein gene product 9.5   Ultraclone Ltd.
                                  Rossiters Farmhouse
                                  Wellow
                                  Isle of Wight PO41 0TE, UK.

Peptides   Amersham International
                                  White Lion Road
                                  Amersham
                                  Bucks HP7 9BR, UK.

References

Avivi C., Rosen, O. and Goldstein, R. S. (1994) New chromogens for alkaline phosphatase histochemistry: salmon and magenta phosphate are useful for single- and double-label immunohistochemistry. *J. Histochem. Cytochem*. **42**, 551–4.

Ballard, S. G. and Ward, D. C. (1993) Fluorescence in situ hybridization using digital imaging microscopy. *J. Histochem. Cytochem*. **41**, 1755–9.

Baskin, D. G. and Stahl, W. L. (1993) Fundamentals of quantitative autoradiography by computer densitometry for in situ hybridization, with emphasis on [33]P. *J. Histochem. Cytochem*. **41**, 1767–76.

Bauman, J. G. and Bentvelzen, P. (1988) Flow cytometric detection of ribosomal RNA in suspended cells by fluorescent in situ hybridization. *Cytometry* **9**, 517–24.

Bolender, R. P. (1993) Software for quantitative immunogold and in situ hybridization. *Microscopy Res. Tech*. **25**, 304–13.

Buttery, L. D. K., Springall, D. R., Andrade, S. P. *et al.*, (1993) Induction of nitric oxide synthase in the neo-vasculature of experimental tumours in mice. *J. Pathol*. **171**, 311–19.

Cohen, A. R., Roysam, B. and Turner, J. N. (1994) Automated tracing and volume measurements of neurons from 3-D confocal fluorescence microscopy data. *J. Micros*. **173**, 103–14.

Coons, A. H., Leduc, E. H. and Connolly, J. M. (1955) Studies on antibody production. I. A method for the histochemical demonstration of specific antibody and its application to the study of the hyperimmune rabbit. *J. Exp. Med*. **102**, 49–60.

Davis, C. A. (1993) Whole-mount immunohistochemistry. In *Guide to techniques in mouse development*, Methods in Enzymology vol. 225 (ed. P. M. Wassarman and M. L. DePamphilis), pp. 502–16. Academic Press, Inc., San Diego.

DeBelleroche, J., Bandopadhyay, R., King, A. *et al*. (1990) Regional distribution of cholecystokinin messenger RNA in rat brain development: quantitation and correlation with cholecystokinin immunoreactivity. *Neuropeptides* **15**, 201–12.

Denk, H., Radaszkiewicz, T. and Weirich, E. (1977) Pronase pretreatment of tissue sections enhances sensitivity of the unlabelled antibody-enzyme (PAP) technique. *J Immunol Methods* **15**, 163–5.

Dirks, R. W., van Gijlwijk, R. P. M., Tullis, R. H. *et al*. (1990) Simultaneous detection of different mRNA sequences coding for neuropeptide hormones by double in situ hybridization using FITC- and biotin-labeled oligonucleotides. *J. Histochem, Cytochem*. **38**, 467–73.

Domoto, T., Bishop, A. E., Oki, M. and Polak, J. M. (1990) An in vitro study of the projections of enteric VIP-containing neurones in the human colon. *Gastroenterology* **98**, 819–27.

Fleming, K. A. (1992) Analysis of viral pathogenicity by in situ hybridization (Editorial) *J. Pathol*. **166**, 95–6.

Gosden, J. and Hanratty, D. (1993) PCR in situ: a rapid alternative to in situ hybridization for mapping short, low copy number sequences without isotopes. *BioTechniques* **15**, 78–80.

Gown, A. M., de Wever, N. and Battifora, H. (1993) Microwave-based antigenic unmasking. A revolutionary new technique for routine immunohistochemistry. *App. Immunohistochem*. **1**, 256–66.

Graham, R. C. and Karnovsky, M. J. (1966) The early stages of absorption of injected horseradish peroxidase in the proximal tubules of mouse kidney: ultrastructural cytochemistry by a new technique. *J. Histochem. Cytochem*. **14**, 291–302.

Guesdon, J. L., Ternynck, T. and Avrameas, S. (1979) The use of avidin biotin interaction in immunoenzymatic techniques. *J. Histochem. Cytochem*. **27**, 1131–9.

Hanker, J. S., Yates, P. E., Metz, C. B. and Rustioni, A. (1977) A new specific, sensitive and non-carcinogenic reagent for the demonstration of horse-radish peroxidase. *Histochem. J*. **9**, 789.

Hannon, K., Johnstone, E., Craft, L. S. *et al*. (1993) Synthesis of PCR-derived, single stranded DNA probes suitable for in situ hybridization. *Anal. Biochem*. **212**, 421–7.

Herrington, C. S., Burns, J., Graham, A. K. *et al*. (1989) Interphase cytogenetics using biotin and digoxigenin labelled probes II: Simultaneous differential detection of human papilloma virus nucleic acids in individual nuclei. *J Clin. Pathol*. **42**, 601–6.

Hsu, S. M., Raine, L. and Fanger, H. (1981*a*) Use of avidin–biotin–peroxidase complex (ABC) in immunoperoxidase techniques. A comparison between ABC and unlabelled antibody (PAP) procedures. *J. Histochem. Cytochem*. **29**, 577–80.

Hsu, S. M., Raine, L. and Fanger, H. (1981*b*) The use of anti-avidin antibody and avidin–biotin–peroxidase complex in immunoperoxidase techniques. *Am. J. Clin. Pathol*. **75**, 816–21.

Huang, W. M., Gibson, S. J., Facer, P. *et al.* (1983) Improved section adhesion for immunocytochemistry using high molecular weight polymers of L-lysine as a slide coating. *Histochem.* **77**, 275–9.

Koch, J. E., Kolvraa, S., Petersen, N. *et al.* (1989) Oligonucleotide priming methods for the chromosome-specific labelling of alpha satellite DNA in situ. *Chromosoma* **98**, 259–65.

Komminoth, P., Merk, F. B., Laev, I. *et al.* (1992) Comparison of ^{35}S- and digoxigenin-labeled RNA and oligonucleotide probes for in situ hybridization. *Histochemistry* **98**, 217–28.

Leong, A. S. Y. and Milios, J. (1993) An assessment of the efficacy of the microwave antigen-retrieval procedure on a range of tissue antigens. *Appl. Immunohistochem.* **1**, 267–74.

Lloyd, R. V., Jin, L. and Fields, K. (1990) Detection of chromogranins A and B in endocrine tissues with radioactive and biotinylated oligonucleotide probes. *Am. J. Surg. Pathol.* **14**, 35–43.

Long, A. A., Komminoth, P., Lee, E. and Wolfe, H. J. (1993) Comparison of indirect and direct in-situ polymerase chain reaction in cell preparations and tissue sections. Detection of viral DNA, gene arrangements and chromosomal translocations. *Histochemistry* **99**, 151–162.

Martinez-Montero, J. C., Herrington, C. S., Stickland, J. *et al.* (1991) Model system for optimising mRNA non-isotopic in situ hybridisation: riboprobe detection of lysozyme mRNA in archival gut biopsy specimens. *J. Clin. Pathol.* **44**, 835–9.

McCabe, J. T. and Bolender, R. P. (1993) Estimation of tissue mRNAs by in situ hybridization. *J. Histochem. Cytochem.* **41**, 1777–83.

McQuaid, S. and Allan, G. M. (1992) Detection protocols for biotinylated probes optimization using multistep techniques. *J. Histochem. Cytochem.* **40**, 949–53.

Montone, K. T., Budgeon, L. R. and Brigati, D. J. (1990) Detection of Epstein Barr virus (EBV) genomes by in situ DNA hybridization with terminally biotin-labeled synthetic oligonucleotide probe from EBV Not I and Pst I tandem repeat regions. *Mod. Pathol.* **3**, 89–96.

Montone, K. T. and Tomaszewski, J. E. (1993) In situ hybridization protocol for overall preservation of mRNA in fixed tissues with a poly d(T) oligonucleotide probe. *J HistoTechnol.* **16**, 315–21.

Nakane, P. K. (1967) Simultaneous localization of multiple tissue antigens using the peroxidase-labelled antibody method. A study on pituitary glands of rats. *J. Histochem. Cytochem.* **16**, 557–60.

Nakane, P. K. and Pierce, G. B. (1967) Enzyme-labelled antibodies for the light and electron microscopic localization of tissue antigens. *J. Cell. Biol.* **33**, 307–18.

Nederlof, P. M., van der Flier, S., Verwoerd, N. P. *et al.* (1992) Quantification of fluorescence in situ hybridization signals by image cytometry. *Cytometry* **13**, 846–52.

Negoescu, A., Labat-Moeur, F., Lorinmer, P. *et al.* (1994) F(ab) secondary antibodies: a general method for double immunolabelling with primary antisera from the same species. Efficiency control by chemiluminescence. *J. Histochem. Cytochem.* **42**, 433–7.

Nunez, D. J., Davenport, A. P., Emson, P. C. and Brown, M. J. (1989) A quantitative 'in situ' hybridization method for using computer-assisted image analysis. Validation and measurement of atrial-natriuretic-factor mRNA in the rat heart. *Biochem. J.* **263**, 121–7.

Pringle, J. H., Primrose, L., Kind, C. N. *et al.* (1989) In situ hybridization demonstration of poly-adenylated RNA sequences in formalin-fixed paraffin sections using biotinylated oligonucleotide poly d(T) probe. *J. Pathol.* **158**, 279–86.

Pryzwansky, K. B. (1982) Applications of double label immunofluorescence. In *Techniques in immunocytochemistry* vol. 1 (ed. G. R. Bullock and P. Petrusz), pp. 77–90. Academic Press, London.

Reading, M. (1977) A digestion technique for the reduction of background staining in the immunoperoxidase method. *J. Clin. Pathol.* **30**, 88–90.

Roysam, B., Ancin, H., Bhattacharjya, A. K. *et al.* (1994) Algorithms for automated characterization of cell populations in thick specimens from 3-D confocal fluorescence microscopy data. *J. Micros.* **173**, 115–26.

Saeger, W., Uhlig, H., Baz, E. *et al.* (1991) In situ hybridization for different mRNA in GH-secreting and in active pituitary adenomas. *Pathol. Res. Pract.* **187**, 559–63.

Sitzmann, J. H. and LeMotte, P. K. (1993) Rapid and efficient generation of PCR-derived riboprobe templates for in situ hybridization histochemistry. *J. Histochem. Cytochem.* **41**, 773–6.

Sternberger, L. A. (1979) *Immunocytochemistry*, 2nd edn. John Wiley & Sons, New York.

Strickler, J. G., Manivel, J. C., Copenhaver, C. M. and Kubic, V. L. (1990) Comparison of in situ hybridization and immunohistochemistry for detection of cytomegalovirus and herpes simplex virus. *Human Pathol.* **21**, 443–8.

Thomas, C. H. and Rose, B. R. (1991) Deleterious effects of formal/acetic acid/alcohol (FAA) fixation on the detection of HPV DNA by in situ hybridization and the polymerase chain reaction. *Pathology* **23**, 327–30.

Thomas, G., Davies, H. G. and Williams, E. D. (1993) Demonstration of mRNA using digoxigenin labelled oligonucleotide probes for in situ hybridisation in formaldehyde-free conditions. *J. Clin. Pathol.* **46**, 171–4.

Tao, Q., Srivastava, G., Loke, S. L. *et al.* (1994) Improved double immunohistochemical staining method for cryostat and paraffin wax sections, combining alkaline phosphate anti-alkaline phosphatase and indirect immunofluorescence. *J. Clin. Pathol.* **47**, 597–600.

Towle, A. C., Lauder, J. M. and Joh, T. H. (1984) Optimization of tryosine hydroxylase immunocytochemistry in paraffin sections using pretreatment with proteolytic enzymes. *J. Histochem. Cytochem.* **32**, 766–70.

Tramu, G., Pillez, A. and Leonardelli, J. (1978) An efficient method of antibody elution for the successive or simultaneous localization of two antigens by immunocytochemistry. *J. Histochem. Cytochem.* **26**, 322–4.

Trembleau, A., Calas, A. and Fevre-Montagne, M. (1990) Ultrastructural localization of oxytocin mRNA in the rat hypothalamus by in situ hybridization using synthetic oligonucleotides. *Brain. Res. Mol. Brain. Res.* **8**, 37–45.

Wakefield, A. J., Dhillon, A. P., Rowles, P. M. *et al.* (1989) Pathogenesis of Crohn's disease: multifocal gastrointestinal infarction. *Lancet* **2**, 1057–62.

Wharton, J., Gordon, L., Byrne, J. *et al.* (1993) Expression of the human neuropeptide tyrosine Y1 receptor. *Proc. Natl. Acad. Sci. USA* **90**, 687–91.

Witkiewicz, H., Bolander, M. E. and Edwards, D. R. (1993) Improved design of riboprobes from pBluescript and related vectors for in situ hybridization. *BioTechniques* **14**, 458–63.

Xerri, L., Monges, G., Guigou, V. *et al.* (1992) Detection of gastrin mRNA by in situ hybridization using radioactive and digoxigenin-labelled probes: a comparative study. *Acta Pathol. Microbiol. Immunol. Scand.* **100**, 949–53.

Further reading

Boehringer Mannheim Biochemica (1992) *Nonradioactive in situ hybridization: Application manual*.

Cuello, A. C. (1993) *Immunohistochemistry II*. John Wiley & Sons, Chichester.

Kiernan, J. A. (1990) *Histological Histochemical methods: Theory and practice* (2nd edn). Pergamon Press, Oxford.

Larsson, L. I. (1988) *Immunocytochemistry: Theory and Practice*. CRC Press Inc, Florida.

Leitch, A. R., Schwarzacher, T., Jackson, D. and Leitch, J. (1994) *In situ hybridization: A practical guide*. (Royal Microscopical Society) BIOS Scientific Publishers Ltd., Oxford.

Maniatis, T., Fritsch, E. F. and Sambrook, J. (1989) *Molecular cloning: A Laboratory Manual* (2nd edn) Cold Spring Harbour Laboratories, Cold Spring Harbour, NY.

Polak, J. M. and Van Noorden, S. (1986) *Immunocytochemistry: Modern Methods and Applications* (2nd edn). John Wright, London.

Wilkinson, D. G. (1992) *In situ hybridization: A practical approach*. IRL Press (Oxford University Press), Oxford.

10

An overview of methodologies for assessing proliferation and cell death

ROBIN DOVER

10.1 Introduction

The clinician would like to know more about the rate of cell growth for a number of reasons. These are primarily to understand the pathology of any given disease, to understand the effects of therapeutic drugs and ultimately to use these to clinical advantage. With such aims in mind it is critical that the information gleaned from proliferative studies is correct. Although there are a number of techniques for studying proliferative activity, few are applicable in clinical material or even particularly informative when used.

The methods can be broadly divided into two types, those that produce state measurements and those that produce rate measurements. State measurements will tell us how many cells are, for example, at a certain stage of the cell cycle or expressing a particular protein; these may be correlated with proliferation. Rate measurements actually measure the production of new cells. As the questions usually being asked involve the issue of how rapidly cells are being made, the rate measurements would seem the best to use. However, rate measurements require multiple samples or biopsies over a period of time and are often technically more difficult to perform. Often the state measurements are chosen for ease of use.

Anyone interested in proliferation should have at least a basic understanding of the cell cycle. In recent years our knowledge of the molecular events in the cell cycle has rapidly evolved. Some of these findings have resulted in cell cycle stage-specific antibodies, which can be very useful. However, these are so far very broad tools and practically a basic knowledge of the cell cycle is all that is required initially. Figure 10.1 shows a diagrammatic representation of the cell cycle. Following its 'birth' at mitosis the cell enters a phase termed G_1. This was originally coined to mean gap 1 because no specific events could be attributed to this phase. In most cells it is the longest and most variable phase. In early embryos there may be either no discernible G_1 or a very short G_1. There is still some debate as to whether events that are now detectable in G_1 are specific to this phase. G_1 is followed by S-phase, where S stands for synthetic. It is in this phase that cells synthesize new DNA to replicate the genome. This was discovered with the advent of radiolabelled nucleotides,

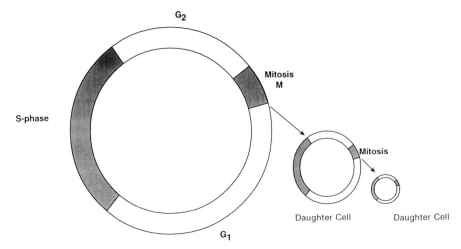

Figure 10.1. A diagrammatic representation of the cell cycle. Following a cell's 'birth' at mitosis it enters G_1, the first so-called gap phase, which is the most variable part of the mammalian cell cycle. At some point the cell becomes committed to replication and specific events in preparation for DNA synthesis occur. This has been termed late G_1 but is difficult to define exactly. The cell then enters S-phase and replicates its entire genome. During this phase DNA precursors will be incorporated and can be used to tag only those cells making DNA. After the S-phase cells enter a second gap phase, G_2, again relatively little is known about this phase, although some G_2-specific gene expression has been reported. At this point the cell has two copies of each chromosome pair, i.e. it has effectively 92 chromosomes. After G_2 the nuclear membrane disappears, the chromosomes condense and the mitotic spindle forms, the chromatids align to form a metaphase plate and then migrate to the poles of the cell. Cytokinesis occurs and two cells are formed which are now in a new G_1 phase. Here the second and subsequent daughter cells are shown smaller, only for clarity. Cytokinesis usually results in equal-sized daughters.

which were found to be incorporated only during this discrete part of the cycle. With the exception of embryonic cells, the S-phase is relatively constant in mammalian cells and is of the order of 6–10 h. Although small amounts of DNA synthesis can occur at other phases of the cycle, this is limited to repair of damage and is termed repair synthesis or unscheduled DNA synthesis to distinguish it from synthesis programmed to occur for cell division. When a cell has completely replicated its DNA it enters a second gap phase, G_2. Although there are specific events detectable in this phase, there are as yet no clinically useful markers available. G_2 is relatively short and may be of the order of 2–8 h. Cells then enter mitosis. Mitosis is usually short, of the order of 1 h. Mitosis is the only histologically visible stage of the cell cycle and hence the original distinction of two stages of the growth cycle into mitosis and inter-phase. The cell cycle time is the time for the entire cycle to be traversed. In practise this will vary between cells in a population.

10.2 Basic methods and technical limitations

10.2.1 Mitotic index

There must be some relationship between the number of mitoses and the rate at which cells are produced. From this assumption one can count the fraction of a population that is in mitosis at any point in time. Technically this is simple although early stages of mitosis can be difficult to distinguish. The fraction is normally expressed as a percentage and is termed the mitotic index (MI). Because the probability of observing an event is related to the duration of the event we can derive,

$$\text{MI} = T_m/T_c$$

where MI is mitotic index, T_m is the duration of mitosis and T_c is the time for the whole cell cycle. However, in the clinical situation it is rare to know these parameters.

Practical considerations include the choice of fixative and size of biopsy. If penetration of fixative is slow then mitoses in the centre of the sample may continue beyond mitosis and thus lead to an underestimate. In clinical practice, *provided that care is taken*, the mitotic index can be a useful indicator of proliferative activity. However, the critical worker will wish to read further about this apparently simple technique (Wright and Alison, 1984). Potential pitfalls include differences between T_m for different populations and changes in the number of cells in the proliferative pool. In the latter case two populations may have exactly the same cycle times but have different numbers of cells engaged in proliferation.

10.2.2 Labelling index

If one can provide the cell population under study with a radioactive precursor that is specifically incorporated into DNA and no other molecule, then one could label the cells in S-phase specifically. Thymidine fulfils these criteria for specificity and is usually used as a tritiated preparation. The isotope is given in experimental animals intraperitoneally or more rarely intravenously. In the clinical situation it is more difficult to apply. In the skin, intradermal injection is possible. Local application may be possible in some other organs. Circulating thymidine is either incorporated or broken down within a short period, certainly within the hour. During this time all cells in S-phase should incorporate the isotope in their DNA. Detection could be by liquid scintillation counting but this approach has serious methodological problems (Wright and Alison, 1984). It is often more informative to have positional information, showing where the proliferative cells are. This is achieved by autoradiography on sections. Basically, after sectioning the sample is covered with a thin layer of photographic emulsion and exposed for a period of days. The radioactive emissions (β particles in the case of tritium) produce a latent image in the emulsion, which when developed is visualized as black grains of silver overlying the site of origin of the radioactivity. Thus clusters of grains appear over cells in S-phase. By analogy with the mitotic index a labelling index can be scored, again this is usually expressed as a percentage. Technically there are many problems including the handling of radioisotope, potential risk to the patient and the availability to all the cells in the population. These problems are discussed in detail elsewhere (Dover and Wright, 1991). Recently it has been demonstrated that incorporation of labelling doses of radioactive thymidine induces p53 expression and transient cell cycle arrest in mammalian cells (Dover *et al.*, 1994), thus the very method we use to measure proliferation may, in itself, invalidate the observation.

Because of the major problems applying this method in the clinical situation it is rarely used. During the last decade an alternative has become more widely available. Bromodeoxyuridine (BrdU) is a structural analogue of thymidine and is only incorporated into DNA. It can be detected using antibodies. Originally a mouse monoclonal was used (Gratzner, 1982) but a range of commercially available polyclonal and monoclonal antibodies are available. These will produce a similar image to autoradiography, with the silver grains replaced by a coloured reaction product. A labelling index can be scored. The technical problems related to thymidine usage also apply to BrdU with the exception of the radioactivity. However, BrdU is a mutagen and light-sensitizing and should not be considered innocuous. Problems such as some cells not incorporating DNA-specific labels during S-phase (Clausen *et al.*, 1983) are as much a problem for workers with BrdU as for those with thymidine. The advantages of BrdU immunocytochemistry is that it is rapid compared with autoradiography and it is ethically more acceptable to administer to patients in some circumstances.

10.2.3 Continuous labelling

This is usually only applied in experimental animals or *in vitro*. This involves continuous delivery of the labelling agent. Theoretically it can generate data on all stages of the cycle but again suffers from a number of technical problems.

10.2.4 Immunocytochemical marker

In recent years a number of antibodies have become available that detect proteins that are expressed for only part of the cell cycle. These have great appeal in the clinical situation as they are simple and rapid to use and are usually inexpensive. This is a rapidly developing field and as space constraints prevent a detailed discussion, the reader is advised to read some basic reviews (e.g. Yu, 1992) prior to attempting such studies.

10.2.5 Ki-67/MIB1

One of the first markers to be used was the Ki67 antigen, which recognizes a nuclear antigen and is expressed in all cycle stages except early G_1 (Gerdes *et al.*, 1984). Originally the antibody was thought to work only in snap frozen material but it can be detected in formalin fixed, paraffin-embedded tissue if microwave antigen retrieval is used. This was realized shortly after a second antibody, recognizing the same antigen, was produced, namely MIB1. This worked on wax sections but improved staining can be obtained with microwave treatment. Ki-67/MIB1 staining correlates well with other indicators of proliferation but almost certainly over-estimates proliferation in tumours (See Figure 10.2).

10.2.6 pcna

Proliferating cell nuclear antigen (pcna) appeared for a time to be a promising candidate for a proliferative marker. However, technical problems beset its use and it can be expressed ectopically in non-proliferating cells surrounding tumours (Hall *et al.*, 1990). The usefulness of pcna as a marker is severely limited by these considerations.

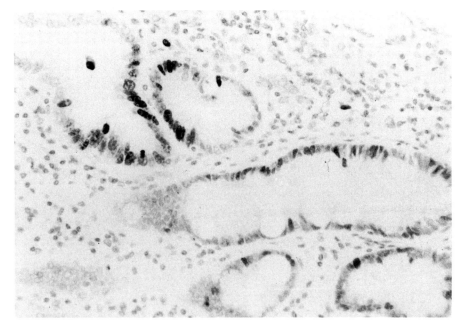

Figure 10.2. This shows a section of stomach stained with an anti-MIB1 antibody. Nuclear staining is evident in the epithelium.

10.2.7 Cytometry

During cell division each cell must replicate its DNA, so direct methods of measuring DNA content on a per cell basis have been developed. Originally this was performed using a microscope on sections. The principle is that a dye is used that binds DNA stoichiometrically. The dye is then excited with either ordinary light of a specific wavelength or with a laser pulse and the amount of light emitted is measured. Thus DNA content can be measured accurately. Using a microscope this is very time consuming and tedious to perform. When cells can be prepared as a single cell suspension, automated techniques exist. Tissue disaggregation remains a problem but for blood cells this method is ideal. The flow cytometer is widely used and can analyse large numbers of cells in a very short time. The principle is the same but a cell suspension is used and organized to pass drop-wise, one cell at a time, past an interrogating laser. Automatic data analysis can be performed and DNA histograms produced rapidly. This method can be combined with other markers such as immuno-histochemistry for multi-variate analysis. One such method uses DNA and BrdU co-measurements (e.g. Wilson and McNally, 1992).

Unless combined with a time element, the cytometer produces only a static measure of proliferation. It can estimate the numbers of cells in each part of the cycle but it can't tell us anything about the rate of movement around the cycle. Time resolved cytometry can be used to produce measurements of the rate of progress around the cycle. Combining a pulse of BrdU with serial samples one can follow the progress of cells that were in S-phase around the cycle (Dolbeare *et al.*, 1983, 1985). This is technically quite difficult and requires a single cell suspension, which can also be difficult to obtain from some tissues. When studying aneuploid populations, as may be the case in tumours, this will affect the DNA distribution and may make analyses more difficult.

10.2.8 DNA analyses

Although probably limited to *in vitro* studies, DNA analysis can be a useful tool. There are a number of methods available, all of which estimate the total DNA content in the sample, using a DNA specific dye that can be excited by an interrogating light beam, emitting a signal that can be machine read (Otto, 1993, 1994). Such assays can therefore be fast and accurate, but again present only a static picture. They are useful for large-scale studies or as a screen before embarking on more arduous studies.

10.2.9 Time lapse

A direct method applicable to cultured cells, time lapse cinematography or videography can be a very powerful technique. The growth rate of individual cells can be determined and relationships between cells studied. Cell lineages can be constructed and cycle times measured (Dover and Potten, 1988), other events such as programmed cell death (apoptosis) can be visualized and quantified. This method is time and labour intensive but can be very useful.

10.2.10 Growth curves

Use of growth curves, in tissue culture studies, is perhaps the simplest method of all. This requires culturing cells at low density in replicate vessels and harvesting and making accurate counts of cell number with time. A classical growth curve with lag, log and stationary phases will result in most cases. It is a quick and simple method for comparing populations. The curve can be read to produce a population doubling time. However, it is insensitive in terms of the detail that can be observed. It can also be a problem deciding where the points of inflection on the curve are. With cell populations that either die or differentiate in culture (e.g. bone marrow, keratinocytes) this method will underestimate proliferation.

10.3 Advanced methods and technical limitations

10.3.1 Stathmokinetics

Whilst the mitotic index can be very useful if carefully performed, the rate of entry into mitosis is a more desirable parameter. It is at mitosis that new cells are produced so a measure of the rate of entry into mitosis should be equivalent to the rate of new cell production.

The mitotic rate (sometimes termed the birth rate) can be measured using the stathmokinetic or mitotic arrest method. This involves the use of a drug to disrupt the mitotic spindle apparatus so that cells can enter mitosis but cannot pass beyond metaphase, where the chromosomes would normally migrate to the poles of the cell. A variety of agents are available and historically colchicine or colcemid were used. More recently the vinca alkaloids have been developed and vincristine and vindesine sulphates are the most common. To be an effective blocking agent the effects of the drug should be limited to mitosis. There is evidence that colcemid can act at other cycle stages and may affect the rate of progression. The principal of the procedure is that the drug is administered locally in human subjects, or systemically in experimental animals. Samples/biopsies are taken at timed intervals over a period of up to 5 h. The samples are fixed and after processing examined microscopically. As the drug will take some time to begin having effects, and as some cells may be able to complete mitosis that they have already entered, there will be a lag before there is any effective arrest. To allow for this, the first sample is normally not taken until an hour after administration. As the drug is toxic to the cells they will die and pyknotic nuclei will develop after extended time periods, thus the experiment is limited to a short 'window' of time following administration of the drug. Cells become blocked in mitosis and new cells enter mitosis with time, thus the number of arrested cells will increase with time. The number of arrested cells is counted and plotted. The slope of the curve gives the rate of accumulation of mitotic cells, which is equivalent to the rate of entry into mitosis. This gives a direct and accurate assessment of the rate of cell proliferation.

Obviously, the need for local drug administration and serial biopsy makes this technique difficult to apply in clinical situations but it is a very powerful method. If it can be demonstrated that all the technical conditions required to ensure a linear arrest have been met and that the appropriate dose of arresting agent has been determined to ensure that there are no cells escaping the block, then it may be more practical to perform an augmented mitotic index. This is simply using an arresting agent but sampling at only one time point, usually 2–3 h later. This removes the problem of possible differences in mitotic duration affecting the result and is a rate-related measurement.

10.3.2 FLM/PLM

A far more complex technique is the percent or fraction labelled mitosis method (PLM/FLM). This method is not for the faint hearted and will be described only briefly here. The principle is that a cohort of cells is labelled and their progress around the cycle followed by a second marker. In practice this usually involves labelling with tritiated thymidine and sampling over an extended period (usually at least a day) and scoring only mitotic cells as either labelled or not labelled. This value will change as cells move around the cycle. At time zero, just after labelling, no mitotic cells will be labelled. After a period equal to the minimum G_2 the first few labelled cells will enter mitosis and the fraction of labelled mitoses (FLM) will begin to rise. The FLM will reach a value of 1 when all the mitotic cells are labelled. As S-phase is much longer than M (typically 6–12 h vs. 1 h) the FLM will remain at 1 for the duration of S-phase and then drop to zero. At this time the last cell that had been labelled in S-phase has passed through the remainder of S, G_2, and M and has entered G_1. All the labelled cells are in G_1 at this time so the FLM will be zero and remain so until the cells have traversed G_1, S (for the second time since the experiment began) and G_2. If a plot is made of the change in FLM with time the durations of all cell cycle phases can be calculated and the overall cycle time inferred. For more details see references by Wright and Alison, 1984; Dover, 1992.

10.3.3 Combined proliferation/immunocytochemistry

The use of BrdU for kinetic studies has highlighted immunological methods and the possibility of combined cytochemistry. Combinations of two proliferative markers, e.g. BrdU and MIB1/Ki-67, theoretically should yield useful information. Combining proliferative methods with markers of differentiation can also yield useful information (Dover and Watt, 1987). The ability to localize cells expressing a range of markers, e.g. differentiation specific antigens, oncogenes, transfected genes or tumour markers, in combination with proliferative estimates, have produced much useful information. This is often limited to geographic descriptions of populations. Although useful in itself it is best used as an adjunct to other techniques.

10.3.4 Combined proliferation/in situ hybridization

In situ hybridization is a powerful technique that has been widely exploited in recent years. Although the processing for *in situ* hybridization often entails procedures that will damage or remove material from the sections, it can still be possible to combine immunocytochemistry and *in situ* techniques. Even where the tissue is damaged or staining removed by tissue preparation methods it can be possible to co-localize signals (Lu and Dover, 1993).

A recent development that shows great promise is the use of *in situ* hybridization to detect proliferation specific histones. Chou *et al.* (1990) used H3 as a marker and found a good correlation with the expected pattern of proliferation. Alison *et al.* (1994) used a cocktail of digoxigenin-labelled probes to H3 and performed quantitative comparisons with BrdU labelling. They found a high correlation between the two methods with H3 consistently detecting a fractionally higher number than BrdU. This is probably attributable to onset of expression in late G_1. The choice of fixative is also important for H3 detection, as for most mRNA studies a cross-linking fixative appears to give the best results. This method can be applied to archive material and seems to be a powerful and very promising technique for the identification of S-phase cells. Again, there are commercially available kits that can be purchased to perform such studies, which may help with reproducibility. When combined with a marker of cells capable of cell division such as Ki-67 or MIB1 this method can be used to investigate archival material, which is often desirable for clinical studies. The histone *in situ* labels a population that is overwhelmingly comprised of S-phase cells and the immunocytochemistry will reveal some information concerning the total population of proliferative cells.

10.4 Cell death

The net cell production in a tissue is governed by a number of factors including the rate of cell proliferation, the rate of differentiation and the rate of cell death. Until relatively recently cell death was thought to be important only in pathological conditions. The description of a specific form of cell death that occurs in a programmed manner and is found in normal development has led to a considerable change in attitudes. Apoptosis is a distinct form of cell death that occurs both as a normal event in the life cycle of a tissue and in response to certain external stimuli. Quantification of apoptosis is still a difficult proposition. The most common methods of detection involve biochemical or molecular biological techniques that destroy all positional information and are difficult if not impossible to quantify. At a tissue level the best method is still electronmicroscopic examination, but this is limited by the small sample sizes and tedious methodology. Histological recognition is possible but is a skilled task and may require thin (plastic) sections. The only other methods available on histological sections exploit a feature of early apoptosis – the induction of strand breaks in DNA. However, it should be considered essential, with our current methodologies, to include a comparison of any methods used against microscopic methods on the same samples.

In experimental situations death in a population may be assessed with gross methodologies such as DNA laddering. This is where DNA from apoptotic cells is revealed by electrophoresis as regular sized fragments (200 bp). This reveals nothing about the type of cells in a mixed population, nor does it allow positional information to be gleaned. Another population-based method is to use cytometry to detect cells with reduced levels of DNA, i.e. less than a G1 complement. As described above, cells are exposed to DNA-binding dyes and individual cells have their DNA content measured. In the resulting distribution cells below the G1 level may be observed. In theory these cells should be apoptotic. Methods may be developed that can determine between apoptosis, necrosis and living cells and can be applied in the cytometer (Echaniz, de Juan and Cuadrado, 1995). Again the problems of obtaining a 'clean' single-cell preparation from solid tissue apply. There is also the possibility of DNA contamination, via DNA released from apoptotic cells or damage caused during the cell preparation (Tounekti, Belehradek and Mir, 1995). Acridine Orange has been used as a dye to differentiate between the denaturability of normal versus apoptotic cells' DNA. This method appears to be very sensitive (Hotz *et al.*, 1994). Other methods used in cytometric analysis rely on differential penetration of the dye into living versus dying cells. In this case unfixed cells would be treated with a non-permeant dye (e.g. propidium iodide or ethidium bromide) – dead/dying cells will take up the dye. Cytometry has also been combined with end labelling (Hotz *et al.*, 1994) (see below).

A number of methods have used essentially nick translation *in situ* to introduce markers into cells with such nicks. These can be biotinylated, radioactive or fluorescent. The method is known by a variety of names, e.g. *in situ* end labelling (ISEL) (e.g. Wijsman *et al.*, 1993), *in situ* nick translation (e.g. Gold *et al.*, 1993), *in situ* tailing (e.g. Gold *et al.*, 1994), *in situ* terminal transferase labelling (tdtl) (e.g. Hayashi *et al.*, 1994). All rely on the same principal, that of exploiting nick translation of whole cells or tissue section; the method will work in paraffin-processed tissue so could be applied to archive material. The method is relatively new but early results were promising (Gavrieli, Sherman and Sasson, 1992) and today commercial kits are available to perform the process. The use of the method is not without some drawbacks

and technical problems. Different tissues may require different enzyme treatment times, tissues where enucleation is part of normal differentiation, e.g.
squamous epithelium, will have partly digested nuclei as part of their normal
repertoire of behaviour. It is obvious therefore that the method is not specific
for programmed cell death. Nicking of DNA is an early event in apoptosis and
thus methods based on nick translation may be prone to underestimation.
Further problems were discussed by Ansari *et al.* (1993). Other markers may
be useful in the future, one such method may be the use of protein kinase C
isoenzymes. Knox, Johnson and Gordon (1993), studied six isoenzymes in cells
undergoing apoptosis; expression of PKC-alpha and -beta were highest in these
cells. It may be possible to develop this observation to produce a method to
detect apoptosis.

As yet there is no really satisfactory method that allows quantification and
positioning information. *In situ* nick translation needs to be carefully controlled
and the results interpreted with some caution but may provide some useful
information. This is an area of intense research interest at present and it seems
probable that new markers or methods will arise in the near future. For
population studies cytometry and end labelling may be a useful combination.
The *Bcl-2* and *Bcl-x* gene products have engendered particular interest as a
body of evidence suggests that they may, at least in part, 'protect' cells from
apoptosis. However, lack of *Bcl-2/Bcl-x* expression does not signify that a cell
will undergo apoptotic death. The molecular events regulating apoptosis are
being unravelled and clinically useful markers may emerge.

References

Alison, M., Chaudry, Z., Baker, J. *et al.* (1994) Liver regeneration: a comparison of in
    situ hybridization for histone mRNA with bromodeoxyuridine labelling for the
    detection of S-phase cells. *J. Histochem. Cytochem.* **42**, 1603–8.
Ansari, B., Coats, P. J., Greenstein, B. D. and Hall, P. A. (1993) In situ end labelling
    detects DNA strand breaks in apoptosis and other physiological and pathological
    states. *J. Pathol.* **170**, 1–8.
Chou, M. Y., Chang, A. L. C., McBride, J. *et al.* (1990) A rapid method to determine
    proliferation patterns of normal and malignant tissues by H3 mRNA in situ
    hybridization. *Am. J. Pathol.* **136**, 729–33.
Clausen, O. P. F., Elgjo, K., Kirkhus, B. *et al.* (1983) DNA synthesis in mouse
    epidermis: S phase cells that remain unlabelled after pulse labelling with DNA
    precursors progress slowly through S. *J. Invest. Derm.* **81**, 545–9.
Dolbeare, F., Breisker, W., Pallavicini, M. G. *et al.* (1985) Cytochemistry for BrdUrd/
    DNA analysis: stoichiometry and sensitivity. *Cytometry* **6**, 521–30.
Dolbeare, F., Gratzner, H., Pallavicini, M. G., Gray, J. W. (1983) Flow cytometric
    measurement of total DNA content and incorporated bromodeoxyuridine. *Proc.
    Natl Acad. Sci. USA* **80**, 5573–7.
Dover, R. (1992) Basic methods for assessing cellular proliferation. In *Assessment of
    Cell Proliferation in Clinical Practice* (ed. P. A. Hall, D. A. Levison and N. A.
    Wright), pp. 63–81. Springer-Verlag, London.

Dover, R., Jayaram, Y., Patel, K. and Chinery, R. (1994) p53 expression in cultured cells following radioisotope labelling. *J. Cell Sci.* **107**, 1181–4.

Dover, R. and Potten, C. S. (1988) Heterogeneity and cell cycle analyses from time-lapse studies of human keratinocytes *in vitro*. *J. Cell Sci.* **89**, 359–64.

Dover, R. and Watt, F. M. (1987) Measurement of the rate of epidermal terminal differentiation: expression of involucrin by S-phase keratinocytes in culture and in psoriatic plaques. *J. Invest. Dermatol.* **89**, 349–52.

Dover, R. and Wright, N. A. (1991) The cell proliferation kinetics of the epidermis. In *Biochemistry and Physiology of the skin* (ed. L. A. Goldsmith), pp. 239–65. Oxford University Press, Oxford.

Echaniz, P., de Juan, M. D. and Cuadrado, E. (1995) DNA staining changes associated with apoptosis and necrosis in blood lymphocytes of individuals with HIV infection. *Cytometry* **19**, 164–70.

Gavrieli, Y., Sherman, Y. and Sasson, S. A. (1992) Identification of programmed cell death in situ via specific labelling of nuclear DNA fragmentation. *J. Cell Biol.* **119**, 493–501.

Gerdes, J., Lembke, H., Wacker, H. H. *et al.* (1984) Cell cycle analysis of a cell proliferation associated human nuclear antigen defined by the monoclonal antibody Ki-67. *J. Immunol.* **133**, 1710–15.

Gold, R., Schmied, M., Giegerich, G. *et al.* (1994) Differentiation between cellular apoptosis and necrosis by the combined use of in situ tailing and nick translation techniques. *Lab. Invest.* **71**, 219–25.

Gold, R., Schmied, M., Rothe, G. *et al.* (1993) Detection of DNA fragmentation in apoptosis: application of *in situ* nick translation to cell culture systems and tissue sections. *J. Histochem. Cytochem.* **41**, 1023–30.

Gratzner, H. G. (1982) Monoclonal antibody to 5-Bromo and 5-iododeoxyuridine: a new reagent for detection of DNA replication. *Science* **218**, 474–5.

Hall, P. A., Levison, D. A., Woods, A. L. *et al.* (1990) Proliferating cell nuclear antigen (pcna) immunolocalization in paraffin sections – an index of cell-proliferation with evidence of deregulated expression in some neoplasms. *J. Cell Path.* **162**, 285–94.

Hayashi, P., Marrone, A., Kleiner, D. *et al.* (1994) Hepatic apoptosis in chronic hepatitis c – analysis by in situ terminal transferase labelling (tdtl). *Hepatology* **20**, A160.

Hotz, M. A., Gong, J., Traganos, F. and Darzynkiewicz, Z. (1994) Flow cytometric detection of apoptosis: comparison of the assays of *in situ* DNA degradation and chromatin changes. *Cytometry* **15**, 237–44.

Knox, K. A., Johnson, G. D. and Gordon, J. (1993) A study of protein kinase C isoenzyme distribution in relation to Bcl-2 expression during apoptosis of epithelial cells *in vivo*. *Exp. Cell Res.* **207**, 68–73.

Lu, Q.-L. and Dover, R. (1993) Computer assisted signal co-localization for simultaneous detection of antigen by immunohistochemistry and DNA by non-isotopic in situ hybridization. *Histochemistry* **99**, 23–7

Otto, W. R. (1993) Fluorimetric and spectrophotometric measures of cell number. In *Cell and tissue culture: laboratory procedures* (ed. A. Doyle, J. B. Griffiths and D. G. Newell), 10E1.1–1.14. John Wiley, Chichester.

Otto, W. R. (1994) Fluorimetric DNA assay. In *Keratinocyte methods* (ed. I. M. Leigh and F. M. Watt), pp. 89–90. Cambridge University Press, Cambridge.

Rao, J. and Otto, W. R. (1992) Versatile fluorimetric assay combining DNA estimation and immunocytochemistry. *Ann. Biochem.* **207**, 186–92.

Tounekti, O., Belehradek, J. Jr and Mir, L. M. (1995) Relationships between DNA fragmentation, chromatin condensation, and changes in flow cytometry profiles detected during apoptosis. *Exp. Cell Res.* **217**, 506–16.

Wijsman, J. H., Jonker, R. R., Keijzer, R. *et al.* (1993) A new method to detect apoptosis in paraffin sections: in situ end labelling of fragmented DNA. *J. Histochem. Cytochem.* **41**, 7–12.

Wilson, G. D. and McNally, N. J. (1992) Measurement of cell proliferation using bromodeoxyuridine. In *Assessment of cell proliferation in clinical practice* (ed. P. A. Hall, D. A. Levison and N. A. Wright), pp. 113–39. Springer-Verlag, London.

Wright N. A. and Alison M. (1984) *The biology of epithelial cell populations*, p. 538. Clarendon, Oxford.

Yu, C. C. W. (1992) The application of immunohistochemistry in assessment of cellular proliferation. In *Assessment of cell proliferation in clinical practice* (ed. P. A. Hall, D. A. Levison and N. A. Wright), pp. 141–59. Springer-Verlag, London.

Chromosomal and molecular assays

11

Assessing proteins and nucleic acids quantitatively and qualitatively by gel fractionation and related techniques

JANUSZ A. Z. JANKOWSKI
and FIONA K. BEDFORD

11.1 Summary

The scope of biological and clinical applications is wide but it must be emphasized that obtaining pure protein, DNA or RNA in sufficient amounts is fundamental to the success of all molecular analysis. This Chapter describes the diversity of techniques available for protein and nucleic acid extraction from clinical tissue. In particular, the advantages of each method are succinctly summarized. Furthermore, some of the aspects of molecular analysis and experimental validation that can be applied to clinical specimens are briefly illustrated.

11.2 Introduction and outline of gel electrophoresis and blotting techniques

In general terms, molecular techniques can be used to assess genomic DNA or gene expression at the RNA or protein level. In most cases this is achieved by gel electrophoresis and the information gained is both quantitative (gene enhancement over control samples) and qualitative (the appropriateness of the gene product's molecular weight).

The techniques of Western blotting (protein), Northern blotting (RNA) and Southern blotting (DNA) have many similarities. The molecules must each be extracted from tissue in the presence of inhibitors that prevent breakdown of DNA, RNA and protein by DNAases, RNAases and proteases respectively (stage of molecular extraction). Subsequently the molecules are electrophoresed on agarose gels or polyacrylamide gels in order to separate the molecules by size (gel fractionation). These gels are rather fragile and therefore the molecules must be transferred (blotted) to rigid permanent supports such as nitrocellulose or nylon membranes. These membranes are subsequently probed by either antibodies in the case of Western blots or nucleic acid probes in Northern and Southern blots (hybridization) (Schagger, 1987).

The extraction step is probably the most important step and may frequently be unsatisfactory. As a result it is usually preferable to assess molecular integrity by using Cam 5.2 (see Figures 11.1 and 11.2), GAPDH or β-actin probes for protein, RNA and DNA respectively (Sambrook, Fritsch and Maniatis, 1989 and Boothwell, Yancopoulos and Alt, 1990).

When separating nucleic acids, DNA is usually linearized in the extraction process, whereas RNA must be denatured so that strands run as separate linear chains in the gel. When separating proteins, β-sheets and α-helices must be

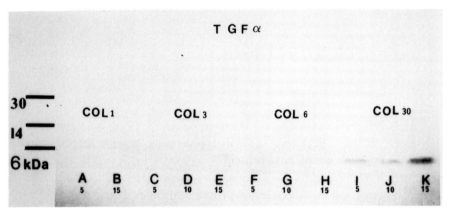

Figure 11.1. Western blots of TGFα (transforming growth factor alpha) expression in colorectal cancer *in vivo* and *in vitro*. The maximal intensity of immunoreactive band is at 6 kDa. Positive control, normal colon and stomach; negative control, lymphoid tissue and muscle tissue (not shown). A: TGFα expression in the HCA-7 cell line – immunoreactivity is maximal in colony 30 and increases with differentiation at 5, 10 and 15 days.

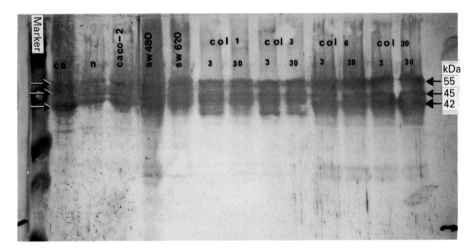

Figure 11.2. Western blot of cytokeratin expression in colorectal tissue. Equal immunoreactivity to the antibody Cam 5.2 is present in most samples at 42, 45 and 55 kDa confirming the equal epithelial content of the protein assessed in previous Western blot gels. Positive control, normal colonic tissue; negative control, neural tissue. The majority of samples had equal loading of protein samples.

disrupted to ensure high resolution of bands during electrophoresis by reducing disulphide bonds and this is achieved by adding sodium dodecyl sulphate (SDS) and β-mercaptoethanol to the gels.

Sometimes very small amounts of either protein or nucleic acids are present in tissue specimens, so that the standard detection methods above may be insufficiently sensitive to identify them. In these circumstances, methods such as ELISA or PCR can be used to identify minute amounts of proteins and nucleic acids respectively.

11.3 Western blots

11.3.1 Protein extraction

The methods used to extract protein from tissue or cells are different but during lysis of the cells it is possible to add one or several proteinase inhibitors to the lysis buffer to block specific degradative enzymes: Aprotinin in PBS, use at 0.1–$2\,\mu g/ml$ and store at $-20\,°C$; leupeptin in distilled water, use at 0.5–$2\,\mu g/ml$ and store $-20\,°C$; pepstatin in distilled water at $1\,\mu g/ml$; and standard trypsin inhibitor (PMSTI) in distilled water at $10\,\mu g/ml$.

1. Wash cells and tissue in cold (0 °C) PBS (phosphate buffered saline) and centrifuge at 1000 *g* for 1–2 min and discard supernatant. When removing adherent cells growing in monolayers, rinse cells in PBS and scrape with a cell scraper or similar device into 15 ml conical tubes (for T75-150 flasks) or Eppendorfs (for T25 flasks) and spin to pellet the cells. Repeat once.
2. Resuspend cells in $1 \times$ SDS gel-loading buffer (without Bromophenol Blue). As a guide, use 1 ml for 5×10^7 cells and place in boiling water for 10 min with periodic vortexing (the sample should become highly viscous due to the release of high molecular weight chromosomal DNA). Place on ice for 5 min. As an alternative fresh tissue can be homogenized in the peptide extraction buffer solution in a fume hood to avoid aerosols (1–2 min) and the supernatant subsequently transferred to $2 \times$ SDS gel-loading buffer.

Peptide extraction buffer

1 M hydrochloric acid, 5% (v/v) formic acid, 1% (w/v) sodium chloride, 1% (v/v) trifluroacetic acid (TFA) and 1 mg per ml ascorbic acid, if oxidation of the peptide is likely.

Peptide SDS gel-loading buffer

100 mM Tris/HCl (pH 6.8), 4% SDS (electrophoresis grade), 20% (v/v) glycerol, 200 mM dithiothreitol (DTT) (add this to required solution

buffer just prior to use) and 0.2% (w/v) Bromophenol Blue, added only to samples already quantitated (see later). (Alternatively, β-mercapto-ethanol 2–10% (v/v) may be a more effective reducing agent than DTT.)

3. The subsequent extract should be centrifuged in a microfuge at top speed for 15–30 min at 20 °C (or in a Beckman at 10 000 g for 10–15 min). Place tube on ice for 5 min. Freeze the resultant (clear) acidic supernatant at −70 °C as 200 μl samples, avoiding any solid matter at the bottom of the centrifuge tube. Prior to running the samples in SDS gels, neutralize with weak (0.1–1 M) NaOH, add 0.2% (w/v) Bromophenol Blue dye and also fresh 2% β-mercaptoethanol or DTT to each sample if they have been frozen more than 1 month.

4. In order to assess the protein concentration in each specimen, one sample is thawed on ice and is diluted into a pre-prepared Bio-Rad protein dye. Depending on the amount of protein present the dye undergoes a gradated colour change and this can be accurately quantified by spectrophotometry:

   (a) Dilute 1 part Bio-Rad dye in 4 parts distilled water and dispense 1 ml into each of the 1 ml polystyrene cuvettes (six cuvettes for calibration and one for each sample).

   (b) The calibration cuvettes receive 0 μl, 2 μl, 4 μl, 6 μl, 8 μl and 10 μl of the BSA protein standard solution (1.54 mg/ml). Place 1 μl of the protein sample into each of the remaining cuvettes. Wearing gloves, hold a clean piece of Parafilm over the top of each cuvette and invert to mix three times.

   (c) Using the first of the calibration cuvettes to zero the spectrophoto-meter, read the absorbance at 595 nm of all the cuvettes.

   (d) Plot A_{595} of the calibration cuvettes on the x axis of a graph and the corresponding protein quantity on the y axis in order to derive the protein concentrations from the A_{595} of your samples. For the Bio-Rad minigel, an identical amount of protein should be loaded to each lane, 15–30 μg, but for a maxigel the amount can be increased to 150 μg.

11.3.2 SDS-PAGE electrophoresis

Use the Bio-Rad protein II apparatus and gel pouring moulds or similar apparatus by Hoeffer or USB.

1. Clean long and short glass plates, plastic spacers and appropriate gel combs with weak detergent solution, then wipe well with absolute alcohol and air dry. Assemble the glass plates according to the manufacturer's instructions and test for leaks by pouring alcohol in between the short and long plates.

2. Make 10 ml of resolving gel mix, but if no leaks are present usually

only 3–5 ml is required per gel. The main variable in these gels is the acrylamide concentration, which varies according to the size of the protein. In general, as the protein size under investigation decreases the resolving ability of the gel and hence the acrylamide concentration should increase; 7.5% (w/v) acrylamide gel is used for proteins over 200 kDa, 10% gel for proteins of 20–200 kDa and 15% gel for proteins < 20 kDa. In this latter case, Tricine can be substituted for glycine to give further enhanced resolution.

7.5%: 5 ml H_2O, 2.3 ml 30% (w/v) acrylamide solution, 2.5 ml 1.5 M Tris (pH 8.8), 0.1 ml 10% (w/v) SDS, 0.1 ml freshly prepared 10% (w/v) ammonium persulphate and 6 μl TEMED (tetramethylethylenediamine).

10%: 3.9 ml H_2O, 3.4 ml 30% (w/v) acrylamide solution, 2.5 ml 1.5 M Tris (pH 8.8), 0.1 ml 10% SDS, 0.1 ml fresh 10% ammonium persulphate and 4 μl TEMED.

15%: 2.3 ml H_2O and 5 ml 30% (w/v) acrylamide solution, 2.5 ml 1.5 M Tris, 0.1 ml 10% SDS, 0.1 ml fresh 10% ammonium persulphate and 2.5 μl TEMED.

Pour sufficient gel mix so as to fill 60–80% of the volume between the glass plates and immediately cover the top of the setting gel with isopropanol and place on a flat undisturbed surface so as to allow the gel to set evenly.

4. After the resolving gel has set (usually 30 min) pour off the isopropanol and wash the gel carefully with distilled water and dry off the last residues by inserting strips of blotting paper between the plates. Then make a 5% acrylamide stacking gel, 5 ml for each gel: 3.4 ml H_2O, 0.83 ml 30% (w/v) acrylamide, 0.63 ml 1 M Tris (pH 6.8), 0.05 ml 10% (w/v) SDS, 0.05 ml 10% (w/v) ammonium persulphate and 5 μl TEMED. Insert the appropriate comb to suit the required number of samples and volume of sample (8 lanes each of 30 μl volume or 15 lanes of 15 μl volume) and pour the stacking gel. Place the remainder of the stacking gel on ice so that shrinking stacking gel can be topped up as it dries (allow 5–30 min to dry). Remove the combs, attach the gel to the electrodes and the transparent electrophoresis tank.

5. Fill the electrophoresis tank with electrophoresis buffer but do not overfill the bottom reservoir (outside tank) as this may aggravate the upward curl of dye front and resolving proteins, due to leakage of current through the edge of the gel.

   Electrophoresis buffer: 25 mM Tris, 250 mM glycine (electrophoresis grade) (pH 8.3) and 0.1% (w/v) SDS (never reuse).

6. Take out one measure of protein sample and add 5 μl of fresh lysis buffer or 10 μl 2% (v/v) β-mercaptoethanol. Partial protein reduction will lead to decreased efficiency of electrophoresis and doublets of soybean trypsin inhibitor and ovalbumin in the marker lanes.

   Boil the protein for 5 min, followed by incubation on ice for 5 min and then load onto the gel. Load markers in an assymetrical position so that the gel can be confidently orientated. Load 5 μl of molecular weight markers with multiple sample proteins: myosin 200 000 kDa, β-galactosidase 116 000, bovine serum albumin 66 200, carbonic anhydrase 31 000, soybean trypsin inhibitor 21 000, lysozyme 14 400 and aprotinin 6500 (markers come as either unstained or prestained) (beware – some dyes bind irreversibly with nylon or PVDF membranes rather than reversibly with nitrocellulose). Always put 5 μl of lysis buffer into empty wells to avoid distorted running of gel.

7. Commence electrophoresis of the gel at low voltage (8 V/cm), 90 V until the protein has run into resolving gel then increase to 100–150 V (15 V/cm) and run for 3 h.

11.3.3 Blotting

1. Prepare 4 l of transfer buffer: 50 mM Tris base, pH 8.3; 380 mM glycine; 0.1% (w/v) SDS; 20% methanol and make up with distilled water.
2. Use Bio-Rad trans-blot apparatus to transfer proteins to nitrocellulose membrane, or other suitable apparatus by Hoeffer or UBS. Cut a paper template to match the size of the resolving gel and use it to make one nitrocellulose membrane (pore size 0.45 μm for most proteins) and four Whatman filter papers for each gel. Soak both the nitrocellulose and Whatman papers in transfer buffer.
3. When electrophoresis is completed cut off the stacking gel and place the gel between the filter paper and nitrocellulose as follows:
   (a) Place backing foam mesh from the trans-blot tank onto the plastic grid (clamping unit) followed by two filters, place on light box which will show up any small air bubbles.
   (b) Gently place gel on the filters and remove air bubbles.
   (c) Place nitrocellulose membrane on the gel and remove air bubbles.
   (d) Place two filters on the membrane followed by the mesh.
   (e) Clamp the whole assembly while submerged in transfer buffer.
   (f) Fill up the tank.
5. Fit the cooling coil, place assembled tank on stirrer and put fast running water through gel so as to bring buffer to optimal temperature of 10 °C.
6. Align the unit with the membrane towards the +ve terminal and gel at −ve terminal and carry out transfer ∼0.5 A for 2–4 h (or 0.02 A overnight). To assess whether transfer is complete, remove the plastic grid and gently peel back the gel from the filter, if prestained markers are used the blue bands will be transferred to the membrane.
7. Following transfer, remove membrane and leave to dry at room temperature for 1 h, ready for immunoblotting. At this stage it is possible to store

the blot in a desiccator between sheets of filter paper at 2–8 °C for up to 3 months.

The gel can be stained with Coomassie Brilliant Blue R250, or the membrane stained with Ponceau S (see later) to check that all the proteins have been blotted from the gel to the membrane.

To stain the gel with Coomassie Blue, place the gel in a shallow container containing the dye for 1–12 h. The gel is easily overstained, so that individual bands can not be identified. Destain in a 20% (v/v) methanol/water destaining solution with a 2 cm diameter ball of tissue paper (used to absorb the dye) and subsequently dry the gel onto Whatman paper and seal it in a bag for a permanent record.

11.3.4 Immunostaining

1. Optional step – membrane fixation. Fix membrane in 0.2% (v/v) glutaraldehyde (made up in PBS) for 45 min at room temperature (RT) for up to 1 h with gentle shaking. Wash membrane 3× in PBS +0.1% (v/v) Tween 20 (PBST). This step may help to fix proteins to membrane but it will also increase protein breakdown with potential loss of the desired epitope.

2. Optional step – Ponceau S staining of the membrane.
   (a) Resoak dry membrane from beneath by capillary action in deionized water then submerge.
   (b) Soak in ×1 Ponceau S stain, incubate for 2–5 min with gentle agitation. (×10 stock solution: Ponceau S 2 g (2%), trichloroacetic acid 30 g (30%), sulfosalicylic acid 30 g (30%), H_2O to 100 ml), use 10 fold-dilution in deionized water.
   (c) When bands are visible, wash the filter in several washes of deionized water (*not* PBS) and mark bands of unstained molecular weight standards with a waterproof pencil.

3. Block the membrane in 3% (w/v) BSA (bovine serum albumin) in PBS (or 5% (w/v) nonfat dried milk (Marvel) or 0.5% (w/v) gelatin) for 30–60 min at RT with gentle shaking. Wash the membrane twice in 1× PBS and once in 1× TST (50 mM Tris, 500 mM NaCl, 0.1% (w/v) Tween 20 and adjust to pH 9), 5 min each wash.

4. Incubate the membrane with the primary antibody diluted in TST for 1–2 h at RT (or overnight) with moderate shaking (20–60 r.p.m.). If the antibody gives high backgrounds, 1% BSA can be added to the primary antibody solution. Subsequently, the membrane should be washed three times in TST, 5 min each wash. Use antibodies at appropriate dilutions but in general use polyclonals at 1/100–1/5000, monoclonals undiluted–1/20 000, hybridoma cells undiluted–1/100, and ascitic fluid at 1/1000–1/10 000.

5. Staining
   (a) *Peroxidase anti-peroxidase (PAP) method*
       Wash membrane in TST × 1 and PBST × 2.
       Add peroxidase conjugated anti-mouse antibody (1/100–1/1000) in
       PBS +2.5% (w/v) BSA, or unconjugated anti-mouse will also do
       (rabbit/goat/sheep) and incubate for 1 h.
       Wash × 3 in PBS/0.1% (v/v) Tween.
       Add mouse PAP complex (1/1000).
   (b) *Avidin–biotin complex (ABC) method*
       Wash membrane in TST × 1 and PBST × 2.
       Add biotinylated antimouse (rabbit/goat/sheep) as above, incubate
       for 30–60 min.
       Wash × 3 in PBS/0.1% Tween (5 min each wash).
       Add ABC peroxidase complex (make up ABC 30 min beforehand)
       (6 μl A, 6 μl B in 20 ml PBS + 0.1% (w/v) Tween).
   (c) *Visualization by diaminobenzidine*
       Use 6 ml of 0.3% (w/v) nickel chloride or nickel ammonium
       sulphate buffered in sodium acetate/acetic acid buffer, 0.1 M pH
       6.0, 1.44 ml DAB (diaminobenzidine) (25 mg/ml) and 54 ml PBS.
       This solution should be filtered and 30–60 μl of H_2O_2 should be
       added immediately before adding it to the blot.
       The reaction should be allowed to progress for up to 10 min so that
       only the bands of interest appear, but as soon as the membrane
       shows background colouration the reaction should be stopped by
       washing 3 × in fresh double distilled water.
       If the band of interest is still faint the colouration can be enhanced
       by washing the gel in $CuSO_4$. Otherwise, semi-dry the membrane
       and seal it in plastic for a permanent record. It is advisable to
       photograph the gel within 6 h, as the chromogen may fade.
   (d) *Chemiluminescence*
       This method is usually 10-fold more sensitive but can be technically
       more difficult for the novice. A highly reliable kit is the
       ECL-Western blotting detection system by Amersham. Briefly, prior
       to applying the chemiluminescence reagents the primary antibody is
       incubated with either horseradish peroxidase labelled antibody or
       biotinylated antibody followed by a streptavidin horseradish peroxi-
       dase layer. The chemiluminescence reagents A and B are mixed
       together in equal volumes and applied to the blot (some workers
       have found that diluting the chemiluminescence reagents with an
       equal part of water allows a more controlled detection step). Subse-
       quently, the peroxidase attached to the antigen will oxidize the thin
       film of overlying luminol reagent generating free radicals, the by-
       product of which is light emission. To make a permanent record a
       photographic film is exposed over the blot in a darkroom for between
       30 s to 10 min.

6. Membrane stripping protocol

   Depending on the reaction used, the initial antibody can be stripped from the membranes and reprobed with other antibodies. This is not recommended, however, when enzyme-linked antibodies have been used, since substrate deposits are difficult to remove.

   Incubate the membrane at 60 °C in the following solution:

   0.05 M sodium phosphate, pH 6.5, 10 M urea and 0.1 M β-mercaptoethanol.

7. Non-denaturing PAGE analysis

   In some cases the addition of SDS, which results in linearization of the protein of interest, may also destroy the antigen of interest as the epitope crosses several helices and therefore can be excluded. The bands will naturally have poor resolution, nonetheless this method will still allow differential quantification of immunoreactivity in comparative specimens.

11.3.5 Troubleshooting

Some protocols suggest the use of proteolytic enzymes such as trypsin or EDTA to disaggregate cells or tissues prior to protein extraction, however this may seriously degrade specific proteins of interest. If lysis and subsequent centrifugation of proteins during extraction are inadequate, the protein solution may be insoluble or run with poor resolution during subsequent electrophoresis. High levels of background staining on the membranes are usually due to inadequate blocking of the antibody. Ensure equal loading by reprobing the membrane with an antibody to a ubiquitous housekeeping protein (Figures 11.1 and 11.2).

11.4 Specialized protein analysis techniques

11.4.1 Outline of immunoprecipitation

Immunoprecipitation (IP), when combined with SDS gel electrophoresis, can help determine the quantity and size of antigens, presence of post-translational modifications, interactions with other proteins and nucleic acids.

The technique does, however, require that there is abundant antigen, that the antigen is not insoluble or highly polymerized and that antibodies are available that bind to the antigen with affinity and specificity to the antigen. Polyclonal antibodies bind antigens best in IP as they bind multiple sites on the antigen by stable multivalent interactions but polyclonal antibodies may also have increased non-specific binding to other antigens. Monoclonal antibodies have specific but weak interactions unless the antigen is multimeric and therefore use of pooled monoclonal antibodies has several advantages.

11.4.2 Metabolic labelling of cells to label antigen (optional)

Radiolabelling followed by immunoprecipitation allows the monitoring of the synthesis of a protein and in pulse chase experiments the monitoring also of the kinetics of processing and turnover of the protein.

1. Aspirate medium from cells; use 9 cm dish with cells sub-confluent.
2. Wash 2 × with 1 ml medium lacking methionine (or cysteine), which both have sulphur.
3. Incubate cells for 20–30 min at 37 °C in 1 ml prewarmed medium minus methionine, or use cysteine-minus medium containing 5% (v/v) dialysed FCS (fetal calf serum).
4. Aspirate and replace with 100–200 μCi of labelled (^{35}S) fresh medium containing appropriate amino acid, i.e. methionine or ^{35}S cysteine.
5. Incubate for 1–5 h (2 h), rocking dish every 30 min.
6. Aspirate the radioactive medium with a Pasteur pipette.
7. Transfer Petri dish to a flat surface chilled to 0 °C.
8. Wash cells 2 × with PBS.

11.4.3 Lysis of cells (to release the antigen)

1. Add lysis buffer to washed cells. Lysis buffer:
   50 mM Tris CL, 150 mM NaCl, 0.02% (w/v) sodium azide, 100 μg/ml PMSF (phenylmethanesulphonyl fluoride) and 1 μg/ml aprotinin.
2. Incubate for 20 min on flat surface on bed of ice.
3. Scrape cells off and transfer to chilled Eppendorf tube.
4. Centrifuge lysate at 12 000 g and transfer to a fresh Eppendorf tube and store on ice.

Optional preclearing of the cell lysate is important when studying minor cellular polypeptides.

1. Add 0.05 vol. of pre-immune serum (normal rabbit serum).
2. Incubate for 1 h at 0 °C.
3. Transfer to Eppendorf tube containing a pellet of killed fixed *Staphylo-coccal aureus* cells Cowan 1 (SAC 1) 500 μl in 10% suspension per 50 μl of serum.
4. Incubate at 0 °C for 30 min.
5. Remove *S. aureus* cells by centrifugation at 12 000 g for 5 min, 4 °C.
6. Transfer supernatant to a fresh Eppendorf.

11.4.4 Immunoprecipitating protein of interest

1. Divide antigen preparation into two equal portions.
2. Add NET-gel buffer to a volume of 0.5 ml.
   NET-gel buffer: 0.1% (w/v) Nonidet P40, 1 mM EDTA, 50 mM Tris/Cl

(pH 7.5), 150 mM NaCl, 0.25% (w/v) gelatin and 0.02% (w/v) sodium azide.

3. Add 5 μl polyclonal serum, 100 μl hybridoma serum or 1 μl ascitic fluid, which will yield 1 μg of antigen. To one tube add the specific primary antibody, to the other tube add pre-immune serum.

4. Put on ice for 30 min–2 h.

5. Add 5–10 μl of secondary antibody.

6. Mix and leave on ice for 30 min.

11.4.5 Purification of immune complexes

Protein A is derived from *Staphylococcus aureus* and is cross-linked to agarose beads. The Fab region of antibodies bind to the protein A leaving the Fc regions available for antigen binding. As a result, unbound proteins are removed by washing the solid phase leaving purified Ab/Ag complexes bound to matrix.

1. Pre-swell a column of protein A-Sepharose (2 mg/ml) with binding buffer (binding buffer: 10 mM sodium phosphate, 0.15 M sodium chloride, pH 8.2) and adjust the pH of the sample to equal that of the binding buffer.

2. Apply the sample to the column.

3. Wash the column with 10 bed volumes (10 ml) or until A_{280} of the elutant reaches zero.

4. Elute the immunoglobulin fraction with 0.1 M sodium citrate pH 3.0. Generally 2–5 column volumes are required for quantitative removal of immunoglobulin.

5. Spin at 12 000 *g* for 20 min at 4 °C and discard the supernatant.

6. Add 30 μl of 1 × SDS loading buffer:
50 mM Tris/Cl, 100 mM DTT, 2% (w/v) SDS, 0.1% (w/v) Bromophenol Blue and 10% (v/v) glycerol.

7. Partially denature proteins by heating to 100 °C for 10 min.

8. Remove protein A-Sepharose by centrifugation at 12 000 *g* for 20 s.

9. Load on gel and proceed as for Western blot.

11.4.6 Outline of ELISA

The vast majority of ELISA performed are available in kit form and therefore only the basic principals rather than a specific protocol are outlined below.

1. Incubate samples (extracted as above for Western analysis), negative controls and serial dilutions of positive controls for 1 h at room temperature by pipetting into wells of 96-well plates (pre-coated with antibody of interest and usually pre-blocked with 5% (w/v) BSA). Specific proteins of interest will be bound by immobilized monoclonal antibody on the plate.

2. Wash in protein buffer to remove components not specifically bound.

3. Add the secondary antibody, usually an alkaline phosphatase-linked polyclonal antibody, to the protein of interest. This will bind to another site of the protein and 'sandwich' the protein between the monoclonal and polyclonal antibodies.

4. After another wash add the substrate, e.g. Formazan or OPD (oxyphenyl diamine), to activate alkaline phosphatase or peroxidase respectively (see below). This will form a colour change, which is amplified according to the amount of protein present (see below for details). Read samples on an automatic ELISA plate counter.

5. Create a standard curve, plotting the changing absorbance on the y-axis against the log concentrations of known amounts of protein of interest on the x-axis. ELISA kits may come with a small amount of sample protein for this purpose.

11.5 Northern blotting

11.5.1 General considerations

11.5.1.1 Tissue yield of RNA

An average mammalian cell has about 1×10^{-5} μg of RNA; 10^8–10^9 cells, approximately 1 g or 1 ml of packed cells, have 5–10 mg RNA. Several tissues, however, have exceptionally high levels of endogenous nucleases, such as the pancreas and prostate, while other tissues, for example, the liver, contain molecules such as glycogen and lipids, which may inhibit the efficiency of nucleic acid purification and lead subsequently to low levels of extraction.

Minute clinical specimens such as endoscopic pinch biopsies and exfoliative cytology will by their very size result in a low nucleic acid yield. In this regard it is advisable to optimize extraction conditions by using fresh solutions, plastic containers to avoid nucleic acid adhesion to glass and to store specimens in liquid nitrogen ($-170\,^{\circ}$C rather than $-80\,^{\circ}$C).

11.5.1.2 Use of sterile diethylpyrocarbonate (DEPC)-treated glassware

DEPC should be used to irreversibly destroy exogenous RNAases. The DEPC-treated glassware can then be autoclaved the following day to inactivate the DEPC as this may otherwise decrease RNA extraction. In addition, it is recommended that gloves are worn throughout RNA experiments, that the work is carried out on an alcohol-cleaned bench and that sterile pipette tips and tubes are always used.

11.5.1.3 Use of RNAase free solutions

Only solutions that have been both DEPC treated (using a final concentration of 0.1–0.05% (w/v)) and autoclaved should be used. There are some exceptions, however. Do not autoclave alcohols and chloroform as they are highly volatile and may completely evaporate and, rarely, explode. In addition, do not add DEPC to those solutions containing Tris or strong SDS ($> 10\%$), sarcosyl or guanidinium. These solutions should be made up in water already DEPC treated, autoclaved and filtered through a 0.22–0.45 μm Nalgene filter, after solubilization under RNAase free conditions. During electrophoresis formaldehyde is added to the gel to inactivate RNAases. Recently, several products have become available such as vanadyl complexes and RNAsin^{TM} (Promega) reagent, which effectively inhibit RNAases. Furthermore, other ribonuclease inhibitors can be added to all RNA solutions during experiments.

11.5.1.4 *Type of RNA required – total RNA, cytoplasmic RNA or nuclear RNA*

There are many techniques available for RNA purification and each varies in complexity, purity of nucleic acid, rapidity and essential apparatus required. One other difference between these methods also exists, namely the type of RNA extracted. In this regard, 'total RNA' extraction methods will purify all RNA (nuclear and cytoplasmic species). Alternatively, either species can be removed separately by different techniques. For the vast majority of purposes, total RNA extraction will be satisfactory, however, in specific cases cytoplasmic extraction should be considered for creation of cDNA libraries, for cloning, micro-injection and *in vitro* expression studies. On the other hand, nuclear extraction is essential in order to study the splicing variants of certain pre-RNA species.

11.5.1.5 *Precipitating RNA*

Do not use potassium acetate for precipitation as traces of SDS (if used) can result in a completely insoluble potassium acetate pellet. Therefore, use 0.5 volumes ammonium acetate or 1/10 volume 3 M sodium acetate, pH 5.2 instead. In addition use greater volumes of ethanol for RNA, 2.5–3 volumes, mix vigorously and precipitate at $-20\,°C$ for 1 h if RNA concentration is 10 μg/ml but overnight if RNA is < 100 ng/ml.

11.5.1.6 *Choice of membrane*

Positively charged nylon filters are more durable, bind nucleic acids more avidly, spontaneously or by cross linking, but may give higher background signals than nitrocellulose membranes.

11.5.2 *Overview of RNA extraction methodologies*

Until recently, the most commonly used techniques relied on rapid dissolution of fresh or frozen cell membranes, either manually by use of a slow circular action in a glass pestle (Dounce homogenizer) or mechanically in a rotary tissue homogenizer (Polytron). The speed at which this initial step is carried out largely ensures the quality of the RNA, which would otherwise be rapidly broken down by endogenous RNAase enzymes. Most RNA extraction solutions contain chaotrophic agents, which bind reversibly to the RNAases, thereby temporarily preventing further enzymatic breakdown.

The next stage is to separate the individual constituents of the cell (DNA, RNA, protein and lipids) by centrifugation. Finally, the RNA is recovered from the pellet and the supernatant can be discarded; alternatively the supernatant can be further purified to extract the DNA also. The RNA requires to be desalted and this is usually achieved by precipitation in 2 volumes of cold ethanol or 1 volume of isopropanol.

There are two basic and well-tested methods commonly used, one involves ultracentrifugation (50 000 r.p.m.) in a caesium chloride gradient and the other requires only moderate-speed centrifugation (10 000 r.p.m.) for 30 min in an 'all-in-one' extraction buffer (Sambrook *et al.*, 1989). While the former method is more time-consuming, requires more reagents and an ultracentrifuge, it gives purer RNA in larger quantities and in addition DNA can also be extracted from the same specimen at the same time (Brown, 1991).

Recently, however, an adaptation of the second method has been described. In this regard, RNAzol^{TM} promotes formation of complexes of RNA with guanidinium and water molecules and abolishes hydrophilic interactions of DNA and proteins (including RNAases), which are removed in the top aqueous phase after centrifugation (Chomczynski, 1993). This method is an adaptation of the popular Chomczynski method (Cinna/Biotecx, Houston, USA).

11.5.2.1 Method using chaotrophic agents (i.e. RNAase loses shape and is denatured) – guanidinium–CsCl method to isolate total RNA from mammalian cultured cells

SDS Proteinase K methods of RNA extraction are efficient but the RNAase inactivation step is slow and therefore not suitable for solid tissues that have rapid RNAase activation, i.e. pancreas. Methods that rely on the denaturing effect of 4 M guanidinium and β-mercaptoethanol and then gradient formation in CsCl give rise to highly intact RNA. Although longer than methods above, they give higher and reproducible yields (30–500 μg according to tissue and differentiation state) and also the RNA is purer, free of DNA contamination and therefore suitable for reverse transcription.

(a) Outline
Homogenize tissue quickly in the presence of lysis buffer.
Ultracentrifuge on CsCl layer: protein on top, DNA in the middle layer and RNA on the bottom.

Add RNA to 0.5% SDS to inhibit RNAases.
Ethanol precipitate to desalt.
Quantitate, assess quality and store.

(b) Method

1. Add 1 cm^3 of tissue frozen on cardice (-70 °C) to 59.5 μl β-mercapto-ethanol (denaturing agent) (700 μl/100 ml) on ice.
2. Add 8.5 ml of RNA extraction buffer (to lyse cells): 0.5% (w/v) SDS, 10 mM Tris HCl (pH 7.4), 1 mM EDTA.
3. Homogenize tissue with Polytron (or use Dounce homogenizer to crush to powder at -70 °C as the cells are not protected until RNA is released from them). In order to break up DNA, mix the suspension well, then centrifuge at 5000 g for 10 min at room temperature.
4. Transfer supernatant to a fresh tube, add 3.4 g of 5.7 M caesium chloride.
5. Place 2.5 ml of caesium chloride RNA cushion in an ultracentrifuge tube via a syringe (23 gauge needle) (mark the position of the RNA cushion on the easy-seal centrifuge tube so as to see where RNA will lie after centrifugation).
6. Add sample homogenate via a syringe.
7. Balance the tubes to 10 mg with guanidinium thiocyanate/extraction buffer mixture.
8. Seal the tops.
9. Centrifuge at 41 000 g overnight at 20 °C min., 50 °C max. acceleration and deceleration.
10. Punch a hole in the top of the centrifuge tube and withdraw the supernatant with a syringe. Cut open the centrifugation tube with a hot blade (not scissors) and decant the remaining supernatant. Wipe away all traces of the supernatant, which contains viscous cellular DNA, and carefully turn the tube upside down and allow the residue to drain for 5 min. Wash with 70–95% (v/v) ethanol at room temperature and invert the tube again to drain ethanol and allow the tube to dry for 5 min.
11. Take up the RNA pellet with 200 μl take-up buffer (Tris HCl, 10 mM, pH 7.4, EDTA 5 mM, SDS 0.1% (w/v), RNAase free). Vortex with the tube covered with parafilm and place in an Eppendorf and re-extract with a further 200 μl of take-up buffer. If the RNA is difficult to dissolve, freeze then thaw the sample twice and then heat to 45 °C. Alternatively, carefully push the pellet (because it is very large) into a 5–50 ml tube containing a larger amount of buffer (1–5 ml) and vortex to dissolve.
12. Extract the RNA with 400 μl chloroform:butanol (4:1), spin out phases.
13. Take off 350 μl of upper phase. Transfer to microfuge tube. If using cells from culture, proceed to step 14.
    (a) If using tissue, then proceed with phenol:chloroform:isoamyl alcohol (P/C/I) ratio 25:24:1 extraction, then C/I, then precipitate.

  (b) If using cells transfected with foreign DNA (virus or otherwise):

Add $MgCl_2$, 10 mM and dithiothreitol, 0.1 mM.

Add placental RNAase inhibitor or vanadyl ribonucleoside complexes to a final concentration of 1000 units/ml or 10 mM respectively.

Add RNAase-free DNAase 1 to a conc. of 2 mg/ml and incubate for 60 min at 37 °C.

Add EDTA to 10 mM and SDS to 0.2% (w/v).

Extract with P/C/I, then C/I, then 3 M sodium acetate and ethanol for 30 min on ice. Centrifuge at 12 000 *g* for 5 min at 4 °C.

14. Precipitate RNA with 1/10 (35 μl) vol. of 3 M sodium acetate, pH 5.2 (DEPC autoclaved) and 2.2 vol. (840 μl) absolute ethanol. Store at 0 °C or −20 °C overnight or longer at −80 °C.

15. Spin at 10 000 *g* (30 min, 4 °C), wash with 70% (v/v) ethanol, recentrifuge and allow the pellet to dry in the air.

16. Redissolve the RNA in a small volume of water. Add 3 vols. of 100% ethanol and store at −70 °C.

17. To recover the RNA, remove a sample, add 1/10 vol. 3 M sodium acetate, pH 5.2 and 2 vol. ethanol, mix well and centrifuge at 12 000 *g* for 5 min at 4 °C.

18. Measure absorbance at 260 nm.

11.5.2.2 Single-step acid–guanidinium–phenol–chloroform method to isolate total RNA (tissue extraction by Chomczynski method)

Although this method has a moderate yield of 5–10 μg RNA per million cells, the RNA is protected from RNAases by 4 M guanidinium and therefore is useful for small precious samples of 5–50 million cells.

(a) Outline

Extract by serial addition of guanidinium, acetate, phenol and chloroform.

Shear DNA by efficient homogenization.

Precipitate RNA/DNA fragments and centrifuge.

Reprecipitate RNA in guanidinium to protect RNA.

Ethanol wash to desalt.

Quantitate, assess integrity and store.

(b) Method

1. Extraction step: Take 100 mg tissue, about 0.25 cm³ (10^8 cells, 1 g tissue = 10^9 cells), homogenize in a 20 ml large-bottomed tube at RT in 1 ml solution D (4 M guanidinium thiocyanate, 25 mM sodium citrate, pH 7 and 0.5% (w/v) sodium lauryl sarcosyl, 0.1 M β-mercaptoethanol) in a glass-Teflon homogenizer. Then add sequentially 0.1 ml 0.2 M sodium

acetate, pH 4, 1 ml of phenol (water saturated) and 0.2 ml chloroform–i-soamyl alcohol mixture (49:1) with thorough mixing by inversion after each reagent. At the stage of adding phenol all froth should disappear. If greater amounts of tissue are used, scale up each volume added. The final suspension is shaken vigorously for 10 s and cooled on ice for 15 min. Samples are centrifuged at 10 000 *g* for 20 min at 4 °C. RNA is present in the aqueous phase whereas DNA and protein are present in the inter-phase.

2. Precipitation step: The aqueous phase is transferred to a fresh tube and mixed with 1 ml of isopropanol and then placed at −20 °C for at least 1 h to precipitate RNA. Then sediment at 10 000 *g* for 20 min.

3. Reprecipitation step: The RNA pellet is suspended in 0.3 ml of solution D in a microfuge tube and left to stand for 30 min on ice to dissolve. RNA is subsequently precipitated with 1 vol. isopropanol at −20 °C for 2–12 h.

4. Wash step: Centrifuge for 20–30 min at 4 °C and resuspend the pellet in 75% (v/v) ethanol. Aspirate on ice, sediment and vacuum dry (15 min).

5. Optional solubilization: Dissolve in 50 μl 0.5% SDS (filtered, DEPC-treated solution) at 65 °C for 10 min. (The addition of vanadyl complexes, to a final dilution of 1:30, has been reputed to help pellet dissolution – personal correspondence from Dr Sukdev Singh, University Dept. of Medicine, Birmingham, UK.)

6. RNA storage: Dissolve RNA in Q water (or 1 mM EDTA, pH 8, or 0.5% SDS) treated with diethyl pyrocarbonate (DEPC). If samples are difficult to dissolve, heat to 37 °C for 20 min and read absorbance at 260 nm.

11.5.2.3 Selective isolation of poly mRNA (cytoplasmic) extraction by the poly(dT) method

Several methods are available but all rely on affinity chromatography with oligo(dT) cellulose or poly(U)-Sephadex.

(a) Outline
Extract total RNA in chaotrophic lysis buffer.
Affinity chromatography with oligo(dT)-cellulose or poly(U)-Sephadex.
Wash RNA in medium-salt solution.
Elute RNA using low-salt buffer (mRNA is 0.5 kb–10 kb).

(Under high-salt binding conditions, 400–500 mM NaCl, poly(A) tails will bind to the column, which can be washed in medium-salt solution and then eluted using low-salt buffer.)

(b) Method

1. Homogenize tissue as above and pellet the cells in RNA lysis buffer ($10 \, ml/10^8$ cells).
2. Wash cells in PBS, re-pellet in 50 ml.
3. Add 10 ml lysis buffer (for 10^8 cells); vortex 15 s at highest setting.
4. Force lysate through a 21-guage needle using a 20 ml syringe, approx. five times until a homogeneously even (light) viscosity lysate is obtained.
5. Incubate 1.5–2 h at 45 °C with light agitation.
6. Weigh 20 mg of oligo(dT) per 10^8 cells into a 50 ml conical culture tube; hydrate with 10 ml binding buffer with light agitation.
7. Pellet oligo(dT), wash in fresh binding buffer (0.5% (w/v) SDS, 10 mM Tris-HCl (pH 7.4), 1 mM EDTA and 400 mM NaCl); dispense into equal ($20 \, mg/10^8$ cells) samples; pellet.
8. Add 600 μl 5 M NaCl to each 10 ml of lysate, mix; add to oligo(dT), incubate with shaking at RT for 20–60 min.
9. Pellet oligo(dT) 3000 r.p.m., wash at least three times with 10 ml of binding buffer, resuspend by pipetting up and down; the oligo pellet is very fragile so aspirate, don't pour.
10. Aspirate all but 0.5 ml binding buffer over the last pellet, transfer oligo(dT) to Millipore filtered tubes in 400 μl samples, rinse tube with 400 μl of binding buffer, wash oligo(dT) three times with 400 μl binding buffer (max. speed in microspin, 25 sec).
11. Transfer filter to a DEPC-treated Eppendorf tube and add 200 μl of elution buffer (0.5% (w/v) SDS, 10 mM Tris-HCl (pH 7.4), 1 mM EDTA), spin, elute once more with 200 μl of elution buffer.
12. Add 1/10 volume (approx. 40 μl) of 3 M sodium acetate and 2 vols. (approx. 880 μl) ethanol, leave at -70 °C overnight.
13. Spin for 5 min at max. speed, decant, wash with 70% ethanol and dry for at least 10 min.

Read the A_{260} of 10 μl of the sample diluted 290 μl of water. Absorbance $\times 30 = $ conc. in μg/ml. DNA Dipsticks^{TM} (Invitrogen) can now be used to assess nucleic acid concentrations of oligonucleotides, RNA and DNA. The procedure is easy and rapid and relies on a colorimetric change induced on the tip of the Dipstick by 5 μl of nucleic acid solution. This technique is also highly sensitive as it can reliably detect concentrations as low as 0.1 ng/μl compared with 5 μg/μl on an agarose gel and 10 μg/μl on a spectrophotometer.

11.5.2.4 Commercial kits for RNA extraction

Qiagen, Promega, Stratagene and many other companies market kits for either total RNA or mRNA extraction. Their advantages are that the novice can relatively easily extract good-quality RNA. These kits are, however, moderately expensive and have enough reagents for only a few specimens. In addition, some kits do not specify the individual constituents clearly and therefore 'trouble-shooting' when errors occur may be empirical. One of the several kits that does have important applications is the Promega 'Fastrack™' kit for rapid and efficient mRNA analysis. When a cDNA library is to be constructed the most important prerequisite is that as many undegraded mRNA species as possible are recovered from the specimen and in this regard these kits utilize either biotinylated oligo(dT) magnetic beads or oligo(dT) resin for ultrafiltration (Chomczynski and Sacchi, 1987). The poly(A) tail of mRNA species will bind avidly to the oligo(dT) and the remaining tissue constituents are discarded. Subsequently, the intact mRNA is recovered in elution buffers in very high purity.

11.5.3 Assessment of RNA quantity and integrity

A sample of nucleic acid (usually 0.5–10 μg of mRNA) is loaded onto a 1–3% agarose gel and integrity of the nucleic acid bands is assessed. Two, possibly three, distinct bands will be visible on the gel corresponding to different molecular weights of mainly ribosomal RNA, 5 S, 18 S and 28 S bands (the latter nearest to the well) (see Table 11.1). Purity (degree to which RNA is free from protein) and quantity of RNA extracted can be assessed by spectrophotometry.

11.5.4 Electrophoresis of Northern denaturing gel

1. The gel tank and gel tray should be soaked in a 5% H_2O_2 solution in DEPC-treated water for 1 h before use.

Table 11.1. *Guidelines for assessment of RNA by spectrophotometry and on agarose gels and RNA storage*

Integrity of RNA (denaturing gel electrophoresis)
  Ratio 28 S (5 kb)/18 S (2 kb) bands should be 2:1
  Reversed (28 S/18 S) ratio or smear suggests moderate and advanced degradation, respectively. Remember poly(A)mRNA appears as a smear at 0.5–10 kb and the median size is 2 kb.

Quantification of RNA (spectrophotometry)
  1 A_{260} unit = 30 μg RNA/ml
  Faint band on gel = 10–20 μg RNA

Purity of RNA (spectrophotometry)
  260/280 ratio < 1.7–2.0 indicates contamination
  260/230 ratio < 2 indicates guanidine contamination

Storage of RNA
  0.5% SDS −20 °C
  1 mM EDTA −20 °C
  75% ethanol precipitate −70 °C

Make up a denaturing 0.8 cm agarose gel using a 200 ml Duran (bottle) containing 108 ml water, 1.5 g agarose (1% (w/v)), 15 ml 10 × MOPS pH 9, heat in a microwave until melted, cool to 50 °C in a water-bath then add 27 ml formaldehyde and allow to set in a fume hood for 30 min. Use broad combs to hold 50 μl of sample and the gel should ideally be 15–20 cm long. Make up 2 l of running buffer containing 1 × MOPS. Antinuclease precautions are not needed. As a safer alternative, glyoxal can be used to denature the RNA strands in place of formaldehyde but this results in less detection sensitivity during hybridization.

2. Precipitate sufficient RNA, usually 5–40 μg per lane, with 0.25 M sodium acetate, 1 vol. ethanol and 20 μg glycogen per tube, or use a lyophilizer if one is available. Mix the sodium acetate and glycogen first then dispense into tubes, 10 μl 3 M sodium acetate and 1 μl 20 mg/ml glycogen per sample. Add enough DEPC-treated water to make the contribution of the RNA up to 100 μl, then add 250 μl ethanol and finally add the RNA. Mix well and spin at top speed in a microfuge for 30 min at 4 °C, wash the pellets with 75% ethanol, remove the supernatant and air dry the pellet on ice.

3. Dissolve the RNA pellet, containing 2 μg poly(A) or 40 μg total RNA, in 5.5 μl water and add 14.5 μl fresh sample buffer containing 200 μl 10 × MOPS, 356 μl 37% (v/v) formaldehyde, 1 ml deionized formamide. Incubate the RNA at 70 °C for 15 min, then place on ice for 5 min and add 2 μl dye loading buffer: 50% glycerol (v/v), 1 mM EDTA, 1% Bromophenol Blue dye solution (w/v) (i.e. 0.01 mg/ml), 1% xylene cyanol (w/v) and 15% Ficoll (optional). Load quickly on a semi-submerged gel and top up the gel tank when both dye bands have migrated into the gel.

4. Prepare commercial RNA markers or use samples of total RNA with clear 28/18 S bands similarly to step 3. DNA markers are not suitable as they run quicker and are more fuzzy than equivalently sized RNA markers. Cut off the marker tracks and stain with Methylene Blue for 15 min in 5% (v/v) acetic acid and 5–10 min in 0.5 M sodium acetate (pH 5.2), 0.04% (w/v) Methylene Blue. Alternatively add 1 μl of 1 mg/ml ethidium bromide to all samples and markers and photograph the gel once electrophoresis is completed.

5. Commence with 100 V (3–5 V/cm) and run the gel in a fume hood until Bromophenol Blue reaches the bottom (running negative to positive for 4–8 h). In a 1% gel the xylene cyanol should run near the 18 S ribosomal band (2 kb). Remember to mix each buffer reservoir after every hour to maintain appropriate pH and do not reuse, as the buffer will be exhausted.

6. Rinse the cut gels three times quickly in water, followed by a wash for 30 min in 0.05 M NaOH and then 30 min in 10 × SSC. Large mRNAs greater than 9 kb may transfer more efficiently to the membrane than similarly sized DNA molecules because the former are already denatured

in the gel and therefore depurination is rarely required for Northern analysis.

7. Optional steps to visualize gels where prestained RNA is faint or not discernible, usually because $< 5 \mu g$ mRNA is present in each lane, are:

    (a) Stain gel in $4 \mu l$ of ethidium bromide, $10 \mu g/ml$ in 300 ml of $1 \times$ MOPS buffer (or 0.5 M ammonium acetate) for 15–30 min, preferably in the dark. Light may make RNA bands fade and disappear.

    (b) Destain the gel by washing in 300 ml of $1 \times$ MOPS buffer (or 0.5 M ammonium acetate) in the dark for 30 min, as residual ethidium bromide will hinder efficient blotting.

    (c) Visualize the gel under UV light and 28 S, 18 S and 5 S bands should be visible in total RNA or a smear in mRNA but long-term staining may inhibit RNA transfer, especially to nylon membranes.

An alternative staining of RNA in a gel uses Acridine Orange as this dye binds more avidly to single-stranded nucleic acid species. In addition, Acridine is not as toxic as ethidium bromide.

    (a) Make up 1 M sodium phosphate (50% monobasic NaH_2PO4 and 50% dibasic Na_2HPO4).

    (b) Make up 100 ml of Acridine Orange buffer: 1 ml of 1 M sodium phosphate, 8 ml of 405 formaldehyde, 90 ml water, 1 mg Acridine Orange.

    (c) Stain for < 2 min and photograph gel.

    (d) Destain for > 20 min in buffer minus Acridine Orange.

11.5.5 Northern blotting

Three alternatives are available for blotting of nucleic acids from gels to membranes. Capillary blotting is simple and requires the use of blotting paper to cause high-salt buffer to move through an agarose gel taking with it all nucleic acids, depositing them on a permanent membrane. Vacuum blotting achieves the same effect but negative pressure is used to effect the removal of nucleic acids from the agarose. Finally, electrophoretic transfer is the method of choice for transferring DNA or RNA from low pore-sized polyacrylamide gels but it may give less satisfactory results in agarose gels.

Denatured gels are more fragile and slippery than ordinary gels, therefore, handle with care. Make up blot as follows:

1. Wet the nitrocellulose filter (cut the filter so that it is 3 mm smaller in both length and breadth than the gel) in distilled water until completely saturated then transfer to $20 \times$ SSC for 5 min. Transfer apparatus consists of, top to bottom: 1 kg weight, flat plastic base support (to give even pressure), 8 cm of paper towels, three sheets Whatman No 3 covering the nitrocellulose beneath, agarose gel, three sheets of Whatman No 3 with

overhangs into the buffer below, plastic table supporting the gel (usually the inverted electrophoresis tray) and a plastic tray filled with $20 \times$ SSC large enough to hold the entire blotting apparatus.

2. Place the inverted gel tray in the tray containing $20 \times$ SSC and use the Whatman No 3 paper as wicks at two sides, making sure there are no bubbles underneath.

3. If gel is more than 0.5 cm thick, contains more than 1% agarose or the RNA to be analysed is greater than 2.5 kb, the RNA should be hydrolysed to improve the efficiency of transfer. This can be achieved by soaking the gel in 0.05 M NaOH for 20 min then in distilled water and then 45 min in $20 \times$ SSC. After this optional pretreatment the gel should be placed on top of the filter paper and covered from its edges to the tray edges with cling film so as to form a seal inhibiting buffer loss from below.

4. Place nitrocellulose on the gel and remove bubbles. If large amounts of plasmid or low complexity DNA are present, rapid transfer can occur, resulting in subsequent artefacts as a result of rearranging the filter during the initial construction of the blot.

5. Lay three sheets of Whatman No 3 on top, pre-soaked in $20 \times$ SSC

6. Place paper towels on top (5–8 cm high) and leave for a minimum of 24 h (RNA may take longer to blot than DNA). When removing the gel mark the membrane carefully with a pencil so that tracks are identified, usually mark top right corner and date of transfer. Soak filter in $6 \times$ SSC then air dry for 30 min and check the gel under UV that all the RNA is blotted.

7. Fixation of the nucleic acids to the blots can be achieved by three methods: alkali treatment of positively charged membranes for 30 min after blotting; baking the membrane between filter papers at 80 °C (preferably in a vacuum oven) for 2–3 h; and UV crosslinking of a damp membrane at 312 nm for 3 min. The latter is the preferred choice for nylon membranes but is not suitable for nitrocellulose membranes. If the filter is not to be used at once it should be stored in aluminium foil, sealed in a plastic bag and stored at room temperature.

11.5.6 *Prehybridization and hybridization*

The hybridization buffers may contain several ingredients including:

1. Rate enhancers such as dextran sulphate to act as volume excluders to aid the kinetics of both rate and final efficiency of hybridization, particularly with long probes (> 500 bp).

2. Blocking agents such as dried milk, Denhardt's (see below) or heparin and SDS as the detergent to inhibit non-specific probe hybridization to the membrane.

3. Heterologous DNA to inhibit non-specific probe hybridization to the DNA on the membrane.

4. Denaturants such as formamide to decrease the melting temperature of the nucleic acids on the membrane.

Use 1 ml of hybridization solution for each 1 cm^2 of filter surface area.
Hybridize at 45–65 °C but this can be calculated exactly (T_m).

1. Add nitrocellulose to the membrane pre-prehybridization liquid: 3 ml 20 × SSC, 8 ml 25 × Denhardt's solution (0.1% polyvinyl pyrrolidine, 0.1% BSA, 0.1% Ficoll) and 9 ml distilled water, placed together into the bottom of the plastic bag. Air bubbles are expelled by pulling back and forwards over the edge of the bench. Seal with a bag sealer. Incubate for 1 h at 65 °C with vigorous shaking.

2. Cut corner of bag pour off solutions and add the prehybridization liquid: 1 ml 20 × SSC, 8 ml 25 × Denhardt's, 0.2 ml 10% SDS (final conc. 0.1%) (or 8 ml 25% dextran sulphate), 1 ml denatured (boiled for 5 min) salmon sperm DNA 1 mg/ml, 10 ml deionised formamide (final conc. 50%), add either 1.1 g NaCl or sodium phosphate pH 6.5 (to final conc. 50 mM). Incubate for 1 h with vigorous shaking at a maximum of 45 °C. It is possible to avoid using formamide, which is toxic, and replace it with 10 ml of DEPC-treated water but higher temperature hybridization (65 °C) is important to avoid non-specific background bands appearing.

3. Hybridization solution is identical to the prehybridization solution but in addition contains 5–10 million c.p.m. of ^{32}P-labelled DNA probe (usually to a final concentration of 10 ng probe/ml, add 500 000 c.p.m. per ml of hybridization solution). Cut the bag, add the labelled probe behind a screen and incubate for 15–18 h at 65 °C (usually same temperature as for prehybridization) with vigorous shaking.
   (Remember to preheat the probe to 100 °C for 10 min with 100 μg non-homologous DNA and snap cool on ice.)

4. Washing the filter: Remove the filter from the hybridization bag and place it in a plastic tray. Rinse briefly in rinse solution (2 × SSC) (1 min) and place the membrane in a fresh plastic bag. Add 250 ml wash buffer (three washes a-a-a, a-b-b or high stringency a-b-c). During the washing stage the temperature of the washes is much less important than the salt concentration in determining the specificity of binding.
   (a) Low stringency wash: 2 × SSC (or 2 × SSPE) and 0.1% SDS at room temperature per 100 cm^2 of membrane for 20 min.
   (b) Moderate stringency wash 1 × SSC (or 1 × SSPE) and 1% SDS at 42 °C for 20 min, i.e. for β-actin probe.
      Check with Geiger counter. If most of the filter reads as background do not continue, but a more stringent wash at lower SSC concentration or higher temperature may be required.
   (c) Very stringent wash: 0.2% SSC and 1% SDS at 42–65 °C for 20 min, used for high homology probes greater than 500 bp.
   (d) While the blot is still wet, scan it with a Geiger counter and if a large amount of radioactivity is detected scattered all over the filter there

may be non-specifically hybridized DNA and steps 3 and 4 may need to be repeated. Use a negative filter with no known RNA and incubate with probe. If counts still exist this suggests a high background, so continue washes.

(e) In the final step semi-dry the membrane and seal in a thin plastic bag to prevent it drying out, which may irreversibly bind the probe to the membrane. Scan the membrane with a Geiger counter to estimate the required autoradiography period, usually 1 day (if > 50 c.p.m.) or 30 days (if < 20 c.p.m.).

5. Stripping the membrane: The membrane should not be allowed to dry before stripping off the probe.

(a) Pour boiling 0.1% (w/v) aqueous SDS solution onto the membrane and place on a shaker for 5–30 min, replacing the solution once. Discard the solution and allow the gel to cool to room temperature.

(b) If in doubt about the efficiency of probe removal, autoradiograph the membrane at this stage.

(c) Prehybridize the membrane again.

11.5.6.1 *Alternative for hybridization ovens*

1. Prehybridize and hybridize as for incubation in a water-bath. 10 ml of prehybridization solution is recommended for flasks larger than 500 ml. Larger volumes are required if there is more than one membrane in the bottle.

2. Rotate the bottles in the rotisserie in the same direction as the membrane was unrolled. Incubate at 42–65 °C for 15–60 min.

3. Remove the bottle and pour off prehybridization fluid. Add heated hybridization fluid (the same prehybridization solution already in flask can be reused).

4. Hold the bottle upright, use a pipette to inject the *preheated* probe into the hybridization solution. Avoid injecting the probe directly onto the membrane as this subsequently causes hot spots on the filter.

5. Replace cap and gently agitate the bottle to distribute the probe evenly in the hybrid solution.

6. Hybridize at 42–65 °C for 2 h to overnight.

7. Washing and hybridization steps are as for conventional incubation.

11.5.7 *Troubleshooting*

11.5.7.1 *No labelled bands*

In order to detect successfully a particular RNA species by Northern analysis using a high specific activity probe, 5 ng of that mRNA must be present on the membrane (0.1% of 5 μg of a mRNA sample or 1% of a total RNA sample).

11.5.7.2 Appropriateness of band size

Assessment of the molecular weight of the RNA transcripts can reveal whether alternative splicing occurs in various disease states. This can best be assessed by Northern analysis.

11.5.7.3 Membrane background effects

Spotted particulate background artefacts spread randomly over the membrane usually indicate unincorporated nucleotides binding non-specifically to the membrane. This latter observation should not be confused with spotted artefacts along the edge of the filter which are relatively commonplace. Patchy diffuse background artefacts spread over the entire membrane suggest too high a probe production.

11.5.7.4 Differential expression of transcripts between sample and controls

The relative amount of each transcript can be assessed by automatic scanning of each band so that full quantitation can be achieved in relation to known controls and other samples.

11.5.7.5 Experimental validation

The use of positive and negative control samples ensures that each experiment fulfils acceptable standards of analysis. In this regard, when performing RNA analysis, to ensure that intact transcripts are present ubiquitous probes/primers to either glucose-3-phosphodehydrogenase (G3PDH) or β-actin are used (see Figure 11.3). Similarly, tissue that does not express a target gene of interest and that does not express the control transcripts either must be deemed a potentially false negative result.

11.6 Specialized PCR techniques

The polymerase chain reaction (PCR) is theoretically a very simple technique in which specific DNA sequences flanked by predesigned primers are amplified. The process of amplification involves three steps which are performed in 'cycles': denaturation step, to separate the DNA strands; annealing step, to allow the primers to stick to the single-stranded DNA; and polymerization step, to form a new complementary strand resulting in double-stranded DNA (Innis *et al.*, 1990). This process doubles the amount of DNA at each cycle and therefore a few strands of a chosen RNA (cDNA) or genomic DNA can be amplified a million times. The discovery of the heat stable enzyme Taq polymerase and subsequent homologues allowed this cycle to be performed in one sample buffer, changing only the temperature (see pages 271–83).

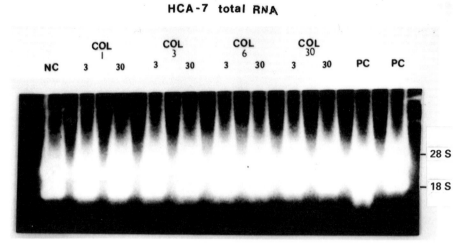

Figure 11.3. Normalization of blots. The intensity of the 18 S and 28 S ribosomal bands were also used as an approximate assessment of the equality and quantity of RNA loading onto gels.

Researchers frequently use PCR to make comparisons of the quantity of gene expression between several clinical samples. It is imperative, therefore, to amplify a control single-copy gene in addition to the test gene in order to ensure normalization (equal loading) of the starting DNA prior to amplification.

Finally, genomic DNA may be harder to amplify than cDNA because the presence of so many repetitive sequences may result in mis-priming. As a result a modification of standard PCR called 'touch-down' PCR may decrease the amplification of artefactual products. Briefly, this method is similar to routine PCR except the anneal temperature of the first cycle is 5 °C higher than that calculated and over the next four cycles the annealing temperature decreases to 'normal'.

11.6.1 General advice for PCR

1. Use master mixes if possible, which will increase accuracy, reduce reagent loss on tips and reduce tube-to-tube variability.
2. To reduce evaporation, overlay the mix with 50–100 μl of mineral oil.
3. Use polypropylene tubes, which respond most rapidly to variable temperatures.
4. Clean the bench top with weak Chloros or other bleach to destroy DNA amplicons.

11.6.2 mRNA preparation – > cDNA, reverse transcription PCR (RT-PCR) – cDNA step by random priming

1. Make up the following buffer: 2–4 μl MgCl$_2$ (25 mM) solution, 2 μl PCR × 10 buffer (10–100 mM Tris/HCl (pH 8.3), 50–500 mM KCl, 15 mM MgCl$_2$), 2 μl (100 mM) of dNTPs, 1 μl RNAase inhibitor (5 units), 2 μl DTT 0.1 M (optional), 0.5 μl random hexamers, 3.5 μl RNA sample (~1 μl total RNA) and add double distilled water to make up to 20 μl.

2. Heat sample with buffers to 65 °C for 5 min to denature RNA and then allow to cool slowly to room temp.

3. Add 1 μl murine Moloney reverse transcriptase (RT), 1 μl avian myeloblastoma leukaemia (AML) RT or SuperScript RT. SuperScript RT is the most efficient, works at higher temperatures and is not inhibited by ribosomal RNA and therefore total RNA also gives excellent yields of cDNA. Incubate the denatured sample with the enzyme at 15–25 °C for 10 min to anneal and then heat the tube to 42 °C (37 °C for AML) for 45 min (or 60 min if the DNA is greater than 10 kb) to extend the cDNA. Do not, however, allow reverse transcription to continue over 2 h as RT has exonuclease actions. Stop the reaction by heating the sample to 95 °C for 5 min to inactivate the RT, which will otherwise inhibit the following PCR.

4. The sample can be purified in Sephadex D50 columns or anion exchange resins to remove unwanted nucleotides but in the majority of cases this is not required for the subsequent PCR. Finally add 1 μl RNAase A to destroy the RNA in the RNA:DNA hybrids.

5. The efficiency of the RT reaction can be tested by checking the A_{260} or alternatively incorporating an α ^{32}P-labelled NTP prior to the RT step and running the products on an agarose gel. After a brief autoradiography period (usually < 1 h) a characteristic smear is discernible in each lane corresponding to cDNA of sizes between 1 and 7 kb.

11.6.2.1 Alternative method using oligo(dT) primers

Use 1 μg RNA in 10 ml water, heat to 75 °C for 10 min, quench on ice for 2 min and add 1 μl of oligo(dT) primer (100 ng), and 8 μl of the following: 4 μl 5 × first strand buffer, 1.35 μl water (DEPC), 0.4 μl dNTP, and last of all add 2 μl 0.1 M DTT and 0.25 μl RNasin. Incubate for 2 min at 37 °C then add 1 μl Superscript-RT (which does not have the exonuclease actions of MML (murine Moloney leukaemia virus) and AML) and incubate for 1 h 37 °C.

This method results in cDNA strands that are all of a similar size but they are usually only derived from the extreme 500 bp of the 3′ end of the RNA. Random primer extension results in unequally sized products from each RNA species but more of the transcript is represented.

11.6.3 PCR step

1. Controls: (a) sterile water (instead of cDNA); (b) positive control of known template cDNA; (c) positive genomic DNA; and (d) extraction control to ensure solutions used for prior DNA purification are uncontaminated.

2. Reaction master mix: volume = no. samples $(n + 1) \times 4$ (i.e. three controls) $\times$ required vol. for each reaction, usually 20–100 μl (below are specifications for 100 μl).

   (a) H_2O (depending on other volumes), e.g. 65–75 μl.

   (b) 10 μl 10 $\times$ PCR buffer: 10–100 mM Tris/HCL (pH 8.3), 15 mM $MgCl_2$, 50–500 mM KCl.

   (KCl helps primers anneal. Can eliminate KCl and increase $MgCl_2$, especially if high concentrations of dNTPs are used. Magnesium is an essential substrate for the enzymatic reaction. 0.01% (v/v) Tween 20 (or 0.1% (v/v) Triton X-100) may help stabilize AmpliTaq. 0.01% (w/v) gelatin (or BSA 100 μg/ml), 0.01% (v/v) NP40 or DMSO are optional additives.)

   (c) 2 μl of each NTP to a maximum final concentration of 1 mM (50–200 μM of each dNTP).

   (d) 1–5 μl forward primer to a final concentration 0.5 μM (0.05–1.0 μM).

   (e) 1–5 μl reverse primer to a final concentration 0.5 μM (0.05–1.0 μM).

   (f) 25 mM $MgCl_2$ solution 4–20 μl (1–5 mM).

   (g) 2 μl cDNA (< 1 μg/100 μl).

   (h) AmpliTaq DNA polymerase (only 0.5 μl (2.5 units)/100 μl required).

3. Add 1 drop mineral oil to each tube (50 μl), vortex and spin for 10 s to separate phases and load onto PCR plate. 'Hot-start' melts all the dsDNA to single strands and will also inactivate any trace proteases that may interfere with the enzymes. This is achieved by heating the sample with or without the enzyme to 95 °C for 5 min.

4. Programming the PCR machine. Below are the typical reaction conditions of many PCR experiments:

   Step 1.0 – 94 °C 5 min.

   Step 2.1 – 94 °C 1 min to *Denature* (increase temp. to 95–97 °C for G-C rich cDNA).

   Step 2.2 – 56 °C 2 min to *Anneal* (increase temp. to achieve more specific products).

   Step 2.3 – 74 °C 2 min to *Extend* (increase time when fragments > 2 kb or in low initial conc.).

   Cycle 30–50 times.

   Step 3.0 – 74 °C 4 min (twice the extension time).

   (a) Number of cycles required: 25 if 100 000 cDNA molecules, 30 if 10 000 and 40 if 1000.

   (b) Setting extension time: At 70–80 °C > 60 nucleotides/s are added

(2000–4000 bases/min), 24 nucleotides/s at 55 °C and 1.5 nucleotides/s at 35 °C.

(c) Setting annealing temperature: higher temperatures, above 55 °C, result in more specific product, increase by 5 °C increments to find appropriate temperature, usually 0–5 °C below T_m (55–72 °C). If the DNA has a very high G-C content the annealing temp. may need to be > 60 °C or use 7-deaza-2′-deoxy-GTP mixed with dGTP to overcome secondary structure.

(d) Setting denaturing temp: 65% of Taq activity is left after 60 cycles, whereas 95% activity is left after 30 cycles. Increase denaturing temp above 95 °C if G-C content is high. (Rough rules: keep the DNA conc. to a minimum, time of cycle also minimum, temp. of cycle maximum.)

5. Once PCR is finished the DNA may be kept at 4 °C. Prior to running of a diagnostic agarose gel (1–1.5% SeaKem agarose gel or 3% NuSieve with 1 μl of 10 mg/ml ethidium bromide added), spin down briefly and add 2 μl stop/loading dye onto parafilm and apply 5–20 μl of the PCR reaction.
To remove mineral oil add 300 μl chloroform to each tube, vortex briefly and centrifuge for 30 s and recover aqueous phase).

6. Choosing primers:
(a) Primer sequences should not compliment within themselves and to each other, especially the 3′ end as this will result in primer-dimers whereas complimentarity at the 5′ will still allow normal extension in the 3′ direction.

(b) Ideally 15–30 bp in length and the primer concentration should be 10–60 pmoles.

(c) Ensure T_ms = 55–80 °C; 2 °C for each A or T base and 4 °C for each G or C.

(d) Ensure DNA to be primed does not have secondary structure, i.e. is G-C rich.

(e) To amplify cDNA rather than genomic DNA ensure that the primers cross intron/exon boundaries.

7. The choice of thermostable DNA polymerase is increasing dramatically:

Taq − good for most uses but leads to 1 in 10 000 base errors (the latter can be minimized by keeping dNTPs to a minimum – half normal)

Taq + pwo − allows efficient amplification of DNA and generation of A overhangs for TA cloning but pwo allows longer products to be generated also

Vent (Biolabs) or Pfu (Stratagene) generate amplified products with few misincorporations but do not generate A overhangs for TA cloning (Invitrogen). In this instance the CloneAmp dU primer cloning system is a good alternative (Life Technologies).

11.6.4 Troubleshooting in PCR

Three types of problems can occur during PCR reactions: (a) failure to amplify any bands; (b) multiple bands; or (c) a DNA smear. The possible explanations are listed in Table 11.2 and see Figure 11.4.

11.7 Southern blotting

11.7.1 DNA extraction

DNA extraction is less fraught than RNA extraction and therefore more methods are routinely used. The method of guanidinium chloride extraction described above is sufficient but the caesium chloride gradient must usually have ethidium bromide (EtBr) added to ensure adequate separation of RNA and DNA (DNA binds to EtBr more avidly and therefore settles at a different height in the spun gradient).

Protocol 1 – DNA from cells in culture or tissue

1. Grow up cells in culture flasks until they are confluent and wash twice with PBS and drain well.
2. Add 4 ml lysis buffer to each 15 cm^2 flask or onto homogenized 1 cm^3 specimens in 15 ml tubes and place in a incubator for 3–15 h. Lysis buffer: 100 mM NaCl, 10 mM EDTA pH 8.0, 10 mM Tris pH 8.0, 0.5% (w/v) SDS and add proteinase K to final conc. 50 μg/ml immediately prior to adding buffer to cells.
3. Scrape lysate into 50 ml conical tube with a cell scraper or spatula.
4. Perform two phenol and two chloroform/isoamyl extractions with gentle mixing on a rotator for 30 min (this prevents shearing of the DNA).
5. Precipitate DNA with 1/10 vol. 3 M sodium acetate pH 5.2 and two

Table 11.2. *Troubleshooting failed PCR experiments*

Problem	Solution
No bands amplified	Repeat and increase DNA conc.
	Decrease annealing temp.
	Increase Mg^{2+}
	Check primers, etc.
	Check enzyme
Background bands	'Hot-start' (increase initial denaturation temperature)
	Add TaqStart antibody to inhibit non-specific DNA polymerase activity at low temperatures
	Decrease Mg^{2+} (5–25 mmol range, 15 mmol is usual)
	Increase annealing temp.
Smear	Decrease starting DNA by a factor of (2–10)

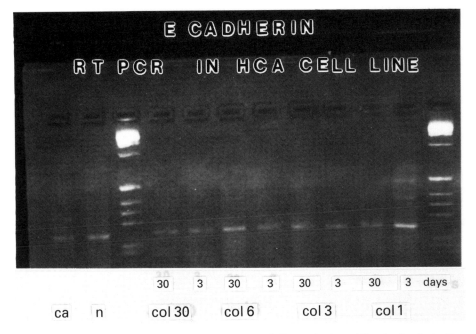

Figure 11.4. RT-PCR of E-cad in the HCA-7 cell line colonies. All colonies including cancer (ca) and normal colon (n) expressed appropriate transcripts of E-cadherin of 220 bp. Interestingly colony 1 at 3 days growth also had weak amplification of larger fragments (370 and 490 bp), consistent with either altered splicing or DNA contamination. Positive control, (normal colon) and GAPDH primers were used and reaction optimized at 30 cycles.

volumes absolute alcohol and hook out precipitate with a sealed, bent glass Pasteur pipette.

6. Wash with 70% alcohol, hook out again, drain alcohol, wash in 100% ethanol briefly and resuspend in 1.5–3 ml TE (10 mM Tris and 1 mM EDTA). It may be necessary to allow the DNA to dissolve fully overnight. DNA may be stored at 4 °C prior to use but ice crystals at −20 °C may smash DNA unless it is lyophilized.

Protocol 2 – CsCl gradient (see RNA and plasmid section)

Commercial kits for DNA extraction

Various kits are sold by all the molecular biology companies but, perhaps, Promega offer the largest range of products called Magic™ or Wizard™ DNA purification systems. In this regard kits that can process various samples sizes from 10 μg, 1 mg and 1 g of total DNA are available (mini, maxi and mega DNA purification systems). These kits eliminate the need for caesium chloride/ethidium bromide gradient centrifugation.

11.7.2 Pre-digestion of large DNA molecules

As genomic DNA is so large it must be pre-digested with enzymes to cut it into smaller pieces for analysis. The choice of enzyme can be determined by assessing the potential cleavage sites in a DNA or cDNA nucleic acid map, however *Eco*RI can be used for most mammalian genomes. Completed digestion should result in an even DNA smear, RI bands (in the case of *Eco*RI) and the absence of high molecular weight material at the top of the gel on agarose gels post-digestion.

1. Digestion
   (a) Add up to 20 μg of DNA (DNA solutions are viscous and should be measured with a pipette tip cut to give a wider bore (2 mm).
   (b) Add 5 μl of 10 × restriction enzyme buffer with the appropriate concentration of 5 M NaCl.
   (c) Add the enzyme to no more than 5% of the total volume as glycerol used to protect the enzyme will inhibit digestion. On average 1 unit of an enzyme cuts 1 μg of DNA in an hour.
   (d) Make up to 50 μl with water and incubate for a minimum 1 h but overnight is best.
2. (a) Add ammonium acetate, followed by 2.5 vols. absolute alcohol and mix.
   (b) Leave on dry ice for 10–30 min until solution becomes viscous but not frozen and spin in a microfuge at 4 °C for 10 min. Pour off alcohol and drain pellet well then add 0.5 ml 70% ethanol and spin at 4 °C for 10 min.
   (c) Again pour off ethanol and add 0.5 vol. 100% alcohol to hasten drying and spin at 4 °C for 10 min.
   (d) Pour off alcohol and air dry the pellet, resuspend in water, measure A_{260} and dilute to 1 μg/μl.

11.7.3 Assessment of DNA purity

DNA, once extracted and cut by restriction enzymes, should be run on a gel to ensure the presence of large molecular weight molecules (> 5–20 kb, but < 50 kb) and in some cases this step will also confirm whether contaminating RNA species are also present (< 5 kb).

The DNA sample can also be quantitated by spectrophotometry. This will also indicate the degree of protein contamination. The spectrophotometer will require a tungsten lamp in addition to the conventional halogen lamp so that infrared readings can be taken. In this regard the 280/260 nm ratio should ideally be 1.7–2.0 (ratios < 1.5 indicate heavy protein contamination, ratios above 2.1 indicate contamination with residual extraction reagents).

11.7.4 Non-denaturing gel electrophoresis

1. Make a 0.6–0.8% agarose gel and add 100 ml 1 × TBE buffer and 50 μl ethidium bromide (10 mg/ml in water) or add 1 μl of 10 mg/ml ethidium bromide to each sample.
2. Heat to 55–65 °C, pour and allow 1 h to set.
3. Make DNA samples up to 20 μl in water and add 8 μl sample buffer (50% glycerol, 10% 5 × TBE, 35% 0.5 M EDTA pH 7, mix thoroughly then add 5% 10% (w/v) SDS last. Add ~100 μg Bromophenol Blue using a pipette tip.
4. Load on gel with markers, either lambda phage 174 *hind*III or *hae*III digest.
5. Run gel:
   (a) Fast test gel at 120 V (50 mA) for 2–4 h of 1 μg (1/20th of the sample).
   (b) If DNA is cut to completion an even smear of DNA should be seen. Run the entire sample, up to 20 μg, at 25–30 V (20 mA) for 18 h overnight. Use wells that will allow 60 μl of sample to be loaded.

Visualization – see RNA section and photograph gel alongside a ruler, but avoid long exposure to UV light.

11.7.5 Southern blotting

To enable the large fragments of DNA to blot from the gel onto the membrane the DNA must be linearized by depurination. Depurinate with 0.25 M HCl for 10–20 min. This is particularly important for restriction fragments greater than 1 kb, which transfer inefficiently, and smaller fragments, which hybridize inefficiently. The depurinated sites are subsequently cleaved during alkali treatment. Wash several times with water afterwards. The Bromophenol Blue tracking dye should change from blue to yellow, indicating completion.

1. Denature DNA in gel into separate strands by using two 15–30 min washes of 250 ml (250 ml 1 M NaOH, 100 ml 5 M NaCl and 150 ml distilled water).
2. Neutralize the gel with 2 × 30 min washes of 250 ml of solution (250 ml 1 M Tris/HCl, pH 7.4 and 250 ml 5 M NaCl). The Bromophenol Blue should change colour again yellow to blue.
3. Soak in 10 × (or 20 ×) SSC for 5 min then prepare blot as for Northern procedure (see Figure 11.5).
4. After blotting for 12–18 hour wash in 2 × SSC for 5 min.
5. Bake nitrocellulose for 2 h at 80 °C, or cross-link nylon for 30–60 s. Nylon is less brittle and binds DNA covalently, so probes can be completely stripped from the membrane after stringent washing.

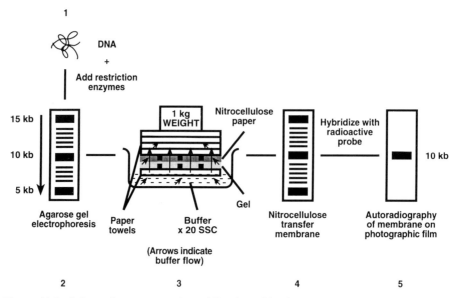

Figure 11.5. Schematic representation of Southern blotting.

11.7.6 Hybridization

The pre-hybridization step may take longer in Southern blotting as there are a lot of repetitive sequences, however the solutions are prepared identically (see RNA section). Prokaryotes or simple eukaryotes with single copy genes of interest may need hybridization for only an hour for 1 μg of DNA whereas single-copy mammalian genes (with bigger genomes on the filter) may need overnight hybridization for 10 μg of sample DNA (Figure 11.6). Use of dextran sulphate is usually essential in Southern blots in order to achieve suitable hybridization kinetics of the probe and high complex genomic DNA. It does, however, in some cases increase the background hybridization. Hybridization in Church buffer (7% (w/v) SDS and 0.5 M sodium phosphate, pH 7.2) tends to give a low background and better signal-to-noise ratio but makes reprobing the filters more difficult.

11.7.7 Troubleshooting

If unexpected bands appear on the autoradiograph at a higher molecular weight this suggests partial cutting of the DNA. If dense smears appear that will not wash out this suggests inadvertent use of repeated sequence in the probe, whereas a weak signal usually indicates a blotting problem.

11.8 Production of double-stranded uniformly labelled probes

Probes can be labelled with either ^{32}P or ECL (enhanced chemiluminescence), however, ^{32}P is more sensitive because as little as 10 fg of target sequence can

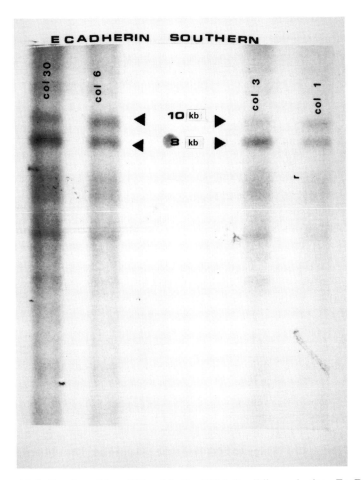

Figure 11.6. Southern blot of E-cad in the HCA-7 cell line colonies. *Eco*RI and *Bam*HI digestion of genomic DNA resulted in two major bands, 10 kb and 8 kb selectively, bound to the E-cad probe. These fragments correspond with those predicted for the E-cadherin gene. A minor increase is seen between colonies suggesting that amplification of E-cadherin may occur between the individual samples.

be detected compared with 500 fg with ECL. The disadvantages of ^{32}P are that probes can be stored for a maximum of 1 week compared with 2 months and that radioactively labelled probes frequently need to be purified and the final band resolution on gels is less good. Aim for a specific activity (dpm/μg) of 5×10^7–2×10^8. Random primer labelling can result in probes of specific activity of 5×10^9 whereas nick-translated probes and PCR-generated probes have a maximum specific activity of 5×10^8. In addition, random priming requires only 25 ng of moderately pure ssDNA template whereas nick translation requires 1 μg of highly purified dsDNA template.

11.8.1 Nick translation

1. (a) Make probe DNA (should be dissolved in distilled water or 10 mM Tris/HCl) aiming for a conc. of 0.1–1 $\mu g/\mu l$, dispense 500 ng–2 μg and place in an ice water bath at 15 °C.
   (b) Add 1 × nucleotide buffer solution – 20 μl.
   (c) Unlabelled dNTP, 20 nmoles of each, 10–50 μl.
   (d) Labelled dNTP 16 pmoles (5–10 μl, spec. activ. > 3000 Ci/mmole, 10 $\mu Ci/\mu l$), usually $\alpha^{32}P$ dCTP.
   (e) Water 10–60 μl (i.e. make final volume to 100 μl).
   (f) BSA, 1 μl of 1 mg/ml to stabilize DNA polymerase (optional).
2. Make enzyme solution up to 10 μl.
   (a) Add pancreatic DNAase I 2–5 μl (10 ng/ml) (produces small nicks in the DNA strands without removing nucleotides, by producing free 3′-OH and free 5′-PO_4 groups), mix by pipetting at room temp.
   (b) Add 2.5 units E. coli DNA polymerase I (removes nucleotides by 5′–3′ exonuclease activity from the free 'nicked' 5′ ends and the polymerase activity adds labelled nucleotides to the free 'nicked' 3′ ends.
3. Incubate the reaction for 1–2 h at 16 °C (if the reaction is conducted at 20 °C or above 'snapback' DNA may be generated with a newly polymerized strand looping back on itself to form elongated and inefficient probes).
4. Stop the reaction by adding 1 μl 0.5 M EDTA (pH 8) or 3.3 μl of 5 M NaOH and leave at room temp. for 10 min.
5. Determine the proportion of labelled dNTPs incorporated into the DNA. Take 1 μl or 2 μl of reaction into the tube containing 200 μl of water or 0.2 M EDTA and mix well.
6. Separate the radiolabelled DNA from unincorporated dNTPs, by either chromatography or spin columns. If chromatography is used, two peaks are formed on the column, the first peak is labelled DNA, the second is unbound ^{32}P. Collect the first peak in a volume of 1–2 ml, count a 5 μl sample and estimate the specific activity (c.p.m./μg). (Use a G-50 Sephadex column preswollen in buffer containing 150 mM NaCl, 10 mM EDTA, 0.1% SDS and 50 mM Tris/HCl, pH 7.5.)

11.8.1.1 Trouble shooting

1. Specific activity lower than expected – carry out test reaction on standard DNA. If this is okay there may be inhibitors in the test DNA sample.
2. Slower reaction progression than expected – try larger volume.

11.8.2 Random priming

1. Random primers, use either
   (a) salmon sperm DNA digested with DNAase I or

(b) random oligonucleotides or

(c) synthetic octamers

2. Denaturation.

In a tube mix 20–200 ng dsDNA (1 μl) with 75 ng random primers (1 μl), boil for 5–10 min then cool on ice.

3. Place in another small microfuge tube:

20 mM dithiothreitol, 1 μl

unlabelled dNTPs, 5 mM, 1 μl

10 × RP buffer (500 mM Tris/HCl, pH 7.5, 100 mM MgCl$_2$), 1 μl

labelled dATP or dCTP (50 μCi[α^{32}]), 3 μl

water, 1 μl

4. Mix DNA/random primers with contents of second tube.

5. Add 5 units (1 μl) Klenow fragment (lacks 5′–3′ exonuclease activity) and mix and then spin at 12 000 *g* for 1–2 sec.

6. Labelling reaction. Incubate for 3 h at 20–37 °C (more rapid reaction at higher temperature).

7. Add buffer A (50 mM Tris/HCl, 50 mM NaCl, 5 mM EDTA, pH 8 and 0.5% (w/v) SDS) and store at −20 °C until needed (but remember that probes decay rapidly).

11.8.3 PCR-generated probes

It is also possible to use a labelled ^{32}P NTP in a PCR reaction, however it is usual to use only half the concentration of the labelled NTP in the reaction compared with unlabelled NTPs.

11.8.4 Non-radioactive labelling and detection methods

1. dUTP nucleotides can be linked to biotin or fluorescein so that they act as reporter molecules when the DNA probe is incorporated into the membrane. Alternatively, enzymes like horeseradish peroxidase and alkaline phosphatase can be cross-linked to nucleic acids. Although the latter is more sensitive the technique can have some limits on the hybridization conditions that can be used (see product information, Amersham Life Science).

2. The above labelling reactions can be detected by use of colorimetric substrates as for immunohistochemistry or by chemiluminescent substrates. The latter substrate relies on HRP (horseradish peroxidase) bound either directly to the nucleic acid or indirectly to it via a biotin–antibody complex catalysing the oxidation of the substrate luminol by H$_2$O$_2$. The luminol immediately following oxidation is in an excited state, which decays slowly to the unexcited state via a photon emitting pathway which can be permanently recorded on X-ray film.

References

Boothwell, A., Yancopoulos, G. D. and Alt, F. W. (1990) *Methods for cloning and analysis of eukaryotic genes*. Jones and Barlett Publishers, Boston.

Brown, T. A. (1991) *Essential Molecular Biology, A Practical Approach*, vol. 1. IRL Press, Oxford.

Chomczynski, P. (1993) *RNAzol B isolation of RNA*. Cinna/Biotecx Bulletin No 3. Houston, USA.

Chomczynski, P. and Sacchi, N. A. (1987) A rapid method for total RNA extraction. *Anal. Biochem.* **162**, 156–9.

Innis, M. A., Gelfand, D. H., Sninsky, J. J. and White, T. J. (1990) *PCR Protocols*. Academic Press, San Diego.

Sambrook, J., Fritsch, E. F. and Maniatis, T. (1989) *Molecular cloning: a laboratory manual*. Cold Spring Harbour Laboratory Press, Cold Spring Harbor, New York.

Schagger, H. (1987) Tricine SDS-PAGE electrophoresis for the separation of proteins. *Anal. Biochem.* **166**, 368–79.

12

Molecular cytogenetic analysis of tumours and cell lines and its application in biology and medicine

SURESH C. JHANWAR

12.1 Introduction

The first major breakthrough relating to chromosome change in primary human tumours was the discovery by Nowell and Hungerford, reported in 1960, that a specific chromosome abnormality (an apparent deletion in one of the G group chromosomes, later designated as the Philadelphia Ph[1] chromosome) was consistently associated with chronic myelogenous leukemia (CML) (Chaganti and Jhanwar, 1985).

The introduction of the banding techniques in the early 1970s constituted a second major catalyst to the study of tumour cytogenetics. The Ph[1] chromosome in CML and the 14q+ marker chromosome observed to be a consistent feature of Burkitt's lymphoma (BL) were shown to be derived from specific translocation. During the past 25 years, a vast literature has accumulated that describes in increasing detail the patterns of chromosome changes that occur in various kinds of primary human tumours (Testa and Misawa, 1985; Heim and Mitelman, 1987; Rowley, 1990).

The third and by far the most exciting development in the study of tumour chromosomes is represented by the successful integration of classic cytogenetics with molecular biology and virology that has been achieved in the early 1980s, which in turn has resulted in identification of genes involved in various histological sub-types of leukemias and lymphomas (Rowley, 1990; Rabbitts, 1994).

During the past few years, fluorescent *in situ* hybridization (FISH) technique has become an indispensable procedure in many areas of research in clinical cytogenetics; gene mapping and identification of gene amplification and analysis of complex chromosomal abnormalities not resolved by conventional cytogenetic methods (LeBeau, 1993; Adinolfi and Crolla, 1994).

The FISH method is currently considered the most advanced and sensitive method for the localization of chromosome-specific sequences and identification of numerical and structural chromosome changes, at both metaphase and interphase stages of the cell cycle (LeBeau, 1993).

Thus, identification of genetic alterations in lymphohematopoietic tumours has led to the description of the molecular mechanisms which underlie malignant transformation (Rabbitts, 1994). In addition to correlations with histological sub-type and immunophenotype, a growing literature has documented correlations between genetic alterations and clinical features such as anatomic site of involvement by disease, probability of response to chromotherapy, and patient survival (Offit *et al.*, 1994).

Until the mid 1980s, tumours of the lymphohaematopoietic system had received the greatest attention from cytogeneticists; solid tumours have received less attention because, technically, they are much more difficult to handle than leukemias and lymphomas. However, more recent technical innovations have enabled many investigators to study in detail these tumours as well.

Cytogenetic analysis of solid tumours, though, provided indications of which chromosomes are likely to be altered in particular cancers, including leads for the location of genetic susceptibility loci and tumour suppressor genes. Cytogenetic analysis alone, however, is only able to detect large chromosomal changes. Recent developments in molecular techniques such as transfection assays and deletion mapping by RFLP (restriction fragment length polymorphism) have provided additional tools that are much more sensitive to detecting genetic changes associated with cancer. The best example of the use of these techniques is the analysis of multistep tumourigenesis in colorectal cancer (reviewed by Jhanwar and Gerdes, 1994).

Amplification of certain cellular oncogenes such as N-*myc*, Her-2/neu and Akt-2 has been identified in a number of tumours, and amplification is even correlated with prognosis in some tumours. In fact, amplification of N-*myc* in neuroblastoma (Seeger, *et al.*, 1985), Her-2/neu in breast cancer (Slamon *et al.*, 1987), and Akt-2 in ovarian cancer has been shown to predict poor prognosis and, therefore, shorter survival in patients with these tumours.

12.2 Methods of analyses

12.2.1 Cytogenetic analysis

Cell cultures and chromosome preparations are obtained according to methods described earlier (Jhanwar *et al.*, 1994). Briefly, tumour tissue is cut and diced with surgical scalpels under sterile conditions in a Petri dish containing RPMI 1640 medium. Enzymatic disaggregation with 0.1% collagenase, 0.01% hyaluronidase, and 0.002% DNAase is performed for 30–60 min. After disaggregation, cells are washed twice with RPMI 1640, and then transferred to culture flasks containing RPMI 1640 growth medium, 15% (v/v) fetal bovine serum, 1% glutamine, 5 μg/ml insulin, 5 μg/ml transferrin, 5 ng/ml sodium selenite, 5 μg/ml fungizone, 50 μg/ml streptomycin, and 50 U/ml penicillin.

Flasks are incubated at 37 °C in an atmosphere of 5% CO_2. The growth medium is changed after 24 h and non-attached cells removed. *In vitro* cell growth is observed with an inverted phase-contrast microscope and proliferating cultures are exposed to colcemid (0.02 μg/ml) for 4–12 h on days 3–7 depending upon the proliferative rate. Cells are then detached with 0.025% (w/v) trypsin-EDTA solution and treated with 0.075 M KC1 for 20 min at 37 °C, fixed in 3:1 methanol–acetic acid fixative and air dried preparations are made following methods routinely employed in the laboratory. Chromosomes are analysed using either Quinacrine or Giemsa banding as previously described (Jhanwar *et al.*, 1994). Detection of either the same structural abnormality and non-random gain of a chromosome in two or more cells or non-random loss of a chromosome in three or more cells is considered a clonal abnormality. In addition, a relative deficiency or excess of specific chromosomes as defined by loss or gain respectively of any given chromosome in relation to the expected ploidy level of the abnormal clone, is also scored to assess non-random involvement of specific chromosomes in numerical aberrations (Mitelman, 1991). For example, less than three copies of a chromosome in a triploid clone is scored as a loss, whereas more than three copies as a gain. Similarly, a missing chromosomal segment due to an unbalanced translocation is also scored as a partial loss in relation to the expected ploidy level of the clone (Figure 12.1).

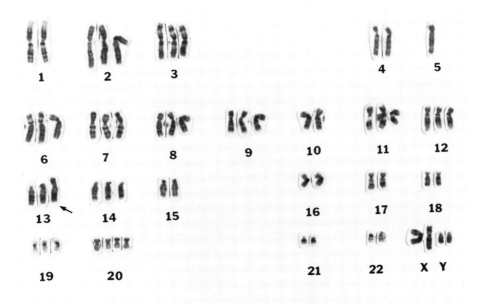

Figure 12.1. A representative G-banded karyotype from a colorectal cancer with chromosome constitution of $60 < 3n \pm >$, XY, +X, +Y, −1, −4, −5, −5, −10, der(13)t(13;17)(p11;q11), −15, −16, −17, −18, +20, −21, −22. Note the relative deficiency of chromosomes 1, 5, 17p, and 18. The chromosome indicated by the arrow is the only unbalanced translocation between 13p and 17q seen in this tumour.

12.2.2 *Restriction fragment length polymorphisms (RFLPs): an overview*

For detailed methods of Southern blotting and hybridizations see Chapter 11. For detailed methods of RFLPs see Chapter 13.

   Subsequent autoradiographs are analysed and scored for loss of heterozygosity in tumours with matched normal tissue DNA used as the control. Cancers which appear to have clearly lost one allele on the autoradiograph are scored as a loss, while those not clearly demonstrating a loss are examined by densitometry to assess for partial loss. Tumours shown to have a greater than 25% reduction in the intensity of one band by densitometry are scored as partial losses (Presti *et al.*, 1991). Only cancers with a matched heterozygous normal DNA for these RFLPs are scored for losses, while those with homozygous normal DNA RFLPs are scored as uninformative. Heterozygous cancers found to have no loss by inspection or densitometry are scored as retained alleles, demonstrating no loss of allele for that segment of the chromosome (Figures 12.2 and 12.3). Scoring of the autoradiographs is performed by one person without the knowledge of the cytogenetic results or the clinical follow-up.

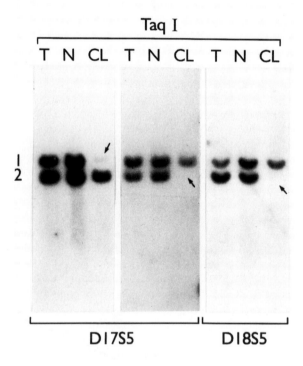

Figure 12.2. Southern blot analysis of DNA from colorectal tumour (T), corresponding normal tissue (N) and cell line (CL) with probes for chromosome 17 and 18 to show partial loss of heterozygosity due to contaminating normal cells in the tumour, whereas complete loss is seen in the cell line. The chromosome locus for each pair is shown at the bottom of the blot with enzyme at the top. LOH is indicated by arrows.

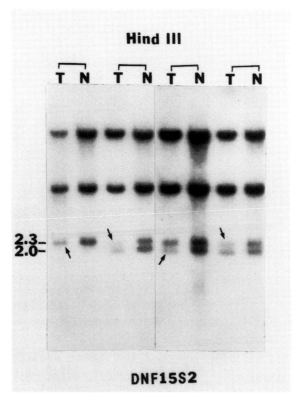

Figure 12.3. Southern blot analysis of DNA from renal cell tumour (T) and corresponding normal tissue (N) with a probe from chromosome region 3p21 to show partial allelic loss due to normal cells in three pairs, whereas a tumour pair on the left was homozygous. LOH in each pair is indicated by an arrow.

12.2.3 Fluorescence in situ hybridization (FISH)

Probes of interest are labelled either by biotin or digoxigenin according to the manufacturer's protocol (Oncor, Inc., Gaithersburg, MD, USA). Molecular hybridization of air-dried slides is performed as follows. Cytological preparations are treated with RNAase A (Sigma) at 100 μg/ml in 2 × SSC at 37 °C for 1 h, rinsed in 2 × SSC (1 × SSC is 150 mM sodium chloride/15 mM sodium citrate), and dehydrated with ethanol and air dried. Preparations are then treated with proteinase K at 37 °C for 5.5 min, washed two times in 2 × SSC, and dehydrated in ethanol. Chromosomal DNA is denatured by treating slides with 70% (v/v) formamide/2 × SSC at 70 °C for 2–3 min, followed by dehydration in ethanol. Hybridization reaction mixture containing hybrid mix (Oncor, Inc.) and 10 μl biotin or digoxigenin-labelled probe is denatured and applied to each cytological preparation and hybridization is carried out at 37 °C for 12–18 h in a closed Petri dish saturated with 50% formamide/2 × SSC at pH 7.0. Post-hybridization washing is performed at 43 °C in 50% formamide/2 × SSC three times (5 min/each washing) followed by four washes in

2 × SSC for 5 min each, and then four washes in 2 × SSC at room temperature. The final two washes are performed in PBD (phosphate buffered detergent) and slides are stored in PBD at room temperature before detection and amplification. Signal is detected by using either the fluorescein-labelled anti-digoxigenin (direct method) or avidin and anti-avidin (indirect method) for 5–30 min at 37 °C or room temperature. Similarly, if required, amplification of signal is accomplished either by rabbit anti-sheep antibody or avidin for 15–30 min at room temperature. Final washes are carried out three times in 1 × PBD for 2 min each and the slides are then stained with propidium iodide for 5 min, once again the slides are washed in 1 × PBD, and counter-stained with DAPI (4,6-diamidino-2 phenylindole) for 5 min to reveal banding patterns.

Microscopy is performed using a Zeiss photomicroscope fitted for epi-fluorescence and a 100× apochromatic oil-immersion objective. Photographs are taken with Kodak Ektachrome 400 film and printed on Kodak 2001 type F paper.

12.3 Multistep tumourigenesis and genetic alterations in solid tumours

12.3.1 Cytogenetic and molecular alterations in solid tumours and cell lines

Currently, solid tumours as a group are subdivided into four major categories based on their cell of origin. Mesenchymal tumours characterized by specific translocations (Jhanwar *et al.*, 1994; Sreekantaiah *et al.*, 1994), epithelial tumours by deletions (Rodriguez, Sreekantaiah and Chaganti, 1994; Tahara, 1995), neurogenic tumours by non-random structural and numerical changes (Jhanwar *et al.*, 1994; Sandberg and Bridge, 1994) and testicular germ cell tumours are associated with isochromosome i(12)(p10) (Chaganti, Murty and Bosl, 1995). Based on the specificity of the particular abnormality, cytogenetics has not only helped in establishing a differential diagnosis, but in some instances provided important prognostic information. For the purpose of this chapter, data generated in the author's laboratory on some of the tumour types will be presented to highlight the biological and clinical implications of genetic alterations seen in solid tumours.

During the past seven years we have concentrated our research efforts to identify non-random chromosome abnormalities that have a possible role in the origin of human solid tumours. For selected solid tumour types, the cytogenetic analysis was complemented by molecular methods (RFLP and FISH) to examine for specific genetic defects and to identify sites of tumour suppressor genes in a large proportion of tumours in any given histopathological group.

Colorectal cancer has served as an excellent model system to study multistep tumourigenesis in epithelial tumours. According to this model, mutation of the second APC allele in FAP patients (Levy *et al.*, 1994) and somatic mutation of APC gene in sporadic cancer (Jhanwar and Gerdes, 1994), *ras*-gene mutation,

DNA hypomethylation, p53 mutation and 5q alterations appear to occur early in colorectal tumourigenesis, while allelic deletions of chromosome arms 17p and 18q usually occur at a later stage of tumourigenesis (Vogelstein *et al.*, 1988; Fearon and Vogelstein, 1990; Solomon, 1990). In addition, the genetic alterations seen in the initiation, neoplastic transformation and progression in hereditary, as well as sporadic, cancer have been shown to be identical in colorectal cancer (reviewed by Jhanwar and Gerdes, 1994).

Hereditary non-polyposis colorectal cancer (HNPCC), on the other hand, is associated with mutations of two mismatch repair genes: *MSH2* on chromosome 2p and *MLH1* on chromosome 3p. A substantial proportion of HNPCC tumours show instability of short tandem repeat sequences (replication errors – RER) in addition to K-*ras*, APC, and p53 mutations (Fishel *et al.*, 1993; Parsons *et al.*, 1993; Aaltonen and Peltomäki, 1994; Nyström-Lahti *et al.*, 1994; Han *et al.*, 1995). A detailed description of such studies is beyond the scope of this chapter, and therefore, will not be addressed further.

12.3.2 Cytogenetic and molecular genetics of colorectal cancer

Colorectal cancer is the second most prevalent malignancy in the United States. Well-described inherited syndromes have been reported in families with a high frequency of colorectal cancer. The hereditary or familial colorectal cancer groups include Gardner's syndrome, adenomatous polyposis coli (APC) or familial adenomatous polyposis (FAP), and hereditary nonpolyposis colorectal cancer (HNPCC), including Lynch syndromes I and II (Gardner, 1983, pp. 39–60). We have performed cytogenetic and RFLP analysis on a cohort of 63 fresh colonic tumours and have shown that allelic deletions of chromosome arms 5q, 17p and 18q were the most frequent primary abnormalities; consistent abnormalities of chromosomes 1 and 7 as additional abnormalities were seen in a subgroup of high grade and metastatic tumours, indicating that gene(s) on these chromosomes may be associated with progression and metastasis of tumours (Figure 12.1). In addition, our combined cytogenetic and RFLP studies have also provided evidence for the role played by hyperdiploidy (uniparental disomies, trisomies or tetrasomies of a given chromosome), as well as relative deficiency of a given chromosomal segment in allowing expression of underlying mutations and, therefore, the inactivation of tumour suppressor genes (Gerdes *et al.*, 1995). Clinicopathological correlations in this cohort demonstrated that four-year survival for patients with normal chromosomes 5, 17 and 18 was 100% compared to 52% in patients with abnormalities of these chromosomes. Moreover, abnormalities of either chromosome 17 or 18 predicted a poor outcome. Four-year survival of patients with normal chromosome 17 was 80% compared to 40% for those with abnormal 17 ($P = 0.01$). Similarly, four-year disease-free survival of patients with a normal chromosome 17 was 60% compared to 19% with an abnormal chromosome 17 ($P = 0.05$) (Gerdes *et al.*, 1995).

While genetic alterations seen in dividing cells in tumours are a true reflection of the abnormalities of the tumour *in vivo*, abundance of normal cells (stomal and contaminating lymphocytes), invariably present in epithelial tumours, most often mask allelic deletions (Figure 12.2). In addition, limited supply of tumour tissues from patients precludes any *in vitro* studies. During the last seven years, we have successfully developed established cell lines from a variety of solid tumours and have shown that the genetic alterations in cell lines are identical to those seen in corresponding tumour tissues (Gerdes *et al.*, 1989). The established cell lines derived from patients with well-defined clinical and genetic history, therefore, not only provide an excellent model system for the precise molecular characterizations but reagents for several biochemical and cellular studies.

Eighteen of the 23 colorectal cancer cell lines studied revealed deletions or rearrangements involving chromosomes 1p, 7, 5q, 17p, 17q and 18q. A close correlation was observed between cytogenetic and molecular deletions. It was of special interest to note that, while the deletions or rearrangements of chromosomes, 5, 17p and 18q were seen as recurrent abnormalities in a majority of cell lines, the abnormalities of chromosomes 1, 17, 17q and 7 were rarely seen without the abnormalities of chromosomes 5, 17p and 18q. These data thus allow us to suggest that the loss of genes on chromosomes 5, 17p and 18q regulating the normal proliferation and differentiation of colonic mucosa may be of primary significance in neoplastic transformation, whereas the rearrangements or losses of genes from chromosomes 1, 17q and 7 appear to be changes related to progression and development of metastatic potential of the tumours (Gerdes *et al.*, in preparation).

In order to identify gene(s) on chromosome 17 that confer more aggressive phenotype, we have performed Northern blotting, immunohistochemical staining (p53 mRNA, protein) and RFLP analysis using several probes including p53 and nm23. Following immunohistochemical staining, 9/23 cell lines showed high levels (25–80%) of mutant p53; 3/23 cell lines expressed intermediate levels (< 25%); and 9/23 cell lines had low levels (< 10%) or complete absence of the protein. Of these, 7/9 cell lines also had complete loss of mRNA expression. 22/23 cell lines had normal germline configuration for the p53 gene, while one showed a rearrangement. RFLP analysis showed allelic losses around the p53 locus in 16/23 cell lines (70%), four of which were losses of the p53 locus itself. In addition, three cell lines (7/11 informative cases) with losses of nm23 gene also showed concurrent losses of p53 locus, while the remaining four were homozygous for p53. Furthermore, 4/6 cell lines with nm23 deletions were derived from metastatic tumours, whereas one cell line was obtained from recurrent tumour. These data suggest that concurrent functional loss of p53 and nm23 genes accomplished by a variety of mechanisms may be associated with poor prognosis and survival. The mechanism involved in the interaction of these genes in tumour progression and resistance to chemotherapeutic drugs, however, remains to be seen (Elahi *et al.*, 1994).

12.3.3 Cytogenetic and molecular analysis of sarcomas

Malignant sarcomas in general are an extremely heterogenous group of neo-plasms composed of various histological subtypes. The histological classifica-tion of soft tissue sarcomas, especially those with high grade, is often difficult. Recent advances in electron microscopic examination and histochemical tech-niques have helped in the classification of certain types of sarcomas, however, a large number of sarcomas still defy a histogenetic classification, which to some extent can be attributed to a common origin of soft tissue sarcomas from primitive mesenchymal cells (Molenaar *et al.*, 1989). It is this feature of sarcomas that makes this group of tumours interesting from the point of understanding natural history and biology of tumours. Furthermore, sarcomas represent a group of tumours with features similar to those described in lymphoma, which are characterized by specific translocations (Figure 12.4) and are mesodermal in origin. The sequence of genetic events associated with the initiation and progression of lymphomas and sarcomas include deregulation of genes intimately involved with normal cell growth followed by loss or inactiva-tion of tumour suppressor genes. Thus, it may be interesting to identify genes involved in pathogenesis of sarcomas and compare their structural and func-tional properties with those involved in pathogenesis of lymphomas. In an effort to identify recurrent genetic alterations associated with various histo-logical subtypes in sarcomas, which may be predictive of clinical and biological behaviour of tumours, we have performed cytogenetic/molecular genetic ana-lyses on a series of 163 sarcomas (Table 12.1) seen at the Memorial Sloan-Kettering Cancer Center between 1987–1993 (Jhanwar *et al.*, in preparation).

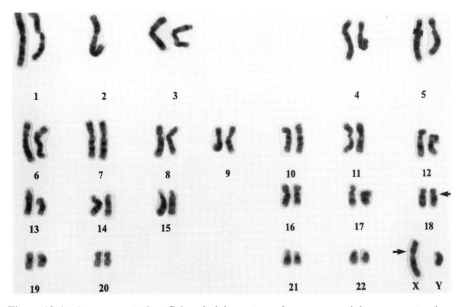

Figure 12.4. A representative G-banded karyotype from a synovial sarcoma to show recurrent chromosome abnormality t(X;18)(p11.2;q11.2) (arrow).

Table 12.1. *Summary of genetic alterations and gene amplification in 163 sarcomas studied at MSKCC (1987–1993)*[a]

Histological type	Cytogenetic abnormality		Molecular abnormality	Gene amplification	
	Primary	Secondary	Gene rearrangement	MDM2	CHOP
Ewing's sarcoma and PNET	t(11;22)	1, 7	EWS 20/23	3/30	–
Liposarcoma	t(12;16)	1, 7, 13, 17	CHOP-TLS 55/76	11/76	5/76
Synovial sarcoma	t(X;18)	7, 12	SSX-SYT 18/21	–	–
Hemangiopericytoma	t(12;19)	–	NA	–	–
Epithelioid sarcoma	t(8;22)	?	NA	–	–
MPNST	17, 22	1, 11, 12, 14	–	–	–

[a] Ladanyi *et al.*, 1990, 1993; Jhanwar *et al.*, 1994; Jhanwar *et al.*, (in preparation); Ladanyi (personal communication).

Cytogenetic analysis of sarcomas revealed a spectrum of clonal abnormalities, which ranged from simple translocations to complex rearrangements and deletions of several chromosomes (Table 12.1). The translocations included t(X;18)(p11.2;q11.2) in synovial sarcoma, t(11;22)(q24;q12) in Ewing's sarcoma and primitive neuroectodermal tumour, t(12;16)(q13;p11) in myxoid liposarcoma, t(12;19)(q13;q13) in malignant haemangiopericytoma and t(8;22)(q22;q11) in epithelioid sarcoma. Malignant peripheral nerve sheath tumours (MPNST), however, exhibited recurrent structural abnormalities of chromosomes 1, 12, 14, 17, and 22. While specific translocations or abnormalities positively correlated with a given histology (Barr *et al.*, 1995; Jhanwar *et al.*, unpublished data), the high grade or metastatic tumours also displayed additional complex abnormalities that are thought to represent secondary abnormalities shared by a variety of cancers. These abnormalities, which included chromosomes 1, 7, 12, and 17, contributed to the inter- and intra-tumouric heterogeneity of individual tumour types (Ladanyi *et al.*, 1990; Jhanwar *et al.*, 1994).

In order to determine molecular markers that may contribute to genetic heterogeneity, we have also performed molecular analysis on a selected group of sarcomas. Of the 23 cases of Ewing's and PNET, including all 14 cases with a classical or variant t(11;22) examined for EWS rearrangement, 20 cases showed rearrangement of the gene. A remarkable correlation between molecular rearrangement and MIC2 expression was also seen in 19/23 cases (Ladanyi *et al.*, 1993). Four cases that showed a discordance may be due to the intratumouric heterogeneity. Similar correlations have been seen between t(X;18) in 18/21 synovial sarcomas studied for the rearrangement of *SSX* (chromosome Xp11.2) and *SYT* (chromosome 18q11.2) genes (Ladanyi, Personal communication). Seventy-six cases of liposarcomas were studied for status of *CHOP* (chromosome 12q13) and *TLS* (chromosome 16p11) genes; all 34 myxoid liposarcomas showed rearrangement of either both or one of the genes. Of the remaining 42 tumours, which included lipoblastic, round cell, well differentiated and pleomorphic liposarcomas, 21 (50%) also showed rearrangement of *CHOP* and *TLS* genes, indicating a possible role of these genes not only in pathogenesis of myxoid but all other histological sub-types of liposarcomas. Amplification of the *mdm2* gene was detected in 3/30 (10%) of cases of ES and PNET (Ladanyi *et al.*, 1995). The three with amplification were primary tumours and were morphologically similar. *mdm2* gene amplification was also detected in 3/11 metastatic, 1/1 recurrent osteosarcomas and 11/76 high grade liposarcomas (Figure 12.5). None of the primary osteosarcomas (16 cases) had detectable *mdm2* gene amplification. Thus the data presented above demonstrate that genetic heterogeneity seen in sarcomas,

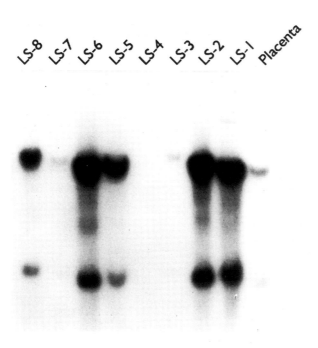

EcoRI / MDM2

Figure 12.5. A Southern blot analysis of DNA from eight liposarcomas to show amplification of MDM2. *Eco*RI-digested DNA was hybridized with the c-DNA MDM2B probe and signal intensities were compared with equal amount of placental DNA digested with *Eco*RI. Several-fold amplification of the *MDM2* gene was detected only in high grade or metastatic sarcomas (LS-1, LS-2, LS-5, LS-6, and LS-8).

acquired during the clonal evolution and progression of the disease, may play a significant role in predicting biological and clinical behaviour of the individual tumours.

12.3.4 Cytogenetic and molecular analysis of neuroblastoma

Neuroblastoma (NB), which originates from the primitive neural crest cells that form the adrenal medulla and sympathetic nervous system, is one of the most common paediatric tumours. The majority of tumours are sporadic; a subset of patients exhibit a predisposition to develop NB, which follows an autosomal dominant pattern of inheritance, these tumours are presumed to be due to a germ-line mutation (Knudson and Strong, 1972).

Neuroblastoma is associated with several cytogenetic and molecular abnormalities (Brodeur and Fong, 1989; Sandberg and Bridge, 1994). Cytogenetic abnormalities reported by several investigators and recently summarized by Sandberg and Bridge (1994) include a total of 130 tumours and cell lines. Cytogenetic abnormalities frequently seen in NB can be divided into two major categories: one category includes tumours with hyperdiploid and near triploid karyotypes, whereas the second category comprises tumours with near diploid karyotype, structural abnormalities of chromosomes 1p, 14q, 17 and 22q, with either double minutes (dmins) or homogenously staining region (HSR). Deletion of 1p involving the region 1p32-pter, first described in NB tumours by Brodeur *et al.* (1988) and subsequently confirmed by several investigators in both tumours and cell lines, has been seen in approximately 70–80% of high stage NBs (reviewed by Sandberg and Bridge, 1994). Several LOH (loss of heterozygosity) studies have identified a common region of deletion at the distal end of chromosome 1 from p36.1 to p36.3. Based on these observations, it has been proposed that a mutation in this region followed by LOH of the same region on the other homologue may be involved in the initiation and progression of the disease (Sandberg and Bridge, 1994). We have performed cytogenetic and molecular analysis on ten high-stage tumours from NB patients and found a remarkable correlation between del(1p) and N-*myc* amplification (Figure 12.6A, B). In one of the metastatic tumours studied, a several-fold N-*myc* amplification was detected by *in situ* hybridization, which correlated with the presence of del(1p), dmin and HSR (Figure 12.7). These observations are consistent with several other studies, reviewed by Sandberg and Bridge (1994). Indeed, the presence of N-*myc* amplification has been shown to correlate strongly with advanced clinical stage and poor prognosis in human NB (Brodeur and Seeger, 1986; Brodeur *et al.*, 1984); some studies even correlated number of N-*myc* copies with the progression-free survival (Seeger *et al.*, 1985) (Table 12.2). In contrast, a recent study on a series of 850 patients with NB reported by Cohn *et al.* (1995) identified six patients with localized NB but with N-*myc* amplification. However, none of these patients, regardless of their histology, did worse than the others in the series. These studies

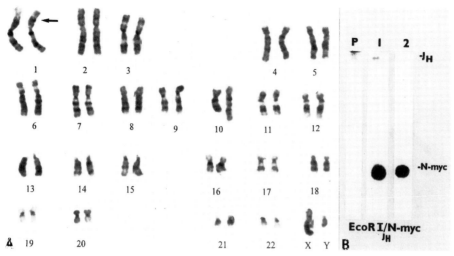

Figure 12.6. A representative G-banded karyotype (A) from a neuroblastoma tumour with chromosome constitution of 46, XY, del(p33), add(10)(q26) and (B) a Southern blot to show amplification of N-*myc* in the DNA from the same patient. *Eco*RI-digested DNA was co-hybridized with JH and N-*myc* probes. Note approximately 20-fold amplification of N-*myc* in tumour DNA (lane 2). DNA from a neuroblastoma cell line (lane 1) serves as positive control.

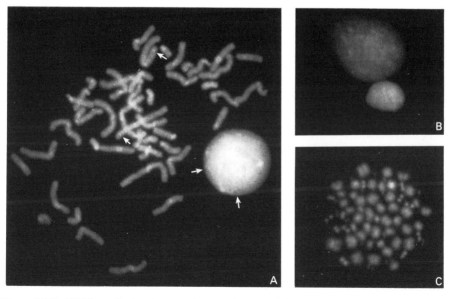

Figure 12.7. FISH analysis of normal metaphase and interphase cells using a N-*myc* cosmid probe (Oncor, Inc.) to show germ line position of N-*myc* (A) and several-fold amplification in interphase cell (B) and metaphase spread with dmins (C) from a high stage neuroblastoma with del(1p). Germ line position at chromosome 2p23 and interphase cell is indicated by arrows in (A).

Table 12.2. *N-myc amplification in neuroblas-*
toma: a study of 89 patients[a]

N-myc copy number	Progression free survival[b]
1	70%
3–10	30%
> 10	5%

[a] Seeger *et al.*, NEJM 313:1111, 1985.
[b] $P < 0.0001$

indicate that the presence of N-*myc* amplification in localized NBs does not
predict poor outcome.

12.4 Clinical and biological implications of genetic alterations in solid tumours and cell lines

Studies summarized above clearly demonstrate that non-random chromosome
abnormalities at cytogenetic and molecular genetic level can be associated with
specific histopathology. It has also been shown that genetic alterations may
predict clinical and biological behaviour of the tumours. Since studies on
sarcomas, renal, colorectal and ovarian tumours have allowed us to define
genetic abnormalities that are primary and those associated with progression of
the tumour, it is not unreasonable to expect that correlations between genetic
alterations and clinical and biological behaviour will be helpful in diagnosis,
prognosis, and understanding the natural history of tumours similar to those
demonstrated in lympho-haematopoietic tumours.

It is also clear from the evidence presented above that the identification of
genetic alterations in human cancers not only helps in our understanding of the
natural history and biological behaviour of cancers but also provides leads for
the diagnosis, successful treatment and management of patients.

Finally, genetically well characterized cell lines derived from tumours such
as Wilm's tumour, colorectal and renal cell carcinomas, have contributed
significantly to the functional characterization of tumour-specific genes (*Wt*,
APC and *VHL*) and the role such genes play in the multistep process of
tumourigenesis (Shimizu *et al.*, 1990; Dowdy *et al.*, 1991; Anglard *et al.*, 1992;
Groden *et al.*, 1995). In addition, cell lines derived from patients with
well-defined clinical and genetic history provide an excellent *in vitro* model
system to investigate chemosensitivity of various chemotherapeutic drugs.

Thus, based on cytogenetic and molecular analysis, a new subclassification of
human solid tumours can be anticipated. To be effective, however, genetic
changes must be correlated with clinical and conventional pathology features,
rate of tumour progression, sites of metastasis, response to therapy, prognosis
and genetic predisposition.

References

Aaltonen, L. A. and Peltomäki, P. (1994) Genes involved in hereditary nonpolyposis colorectal carcinoma. *Anticancer Res.* **14**, 1657–60.

Adinolfi, M. and Crolla, J. (1994) Nonisotopic *in situ* hybridization clinical cytogenetics and gene mapping applications. In *Advances in Human Genetics*, vol. 22 (ed. H. Harris and K. Hirschhorn), pp 187–235. Plenum Press, New York and London.

Anglard, P., Trahan, E., Liu, S. *et al.* (1992) Molecular and cellular characterization of human renal cell carcinoma cell lines. *Cancer Res.* **52**, 348–56.

Barr, F. G., Chatten, J., D'eruz, C. M. *et al.* (1995) Molecular assays for chromosomal translocations in the diagnosis of pediatric soft tissue sarcomas. *J. Am. Med. Assoc.* **273**, 553–7.

Brodeur, G. M. and Fong, C. T. (1989) Molecular biology and genetics of human neuroblastoma. *Cancer Genet. Cytogenet.* **41**, 153–74.

Brodeur, G. M., Fong, C. T., Morita, M. *et al.* (1988) Molecular analysis and clinical significance of N-*myc* amplification and chromosome 1 abnormalities in human neuroblastomas. *Progr. Clin. Biol. Res.* **271**, 3–15.

Brodeur, G. M. and Seeger, R. C. (1986) Gene amplification in human neuroblastomas: basic mechanisms and clinical implications. *Cancer Genet. Cytogenet.* **19**, 101–11.

Brodeur, G. M., Seeger, R. C., Schwab, M. *et al.* (1984) Amplification of N-*myc* in untreated human neuroblastomas correlates with advanced disease stage. *Science* **224**, 1121–4.

Chaganti, R. S. K. and Jhanwar, S. C. (1985) Chromosomal mechanisms of cancer etiology. In *Genetics in clinical oncology* (ed. R. S. K. Chaganti and J. L. German), pp. 60–79. Oxford University Press, New York.

Chaganti, R. S. K., Murty, V. V. V. S. and Bosl, G. G. (1995) Molecular genetics of male germ cell tumors. In *Comprehensive text book of genitourinary oncology* (ed. N. J. Vogelzang, P. T. Scardino, W. O. Shipley and D. S. Coffey), Williams & Wilkins Press, Baltimore (In press).

Cohn, S. L., Look, A. T., Joshi, V. V. *et al.* (1995) Lack of correlation of N-*myc* gene amplification with prognosis in localized neuroblastoma: a pediatric oncology group study. *Cancer Res.* **55**, 721–26.

Dowdy, S. F., Fasching, C. L., Araujo, D. *et al.* (1991) Suppression of tumorigenicity in Wilms tumor by the p15.5-p14 region of chromosome 11. *Science* **254**, 293–95.

Elahi, A., Gerdes, H., Chen, Q. *et al.* (1994) Biological and clinical implication of allelic deletions of chromosome 17 in colorectal cancer cell lines. *Proc. Am. Assoc. Cancer Res.* **35**, 601.

Fearon, E. R. and Vogelstein, B. (1990) A genetic model for colorectal tumorigenesis. *Cell*, **61**, 759–67.

Fishel, R., Lescoe, M. K., Rao, M. R. S. *et al.* (1993) The human mutator gene homolog MSH2 and its association with hereditary non-polyposis colon cancer. *Cell* **75**, 1027–38.

Gardner, E. J. (1983) Familial polyposis coli and Gardner Syndrome. In *Prevention of Hereditary Large Bowel Cancer*. Alan R. Liss, Inc., New York.

Gerdes, H., Tops, C., Chen, Q. *et al.* (1989) Genetics of colon cancer: abnormalities of chromosomes 5 and 17 correlate with the loss of alleles on these chromosomes in colorectal cancer cell lines. *Cancer Genet. Cytogenet.* **41**, 221.

Gerdes, H., Chen, Q., Elahi, A. *et al.* (1995) Recurrent chromosome deletions involving chromosomes 1, 5, 17 and 18 in colorectal carcinoma: possible role in biological and clinical behaviour of tumors. *Anticancer Res.* **15**, 13–24.

Groden, J., Joslyn, G., Samovitz, W. *et al.* (1995) Response of colon cancer cell lines to the introduction of APC, a colon-specific tumor suppressor gene. *Cancer Res.* **55**, 1531–39.

Han, H. J., Mazuyama, M., Baba, S. *et al.* (1995) Genomic structure of human mismatch repair gene, hMLH1, and its mutation analysis in patients with hereditary non-polyposis colorectal cancer (HNPCC). *Hum. Mol. Genet.* **4**, 237–42.

Heim, S. and Mitelman, F. (1987) *Cancer cytogenetics*. Alan R. Liss, Inc., New York.

Jhanwar, S. C., Chen, Q., Li, F. P. *et al.* (1994) Cytogenetic analysis of soft tissue sarcomas: recurrent chromosome abnormalities in malignant peripheral nerve sheath tumors (MPNST). *Cancer Genet. Cytogenet.* **78**, 138–44.

Jhanwar, S. C. and Gerdes, H. (1994) Molecular biology of familial cancer and polyps. In *Cancer of the colon, rectum and anus* (ed. A. M. Cohen, S. J. Winawer, M. A. Friedman and L. L. Gunderson), pp. 137–48. McGraw Hill Press, New York.

Knudson, A. G., Jr. and Strong, L. C. (1972) Mutation and cancer: neuroblastoma and pheochromocytoma. *Am. J. Hum. Genet.* **24**, 514–22.

Ladanyi, M., Heinemann, F. S., Huvos, A. G. *et al.* (1990) Neural differentiation in small round cell tumors of bone and soft tissue with the translocation t(11;22)(q24;q12). An immunohistochemical study of 11 cases. *Hum. Pathol.* **21**, 1245–51.

Ladanyi, M., Lewis, R., Gerin-Chesa, P. *et al.* (1993) EWS rearrangement in Ewing's sarcoma and peripheral neuroectodermal tumor, molecular detector and correlation with cytogenetic analysis and MIC2 expression. *Diagn. Mol. Pathol.* **2**, 141–6.

Ladanyi, M., Lewis, R., Jhanwar, S. C. *et al.* (1995) MDM2 and CDK4 gene amplification in Ewing's sarcoma. *J. Pathol.* **175**, 211–17.

LeBeau, M. M. (1993) Detecting genetic changes in human tumor cells: Have scientists gone fishing? *Blood* **81**, 1979–83.

Levy, D. B., Smith, K. J., Beazer-Barclay, Y. *et al.* (1994) Inactivation of both APC alleles in human and mouse tumors. *Cancer Res.* **54**, 5953–8.

Mitelman, F. (1991) In *International system for human cytogenetic nomenclature: guidelines for cancer cytogenetics. Supplement to an International System for Human Cytogenetic Nomenclature* (ed. F. Mitelman). S. Karger, Basel.

Molenaar, W. M., Dejong, B., Buist, J. *et al.* (1989) Chromosomal analysis and the classification of soft tissue sarcomas. *Lab. Invest.* **60**, 266–74.

Nowell, P. C. and Hungerford, D. A. (1960) A minute chromosome in human chronic granulocytic leukemia. *Science* **132**, 1497.

Nyström-Lahti, M., Parsons, R., Sistonen, P. (1994) Mixmatch repair genes on chromosomes 2p and 3p accounts for a major share of hereditary nonpolyposis colorectal cancer families evaluable by linkage. *Am. J. Hum. Genet.* **55**, 659–65.

Offit, K., Lo Coco, F., Louie, D. C. *et al.* (1994) Rearrangement of the *bcl*-6 gene as a prognostic marker in diffuse large-cell leukemia. *New Engl. J. Med.* **331**, 74–80.

Parsons, R., Li, G.-M., Longley, M. J. *et al.* (1993) Hypermutability and mismatch repair deficiency in RFR tumor cells. *Cell* **75**, 1227–36.

Presti, J. C., Jr., Rao, P. H., Chen, Q. *et al.* (1991) Histopathological, cytogenetic, and molecular characterization of renal cortical tumors. *Cancer Res.* **51**, 1544–52.

Rabbitts, T. H. (1994) Chromosomal translocations in human cancer. *Nature* (London) **372**, 143–49.

Rodriguez, E., Sreekantaiah, C. and Chaganti, R. S. K. (1994) Genetic changes in epithelial solid neoplasia. *Cancer Res.* **54**, 3398–406.

Rowley, J. D. (1990) Molecular cytogenetics: Rosetta Stone for understanding cancer – twenty-ninth G. H. A. Clowes Memorial Award lecture. *Cancer Res.* **50**, 3816–25.

Sandberg, A. A. and Bridge, J. A. (1994) The cytogenetics of bone and soft tissue tumors. R. G. Landes Company, Austin, USA.

Seeger, R. C., Brodeur, G. M., Satter, H. *et al.* (1985) Association of multiple copies of the N-*myc* oncogene with rapid progression of neuroblastoma. *N. Engl. J. Med.* **313**, 1111–16.

Shimizu, M., Yokota, J., Mori, N. *et al*. (1990) Introduction of normal chromosome 3p modulates the tumorigenicity of a human renal cell carcinoma cell line YCR. *Oncogene*, **5**, 185–94.

Slamon, D. J., Clark, G. M., Wong, S. G. *et al*. (1987) Human breast cancer: correlation of relapse and survival and amplification of the HER-2/neu oncogene. *Science* **235**, 177–82.

Solomon, E. (1990) Colorectal cancer genes. *Nature (London)* **343**, 412–14.

Sreekantaiah, C., Ladanyi, M., Rodriguez, E. and Chaganti, R. S. K. (1994) Chromosomal aberrations in soft tissue tumors: relevance to diagnosis, classification and molecular mechanisms. *Am. J. Pathol.* **144**, 1–4.

Tahara, E. (1995) Genetic alterations in human gastrointestinal cancers: the application to molecular diagnosis. *Cancer* **75**, 1410–17.

Testa, J. R. and Misawa, S. (1985) Chromosome changes in leukemia. In *Genetics in clinical oncology* (ed. R. S. K. Chaganti and J. L. German), pp. 159–84. Oxford University Press, New York.

Vogelstein, B., Fearon, E. R., Hamilton, S. R. *et al*. (1988) Genetic alterations during colorectal tumor development. *N. Engl. J. Med.* **319**, 525–32.

13

Analysis of DNA sequence polymorphisms and screening for point mutations

JANE PROSSER and ALAN F. WRIGHT

13.1 Introduction

Outbred organisms show high rates of genetic variation particularly within non-coding regions of the genome. Current estimates suggest that about 98% of mammalian DNA does not code for proteins. Average rates of common nucelotide variation within non-coding regions are in the range of 1 in 200 to 300 base pairs (Cooper *et al.*, 1985). The definition of a polymorphic DNA variant, as opposed to a rare variant, hinges on its frequency in the general population. Polymorphic variants are arbitrarily defined as those with a gene frequency of at least 1%; at least 2% of the population will therefore be heterozygous for the variant. Genetically useful polymorphisms have considerably higher gene frequencies and/or multiple alleles, resulting in average heterozygosity values above 50%. A high heterozygosity value is important for genetic studies since a genetic locus is only 'informative' if the individual is heterozygous at that locus, each allele 'tagging' the corresponding regions from a pair of chromosomal homologues. In this way, the segregation of chromosomal regions can be followed in the offspring by genetic linkage analysis, and loss of heterozygosity in tumour analysis. The expression 'polymorphism information content' (PIC), which is sometimes used, takes into account the total number of informative matings for linkage analysis based on allele frequencies and the frequency of different matings in a random breeding population.

What type of DNA sequence variation is present in the general population? Single base pair substitutions are the most common, but small deletions or insertions (1–5 base pairs), inversions, duplications and expansions also occur. Certain types of base pair substitution are over-represented both in polymorphic variants and in disease-causing mutations, for example, the C to T transition occurring within CpG dinucleotides. The majority of such cytosines in the human genome are methylated and through deamination of methyl-C are susceptible to change to thymine (Duncan and Miller, 1980). The common polymorphisms are generally thought to be selectively neutral, that is, to have no phenotypic effect on reproductive fitness, although this may not always be the case.

13.2 Restriction fragment length polymorphisms

The first common sequence variants found were restriction fragment length polymorphisms (RFLPs), detected in the early 1980s as a result of 'probing' the DNA from different individuals after digestion with a variety of restriction endonucleases. Variants that alter the recognition sequence of a particular restriction endonuclease result in failure of cleavage and altered restriction fragment length (Figure 13.1). The 'probe' is a short specific DNA sequence, which is generally radiolabelled by incorporation of α-^{32}P-dCTP (Feinberg and Vogelstein, 1983) and then hybridized to the filter-immobilized digests by the method of Southern (1975). The technique is exquisitely sensitive, allowing detection of 1–10 picograms of target DNA. Problems with RFLPs include (i) the need for relatively large amounts of genomic DNA for Southern analysis, (ii) the labour-intensive nature of the procedure and (iii) the relatively low heterozygosity of most such polymorphisms.

13.2.1 Protocol (RFLP analysis)

See the Southern blot section (p. 199) in Chapter 11.

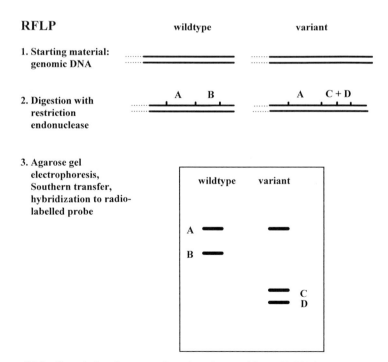

Figure 13.1. Restriction fragment length polymorphisms (RFLP).

13.3 Variable number tandem repeats

The next advance in the identification of polymorphic genetic markers for genome analysis was the discovery of variable number tandem repeat (VNTR) loci, sometimes called 'minisatellites' by analogy with the longer satellite sequences found in chromosomal C-bands. VNTRs are dispersed repetitive sequences containing specific 'core' motifs of 11–60 bp in length, tandemly repeated up to several hundred times, which tend to show high levels of length variation in the general population (Nakamura *et al.*, 1987) (Figure 13.2). They can be distinguished from RFLPs by the ability to detect the variable repeat with a variety of different restriction endonucleases, provided that they cut outside the repeat sequence. VNTRs were initially detected by probing digested genomic DNA with a radiolabelled minisatellite 'core' sequence that detected a family of related VNTR sequences at different sites scattered throughout the genome (Jeffreys, Wilson and Thein, 1985). The resultant gel patterns were highly complex and provided 'genetic fingerprints' with unique patterns for each individual. The minisatellite fingerprints were ideal for forensic studies but too complex for genetic linkage or tumour analyses. Part of

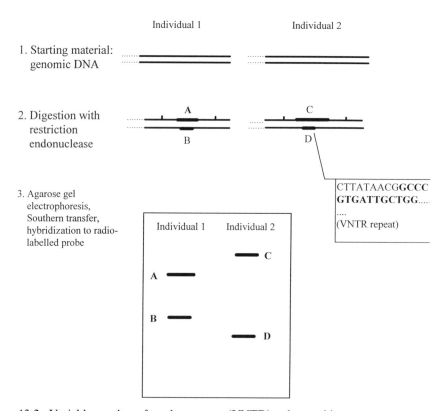

Figure 13.2. Variable number of tandem repeat (VNTR) polymorphisms.

the solution to the complexity was to clone individual VNTR sequences so that they could be analysed separately with less complex gel patterns (Nakamura *et al.*, 1987). The heterozygosity of VNTRs is extremely high (> 70%), making them ideal for genetic linkage or tumour analyses. However, one drawback became apparent when it emerged that they are not evenly distributed throughout the genome and tend to be concentrated towards the ends of chromosome arms (Royle *et al.*, 1988; Wells, Green and Reeders, 1989). Secondly, like RFLPs, they required labour-intensive methods for detection and large amounts of template DNA. This problem was partially solved by the ability to amplify certain individual VNTR repeats by polymerase chain reaction (PCR) (Jeffreys *et al.*, 1988). Their informativeness was improved by identifying sequence variants within individual repeat motifs, which provide almost unlimited allelic variety (minisatellite variant repeats, MVRs) (Jeffreys *et al.*, 1991). MVRs, then, are detected by amplification between a fixed primer on one side of the repeat and primers specific to one or other variant motif, resulting in a 'sequence ladder' of fragments that can be summarized in a unique binary code. However, amplification of VNTR alleles, some of which are 18 kilobases in size, remains difficult, so that probing is still required to detect large or poorly amplified fragments. While VNTRs remain useful and highly informative markers, they have been superseded in non-forensic applications by another class of tandemly repetitive DNA, microsatellite repeats.

13.3.1 Protocol (VNTR analysis)

The procedure is exactly as described for RFLP analysis except that the choice of enzyme is wider, since any enzyme cleaving outside the VNTR repeat is suitable. There are potential problems with the analysis of bands, which are extremely variable, necessitating categorization: usually bands are 'binned' into size categories differing by ≤ 0.5 kb. PCR amplification of VNTRs is frequently difficult because of the large repeat size but it can be achieved with some loci (Jeffreys *et al.*, 1988). Minisatellite variant repeat (MVR) analysis involves PCR amplification using a fixed primer flanking one end of the repeat and alternative primers for annealing to variant forms of the repeat motif itself (Jeffreys *et al.*, 1991). A ladder of amplified repeats at increasing distance from the fixed primer is detected after electrophoresis and Southern transfer by hybridization to a probe from the MVR locus. The sequence of alternative forms of the repeat can then be derived, although as with other genetic markers, complete haplotypes can only be derived using family data. The method is useful for forensic analysis, linkage mapping, loss of heterozygosity studies on tumours, paternity testing and exclusion of sample mix-ups, since a virtually unique 'fingerprint' is obtained from each individual. MVR analysis kits are now commercially available (Cellmark Diagnostics). For determining zygosity, a combination of four or more VNTR or MVR loci is still required, however.

13.4 Microsatellite repeats

Microsatellites are an abundant class of tandemly reiterated simple sequence repeats consisting of 1–5 bp motifs (e.g. CA, CAC, AAAT) (Weber and May, 1989; Weber, 1990). The repeat unit varies in length between individuals in the general population so that several alleles differing in the number of repeat units occur at any one locus. Microsatellites present several advantages over RFLP and VNTR markers. First, they are ubiquitously distributed and occur on average every 30 kilobases (kb) (Weber, 1990). Secondly, while they are less informative than VNTRs, they are more informative than conventional RFLPs, and several thousand markers with average heterozygosity above 50% have now been isolated from the human genome (Gyapay *et al.*, 1994). Thirdly, microsatellites are readily detected even with very small amounts of DNA by PCR amplification (Figure 13.3). The small size of the repeat block makes for very efficient amplification since most fragments are < 350 base pairs in size. High resolution maps of the human and other genomes are now available based on microsatellite markers (Gyapay *et al.*, 1994). Tri- and tetranucleotide repeats are less abundant than CA/GT repeats but are easier to type and so are increasingly used in genetic analysis (Buetow *et al.*, 1994).

13.4.1 Protocol (microsatellite analysis)

Primer design for microsatellite analysis is relatively straightforward and follows the same principles as for other types of PCR amplification. The sequences flanking the repeat are scanned for suitable matching primer pairs, usually about 20 nucleotides in length, with 40–50% GC content, absence of

Microsatellite repeat analysis

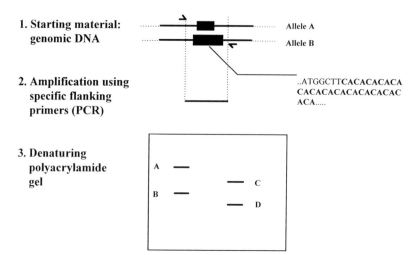

Figure 13.3. Microsatellite repeat analysis.

continuous stretches of four or more G or C residues and of self-annealing or palindromic sequences, and for the presence of a 3′ terminal G or C. Successful design remains a somewhat empirical exercise. The melting temperatures are estimated and matched to within a few degrees, if possible. It remains the case that some apparently well-designed primers work poorly and others that appear unpromising work well. Microsatellites can be analysed by PCR amplification with incorporation of $[\alpha\text{-}^{32}\text{P}]\text{dCTP}$ into both strands during amplification, although this gives complex banding patterns due to differing mobilities of the [CA]-rich and [GT]-rich strands and the presence of progressively weaker 'stutter' bands arising from polymerase slippage that are 2, 4 or 6 nucelotides shorter than the true allelic peak (Murray, Monchawin and England, 1993). End-labelling of one primer with T4 polynucleotide kinase gives simpler banding patterns and this can be rapidly and conveniently carried out as shown below.

13.4.2 Analysis [γ-^{32}P] ATP end-labelled primer

The end-labelling reaction is carried out using the forward reaction of bacterio-phage T4 polynucleotide kinase, which transfers the [γ-^{32}P]phosphate of ATP to the 5′-OH terminus of dephosphorylated primer DNA (Hughes, 1993). The labelling protocol is as follows:

Mix together 2.5 μl 10 × kinase buffer (0.5 M Tris/HCl pH 7.4, 0.1 M $MgCl_2$, 50 mM DTT), 2.5 μl (6 pmol) primer DNA (diluted from the stock solution to 2.5 μM), 1.5 μl T4 polynucelotide kinase (10 U/μl), 1 μl (10 μCi) [γ-^{32}P]ATP (> 5000 Ci/mmol). Incubate for 30 min at 37 °C. Stop the reaction by heating to 65 °C for 5 min to inactivate the enzyme. Remove unincorporated label using a Mermaid kit (Bio101) according to the manufacturer's instruc-tions, with adsorption of the primer onto 5 μl of glass fog and eluting with 2 × 15 μl of water. Add 10 000–20 000 c.p.m. to a standard microsatellite PCR reaction. For example, template DNA (25 to 75 ng), 200 μM dNTPs, 10 mM Tris HCl, pH 9 (at 25 °C), 1.0 mM $MgCl_2$, 50 mM KCl, 0.1% Triton X-100, 250 nM unlabelled primer, 1.5 units Taq polymerase (Promega) in a 25 μl volume, overlaid with mineral oil. Suitable amplification conditions are de-pendent on the primers but a generic protocol has been recommended by Hughes (1993) as follows:

Denaturation for 3 min at 94 °C, amplification for 30–35 cycles of 1 min at 94 °C, 2 min annealing at 50 °C and 2 min extension at 72 °C, with a final extension of 5 min at 72 °C. After amplification, 100 μl of chloroform is added and the products recovered and diluted with 3 volumes of loading buffer (95% (v/v) formamide, 1 × TBE, 0.01% (w/v) xylene cyanol, (0.01% Bromophenol Blue). Samples are then denatured for 3 min at 95 °C and snap-cooled in a water–ice bath. Electrophoresis is carried out after loading 6–8 μl on a denaturing 6% polyacrylamide sequencing gel with sufficient power to maintain the gel temperature in the range 45 °C–55 °C. The gels are then removed from the plates either onto used X-ray film, covered in clingfilm and placed in a cassette with X-ray film directly, or first dried down on a gel dryer at 80 °C for 60 min after transfer to a sheet of Whatman filter paper. The critical variables for microsatellite amplification are the same as those for other types of PCR, namely the annealing temperature, magnesium concentration and cycling conditions. In some cases, increasing the primer concentration may help to amplify fragments when the primers are less than optimal.

13.4.3 Analysis of fluorescence-labelled microsatellites

Non-radioactive labelling methods are increasingly being used for microsatellite analysis. One method is to use fluorescent labelled primers and analysis on automated sequencing machines (Read *et al.*, 1994; Mansfield *et al.*, 1995). This has the advantages of high resolution, quantification of peaks and software that has been developed for accurate sizing of alleles and for direct input to linkage analysis programmes. Alternatively, a chemiluminescence-labelled [CA] repeat oligonucleotide can be used to sequentially probe microsatellites separated on multiplex gels (i.e. the products of several different amplification reactions run simultaneously) after transfer to filters (Weissenbach *et al.*, 1992).

Single fluorescent dye-labelled primers can be used in amplification reactions and analysed using an automated sequencing apparatus (Mansfield *et al.*, 1995). Fluorescein labelling is carried out during primer synthesis by a phosphoramidite addition reaction, which is relatively efficient (90–95%). To achieve the highest sensitivity, removal by HPLC purification of shorter oligonucleotide products, which are unlabelled but compete with the labelled product, is desirable. After deprotection, each newly synthesized primer in ammonium hydroxide solution is passed through a gel filtration column (e.g. NAP-10™, Pharmacia) and eluted in 10 mM sodium phosphate, pH 6.8. Typical reaction mixtures consist of diluted template DNA (25 to 75 ng), 200 μM dNTPs, 10 mM Tris HCl pH 9, 1.0 mM $MgCl_2$, 50 mM KCl, 0.1% Triton X-100, 400 nM primers (one of which is fluorescein labelled), 1.5 units Taq polymerase in a 25 μl volume, overlaid with mineral oil. For large-scale microsatellite analysis projects, reaction volumes of 10 μl or less are increasingly used.

Amplification reactions are set up in 0.5 ml microcentrifuge tubes or in 96-well plates, which are placed on thermal cyclers. PCR conditions suitable for analysis of the majority of the Genethon microsatellites (Weissenbach *et al.*, 1992; Gyapay *et al.*, 1994) are as follows: 4 min at 94 °C (1 cycle); 1 min at 94 °C, 2 min at 55 °C, 2.5 min at 72 °C (26 cycles); 7.5 min at 72 °C (1 cycle).

After amplification, samples are cooled to 4 °C and 'stop' solution (95% formamide, 0.8% Dextran Blue and 25 mM EDTA, pH 9) is added. Samples can be stored at −20 °C until ready for analysis, at which time internal size standards are added and the samples heat denatured at 95 °C for 3 min, then rapidly cooled on ice prior to loading onto the gels. Internal standards can be purchased or prepared by amplification from single stranded (+)DNA of the M13mp18 vector using a variety of 20- to 24-mer primers and the above reaction conditions. Electrophoresis is carried out using an automated single-dye laser fluorescence sequencer. Gels are made up with Hydrolink (Long Ranger™, AT Biochem Ltd) or 6% acrylamide/bis-acrylamide (29:1) in $0.6 \times$ TBE (0.06 M Tris base, 0.05 M boric acid, 0.6 mM EDTA) containing 7 M urea and run in the same buffer at constant power (55 W) and constant temperature (50 °C). Up to 10 sets of reaction products containing internal standards can be loaded per lane without prior concentration. Each gel can be reloaded at least once.

An alternative fluorescence-based system for microsatellite analysis uses a three-coloured dye labelling system (Read *et al.*, 1994). Again, each primer is synthesized and labelled with fluorescent dyes (e.g. HEX, 6-FAM, JOE) and internal standards labelled with a fourth dye (ROX) are added. The advantage of this system is the ability to 'multiplex' to a greater extent than with single dye systems – i.e. run many different sets of markers on the same gel, the different emission spectrum of each laser-excited dye differentiating even overlapping markers. The main disadvantages are the much slower run time due to the need to re-scan the gel at different wavelengths and the lack of temperature-controlled gels. Each system has software packages suitable for automated sizing of alleles, removal of non-allelic peaks and formatting the data for further analysis.

13.5 Unstable trinucleotide repeats

Unstable trinucleotide repeats are a relatively new class of molecular aberration associated with certain chromosomal 'fragile sites' and a number of degenerative neurological diseases. The first to be discovered was a [CGG] trinucleotide repeat at the FRAXA locus in fragile-X syndrome patients, expanded to 200–1000 repeats compared with < 50 in the normal population (Fu *et al.*, 1991; Verkerk *et al.*, 1991). A folate-sensitive fragile site or cytogenetically visible constriction in Xq27.3 is associated with the full expansion and there is increased methylation of a CpG island, located 250 base pairs upstream and extending into the CGG repeat, which appears to regulate expression of the *FMR-1* gene, resulting in diminished or absent expression of *FMR-1* product. In contrast, there are no clear phenotypic effects associated with the [CGG/CCG] expansions associated with the FRA16A and FRAXF fragile sites (Parrish *et al.*, 1994; Nancarrow *et al.*, 1994). Over 100 chromosomal fragile sites have been described in man but only two (FRAXA, FRAXE) have been associated with disease.

The second form of unstable trinucleotide repeat is associated with expansion of polyglutamine tracts at [CAG] or [CTG] codons within expressed regions of genes (Willems, 1994). In these cases, there are never associated fragile sites, the expansions are less extreme (< 150) and, with one exception, they occur within the open reading frame of a gene. Six late-onset neurological diseases have been associated with this type of expansion. The exception is myotonic dystrophy (DM), which has very large expansions (200–4000, cf. $\leqslant 50$ in normals) of [CTG/CAG] tracts within the 3' untranslated region of the *DM* gene. The latter appears to cause reduced expression of the gene, while the expansion of polyglutamine tracts may cause deleterious gain-of-function effects. Another feature of both types of trinucleotide expansion is the tendency for disease expression to increase with increased expansion. In some cases this is detectable clinically as 'anticipation' – the tendency for the disease to worsen in successive generations – and in others as the appearance of severe juvenile or congenital forms of the disease. Disorders associated with untranslated extreme repeat expansions show somatic instability and tissue mosaicism, which is not evident with the expressed expansions. This complicates their analysis.

13.5.1 Protocol (amplification of trinucleotide repeats)

The smaller expansions, as in expressed polyglutamine tracts, are now detectable in a relatively straightforward manner by PCR amplification of the repeat segment and visualization using either fluorescent labelled products on automated sequencers or radioactive labelling of one or both strands and autoradiography (Orr *et al.*, 1993; Barron *et al.*, 1994). The main problems with this type of expansion relate to (i) the relatively narrow cut-off between the normal and pathological range of repeat sizes, necessitating accurate sizing of PCR products; (ii) non-allelic bands generated either during amplification due to strand slippage or as a result of the differing mobility of the two strands; (iii) the less efficient amplification of larger alleles compared with small ones. In the case of the larger expansions in untranslated regions of genes such as DM and FRAXA, the major problems are the somatic instability of the larger expansions, which result in somatic mosaicism, and the technical difficulty of amplifying large expansions. In many cases, instability is so extreme that only a smear of fragments can be detected on probing PCR products or genomic digests. In this situation, radioactive labelling during amplification or probing of PCR products after Southern transfer is probably essential. This may be helped by the recent availability of conditions for amplication of extended fragments $\geqslant 20$ kilobases in size (Barnes, 1994).

A typical protocol for detecting CGG expansions is that of Fu *et al.* (1991), in which 100 ng of genomic DNA is mixed with 3 pmol of each primer flanking the repeat in a total volume of 15 μl, containing PCR buffer, 2 mM $MgCl_2$, 200 μM dATP, dTTP, dCTP and 150 μM 7-deaza-dGTP plus 50 μM dGTP, 10% (v/v) DMSO, 2–4 μCi[^{32}P-α]dCTP, 0.75 Units Taq polymerase. The reaction mix is heated to 95 °C for 10 min and given 25 cycles of annealing (65 °C, 1 min), extension (72 °C, 2 min) and denaturation (95 °C, 1.5 min). The inclusion of 7-deaza-GTP and 10% DMSO and long denaturation steps are necessary to reduce secondary structure formation and because of the stability of high GC content repeats. After addition of formamide loading dye, the products are heat denatured (95 °C, 2 min) and loaded for electrophoresis through a 5% denaturing DNA sequencing gel, followed by autoradiography.

As a result of somatic mosaicism and extreme instability of the larger expansions, detection of CTG repeats in the 3'-untranslated region of the DM gene is best carried out by amplification followed by probing of the PCR products after Southern transfer, using a radiolabelled probe including both the CTG repeat and flanking unique sequence from the DM locus (Monckton *et al.*, 1995).

13.6 Allele loss and instability in tumours

Loss of chromosomal DNA on at least one chromosomal homologue is a frequent occurrence in tumour tissue (Vogelstein *et al.*, 1989). Occasionally both alleles may be lost, the resultant cells undergoing selection for increased cell division if there is concomitant loss of a tumour suppressor gene. Rare events of this sort may undergo strong cellular selection within a dysplastic or neoplastic tissue. The classic paradigm of inherited retinoblastoma, for which the 'two-hit hypothesis' of Knudson (1971) was proposed, is associated with loss of function within the RB1 tumour suppressor gene as a result of germline mutation, with a later somatic event (deletion, chromosome loss, etc.) resulting in loss of the remaining normal gene copy. In the majority of non-heritable tumours, both events leading to allele loss are somatic. Allele losses at different sites are characteristic of different tumours and even of the same tumour at different stages of progression. There appear to be a wide variety of genes capable of altering the growth or metastatic potential of different tumours as a result of loss of function within one or both copies. In some cases, chromosome loss may be a sporadic event within a rapidly dividing tissue conferring neither an increase nor a decrease in growth rate or invasiveness. Background loss of heterozygosity is found in ≤ 20% of breast tumours, for example, and is probably of no significance but there are at least 12 sites that show excessive loss of heterozygosity in these tumours.

13.6.1 Protocol (loss of heterozygosity)

The procedures for detecting loss of heterozygosity in tumour samples are essentially identical to those for RFLP, VNTR or microsatellite typing of genomic DNA. Samples of tumour and non-tumour DNA from the same individual are analysed in order to detect loss of an allele that is present in the non-tumour DNA. Problems arise if the 'tumour' sample also contains non-tumour cells that give a faint background allele, partially obscuring its loss in the tumour. The amount of DNA available from small tumour samples or extracted from paraffin-embedded sections (Jen *et al.*, 1994) may be very limited so that microsatellite analysis is now the method of choice, since amplification requires only a few nanograms of DNA and is very efficient with alleles that are less than about 200 nucleotides in length.

13.6.2 Allele instability

Another mechanism that is common in at least certain tumours leads to a generalized allele instability and hypermutability (Aaltonen *et al.*, 1993, 1994). Mutations within the human homologues of bacterial DNA mismatch repair genes, *mutS* and *mutL*, have been identified in hereditary non-polyposis colon cancer (HNPCC), leading to length variation within tumour microsatellite repeat sequences (see below) scattered throughout the genome (Bodmer, Bishop and Karran, 1994). The tumour cells differ in repeat length from normal somatic cells from the same patients. This type of tumour phenotype has been designated RER$^+$ (Replication ERrors), which is indicative of a potential germline mismatch repair defect (Modrich, 1994, 1995). It is relatively easily identified by demonstrating mutation in at least two microsatellite loci (i.e. the appearance of new alleles in the tumour). Several *mutL* homologues appear to be present in man, in which germline mutations may be associated with particular forms of cancer or somatic mutations with progression to increased malignancy. Sporadic colon cancer cells also show genetic changes within mismatch repair genes (Thibodeau *et al.*, 1993; Ionov *et al.*, 1993). Complementation tests and *in vitro* experiments on post-replication repair suggest that at least seven different genes participate in this process and are therefore potential sites of allele loss and contributors to hypermutability in other cancers (Modrich, 1994, 1995).

13.6.3 Protocol (microsatellite instability)

Extraction of genomic DNA is carried out by standard methods (Sambrook, Fritsch and Mariatis, 1989) unless archived material from paraffin-embedded blocks is required (Jen *et al.*, 1994). Amplification can be carried out using extremely small amounts of DNA from single cells upwards. The major problem with amplification of very small amounts of DNA (< 1 nanogram) is cross-contamination with air or dust-borne DNA from cells or, worse, amplified or cloned material. Several ways of reducing cross-contamination have been proposed. These range from simple precautions advisable in all PCR work, such as setting up reactions away from areas where cloned or amplified material is being analysed, to the use of ultraviolet irradiation, deoxyribonuclease I or uracil-N-glycosylase to remove potential contaminants from reaction mixes prior to addition of DNA template (Kwok and Higuchi, 1989; Furrer *et al.*, 1990). In the case of archived material, it is safest to choose loci with alleles less than about 180 nucleotides in length since these amplify more reliably in DNA that may be of poor quality. The protocols are otherwise identical to those discussed above.

13.7 Amplification

Genomic instability is characteristic of malignant cells and is manifest in a number of processes: not only the deletion of DNA sequences and the altered repeat length of microsatellites, as already discussed, but also amplification of sequences, chromosomal translocations and inversions and increased numbers of small changes in the DNA. Amplification is frequently associated with oncogenes (often cytogenetically visible as double minute chromosomes or heterogeneously staining regions) while deletion often involves tumour suppressor genes (usually detected by loss-of-heterozygosity studies). Knowledge of the specific, and sometimes characteristic, aberrations that arise as a consequence of instability is important to understanding the biological basis of cancer as a whole and of certain cancers in particular.

One of the earliest ways of detecting genetic abnormalities in cancer, as well as in other syndromes, was by cytogenetic analysis of metaphase chromosomes and the karyotypically identified Philadelphia chromosome was the first specific chromosomal abnormality associated with a particular malignancy. To date, many specific cytogenetic rearrangements have been characterized in haemato- poietic tumours, and a number in solid tumours (Rabbitts, 1994). These frequently involve translocations and sometimes inversions, and cloning of the breakpoint regions has led to the elucidation of the specific genetic rearrange- ments involved. Sequences involved in amplification and deletion have been identified in different ways: amplified sequences through physically isolating the DNA for characterization, often hybridizing it back to tumour metaphase chromosomes; deleted sequences by loss of heterozygosity studies of one sort or another. Fine mapping of specific sequences involved in gains and losses is possible with FISH analysis, covered elsewhere in the volume. The methods for determining amplification and deletion have essentially involved site-by-site analysis and required good metaphase preparations. A recent technique, called comparative genome hybridization (CGH), permits scanning of the entire genome for sequence copy number changes. CGH, furthermore, does not require metaphase chromosome preparations from the tissue being analysed and is, consequently, particularly useful in the analysis of solid tumours where good metaphase preparations are difficult to obtain (Kallioniemi *et al.*, 1992).

Labelled genomic test DNA prepared from tumour specimens, or any other sample to be tested, is mixed with differently labelled control DNA prepared from cells with a normal chromosome complement. By *in situ* hybridization the mixed probe is hybridized to normal metaphase spreads. Hybridized test and control sequences are detected via different fluorochromes and the relative intensities of the two dyes reflects the copy number in the test genome compared with the normal genome. A visual inspection of the fluorescent images determines gains and losses. A protocol is offered for this technique. A comprehensive analysis of CGH data can be achieved through digitized analysis of the results, leading to primary data in the form of profiles of the ratio of the intensities of the two fluorochromes along all target metaphase chromosomes. A ratio of 0.5 would represent a monosomy, 1.0 a disomy, 1.5 a trisomy, etc. This 'copy-number-karyotype' is a quantitative display of the variation in DNA copy number throughout the test genome. It requires sophisticated computer facilities and analysis and is not accessible to most researchers (Kallioniemi *et al.*, 1994). Pinkel (1994) commented on the results obtained with the two techniques, FISH using probes from microdissected homogeneously staining regions (hsr) in breast tumours and CGH in breast tumours, and concluded that the two methods are complementary. Each identified regions missed by the other: microFISH identified regions of low copy number missed by CGH, and CGH identified amplifications of sequences not contained in the dissected hsr.

13.7.1 Protocol (CGH protocol using directly labelled nucleotides)

1 μg aliquots of test (tumour) and control (normal) DNA are labelled by nick-translation with Fluorogreen (Fluorescein-11-dUTP, Amersham) and Fluorored (Rhodamine-4-dUTP) respectively as described elsewhere in this volume. The size of the probe DNA is checked by electrophoresis on a 1% agarose denaturation gel. The optimal size of the probe should be between 700 bp and 2000 bp and it is important that the size of both test and normal DNA should be similar.

Good quality slide preparations rich in normal metaphase chromosome spreads are treated with RNAase A (100 μg/ml in 2 × SSC) for 1 h at 37 °C and dehydrated by washing for 2 min in baths of 70%, 90% and 100% (v/v) ethanol. The preparations are then denatured at 70 °C for 3 min in 70% (v/v) formamide/2 × SSC, quenched in cold 70% alcohol for 3 min and dehydrated through 90% and 100% ethanol for 2 min each.

200 ng of Fluorogreen-labelled test DNA is added to 200 ng of Fluorored-labelled control DNA and 10 μl (1 μg/μl) of Cot-1-DNA. Double the volume of ethanol is then added and the mixture is vacuum centrifuged to concentrate the DNA into a pellet. 10 μl of hybridization mix (50% (v/v) deionized formamide, 10% (w/v) dextran sulphate, 2 × SSC, 1% (v/v) Tween 20, distilled water to make up the volume) is added and the DNA is allowed to dissolve at room temperature for 1 h. The probes and competitor DNA are then denatured by incubation at 70 °C in a water bath for 5 min immediately before hybridization to the slide preparation.

Hybridization mix (10 μl) containing denatured DNA is placed on a pre-warmed 22 × 22 mm coverslip. The metaphase slide is carefully lowered on top of this and the edges of the coverslip are sealed with rubber cement. The slide is then placed on a metal tray floating in a waterbath at 37 °C overnight. After overnight hybridization the rubber cement is removed and the slides are washed four times in 50% (v/v) formamide/2 × SSC at 45 °C for 3 min, followed by four washes in 2 × SSC at 45 °C and four washes in 0.1 × SSC at 60 °C. The cover slips will fall off in the first wash. The slides are finally transferred to 4 × SSC/0.1% Tween 20 at room temperature and then mounted in 40 μl Vectorshield (Vector Labs) containing 0.3 μM DAPI (4',6-diamidino-2-phenylindole, from Sigma) as a chromosome counterstain. The cover slips are sealed in place with rubber solution to prevent evaporation of the mountant.

The preparations are viewed with a fluorescence microscope equipped with filter sets suitable for dual excitation/emission of Fluorogreen (excitation max 490 nm, emission 520 nm) and Fluorored (excitation max 545 nm, emission 575 nm). Regions of the genome where the copy number is similar will show approximately equal red and green emission (this will appear orange/yellow), whereas regions of amplification will show an enhanced green signal, depending on the copy number. It may also be possible to detect large deletions by the reduction in green signal, but an imaging system with software capable of performing a quantitative comparison of fluorescence ratios will considerably increase the sensitivity of the analysis. Several commercial fluorescence imaging systems now offer CGH analysis.

13.8 Screening for point mutations

Detecting point mutations is important in human genetics in two main areas: pinpointing the involvement of a particular gene in a known genetic disorder and determining the mutation status of patients and potential carriers when a gene implicated in disease has been identified. In the past decade many techniques have been developed; no single one is unquestionably the best. For this reason, it is important to choose a method suited to a particular circumstance. Practical considerations, like availability of resources and expertise, are best decided by the individual user. The concerns of this Chapter are the intrinsic merits of the techniques used in screening for previously undetected changes, and their suitability for particular situations.

The most useful scanning methods for unknown mutations, and the ones we will describe, include denaturing gradient gel electrophoresis (DGGE) (Myers *et al.*, 1985*b*; Fodde and Losekoot, 1994), chemical cleavage of mismatch (CCM), also known as HOT (for hydroxylamine and osmium tetroxide used in the modification of the mismatch) (Cotton, Rodrigues and Campbell, 1988; Montandon *et al.*, 1989; Condie *et al.*, 1993), single-strand conformation polymorphism (SSCP) (Orita *et al.*, 1989; Glavac and Dean, 1993; Ravnik-Glavac, Glavac and Dean, 1994) together with the related technique of heteroduplex analysis (Keen *et al.*, 1991; White *et al.*, 1992), and most recently, the protein truncation test (PTT) (Powell *et al.*, 1993; Roest *et al.*, 1993; Prosser *et al.*, 1994). Techniques much less frequently used in this context, and not to be discussed here, are RNAase protection assay (Myers, Larin and Maniatis, 1985*a*) as well as one laboratory's carbodiimide alternative to HOT (Ganguly and Prockop, 1990). Two promising new methods, still in development, involve identification of mismatched heteroduplexes with bacterial mismatch repair proteins prior to nuclease treatment (Ellis *et al.*, 1994) or gel retardation assays (Lishanski, Ostrander and Rine, 1994) and the identification and cleavage of mismatch by bacteriophage T4 and T1 endonucleases (resolvases) (Mashal, Koontz and Sklar, 1995; Youil, Kemper and Cotton, 1995). There are several reasonably recent reviews on mutation detection (Cotton, 1993; Grompe, 1993; Prosser, 1993).

The basis of all the scanning techniques to be described is PCR amplification of the starting material and protocols for this are given elsewhere in the volume. Furthermore, inherent in the aim to identify precisely all mutations is the need subsequently to sequence the mutant templates in order to determine the exact nucleotide(s) involved and to verify the scanning results. Two protocols are offered.

13.8.1 *Denaturing gradient gel electrophoresis (DGGE)*

DGGE depends on the electrophoretic differences of partially denatured DNA molecules. The technique is based on the differences in the rate of DNA melting between molecules that differ by as little as a single nucleotide (between homoduplex molecules) or a single mismatch (in heteroduplex molecules). When DNA denatures, it opens over domains of nucleotides, rather than one nucleotide at a time, so that fragments change abruptly from being double-stranded to being branch-shaped or bubble-shaped. When the first domain melts, the rate of migration of the fragment in the gel slows abruptly (Figure 13.4). A single base difference or mismatch in this domain alters the rate of change. Base changes in subsequent, more-slowly-melting domains may not be detected. DNA molecules contain different melting domains, and, ideally for DGGE, will contain a low melting domain at one end of the molecule and a high melting domain at the other. In order to achieve the most suitable melting profile for a PCR amplified fragment, a computer program is used to best place PCR primers from theoretically melted DNA regions of interest. Furthermore, in the actual PCR amplification, one primer incorporates what is called a 'GC-clamp' of about 40 bases of GC rich sequence to assure that each amplified fragment has a very high-melting domain at one end of the molecule.

The success of DGGE depends on the careful choice of denaturing conditions and these can be selected empirically (requiring special apparatus) or theoretically (by computer algorithm). For the best application of DGGE it is important to use the available software (MELT87 and SQHTX available upon request from L. S. Lerman, see below) to position the potential PCR fragments

DGGE

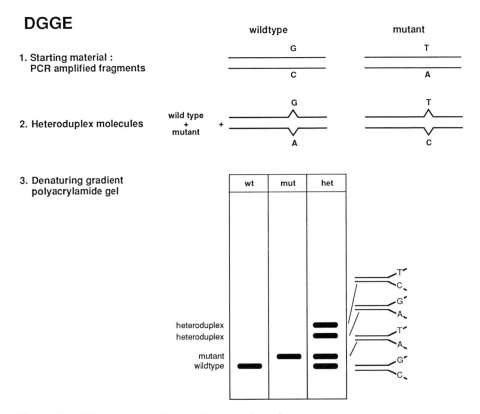

1. Starting material :
   PCR amplified fragments

2. Heteroduplex molecules

3. Denaturing gradient
   polyacrylamide gel

Figure 13.4. Denaturing gradient gel electrophoresis.

along the gene, to check the shape of the melted molecule, to determine the optimal end for adding the GC clamp, and to identify the particular conditions for partial melting. Finally, the melting of each fragment needs to be tested empirically, and special apparatus is required. This is commercially available but can be assembled in laboratory workshops. Once this is done, the screening of samples is easy. And it is for the routine screening of fragments not larger than 400–500 bp that DGGE is most effective. With carefully chosen conditions, efficiency reaches 100% and samples in which the mutant DNA constitutes only a proportion of the total (down to 10%, Hovig *et al.*, 1992) can be effectively analysed.

13.8.1.1 Protocol (DGGE analysis)

The DGGE technique uses ethidium bromide to stain acrylamide gradient gels. Computer programs used to model melting behaviour may be obtained from: Dr. L. S. Lerman at Massachusetts Institute of Technology, 77 Massachusetts Avenue, Cambridge, Massachusetts 02139-4307, USA. They run on an IBM PC or compatible machine with source code and manual included.

Primers are placed according to the 50% melting pattern predicted by MELT87 for the target DNA. The melting contour should be in a stepwise fashion, increasing from the low-melting point end. A GC-clamp may have to be added if the melt pattern is non-stepwise. The GC-rich end primer should be at least 30 bp upstream to the sequence of interest as mismatches in this area are less easy to determine. (Melting behaviour may also be determined empirically by running a transverse gradient gel with 1 well the width of the gel, having previously selected suitable primers.) Primers are synthesized, one, optionally, with a 5′ 40 bp GC-clamp. The sequence of the clamp is as follows: CGC CCG CCG CGC CCC GCG CCC GTC CCG CCG CCC CCG CCC C. A further computer model (SQHTX) may be used to optimize the electro-phoresis run time based on the predicted band separation of a wild type homoduplex and a heteroduplex. After a 30 cycle PCR, products are heat-denatured at 94 °C for 2 min and allowed to cool slowly on the thermal cycler block for 30 min to maximize double-stranded product and hence heteroduplex formation. Samples (15 μl) of the PCR are then mixed 10:1 with Orange G in 50% sucrose and subjected to electrophoresis. Gels are cast in a 15 × 15 × 0.1 cm Hoeffer SE600 vertical apparatus. A Hoeffer SG50 gradient maker attached to a peristaltic pump is used to pour the gels over a period of a few minutes to ensure gradient linearity. The gradient should be poured with the most dense at the bottom. The gel is immersed in 1 × TAE buffer in a custom-built tank with a Grant recirculating immersion heater. A 30% gradient is used (i.e. 15% either side of the 50% melt temperature). Gels are run at 60 V, 60 °C with buffer recirculation between bottom and top tanks for ~ 18 h (exact times predicted by computer model). Gels are then stained with ethidium bromide and photographed under UV light. Mismatches should show up as four bands corresponding to the two homo and two heteroduplexes. Occasionally only three bands are visible when both homoduplexes co-migrate or two bands when both heteroduplexes and both homoduplexes co-migrate.

> 100% denaturant = 7 M urea, 40% (v/v) deionized formamide, 8% (w/v) acrylamide in 1 × TAE
> 0% denaturant = 8% (w/v) acrylamide in 1 × TAE.

13.8.2 Chemical cleavage of mismatch (CCM) or HOT (for hydroxylamine and osmium tetroxide)

CCM or HOT is labour intensive but efficient (around 100%) and has the added advantage of screening large fragments, certainly up to 1 kb and possibly larger. Furthermore, the detected alteration is precisely located relative to one or other end of the molecule scanned. It, too, is capable of detecting mutations in samples in which the mutant DNA constitutes substantially less than 100% (at best in the region of 5–10%) (J. Prosser, unpublished). The technique is based on the principles of Maxam and Gilbert sequencing and involves the chemical cleavage of modified mismatched base pairs in heteroduplex DNA.

Two modifying chemicals (hydroxylamine and osmium tetroxide) react more strongly with mismatched, or unmatched, Cs and Ts respectively, than with these nucleotides when they are correctly base-paired. Modification is followed by piperidine cleavage and the products are analysed on denaturing acrylamide gels (Figure 13.5). Being able to detect all mismatched Cs and Ts in a

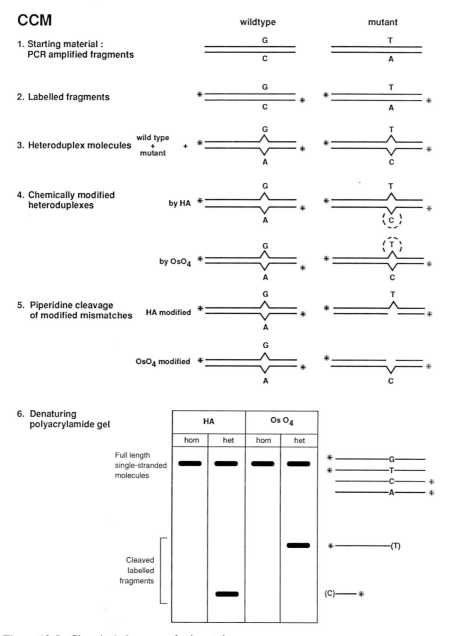

Figure 13.5. Chemical cleavage of mismatch.

heteroduplex molecule (with only one allele labelled) should ensure comprehensive mutation detection, but because some GT mismatches are missed, it is
essential to label both DNA alleles for 100% detection. HOT would be first
choice in analysing large RT-PCR fragments of transcribed genes not especially
prone to protein truncating mutations (see below). It would also be first choice
in the exhaustive analysis of any gene in which implicating mutations are
sought in order to correlate genotype and phenotype. Because it is labour
intensive and uses both radioactivity and hazardous chemicals, it would not be
first choice for routine screening.

13.8.2.1 Protocol (CCM or HOT analysis)

1. Prepare the DNA samples: Both target and wildtype DNAs are used for
   PCR. The DNA to be kinase-labelled is run in low melting agarose,
   excised and Genecleaned, prior to labelling. ($10 \times$ kinase buffer: 0.7 M
   Tris HCl, pH 7.6, 0.1 M $MgCl_2$, 50 mM DTT.) Kinase reaction: (DNA
   excised from low melting agarose and Genecleaned) 100–200 ng DNA,
   2 μl (20 μCi)γATP, 2 μl $10 \times$ buffer, distilled water to 20 μl. 37 °C for
   30 min. At finish, add 25 μl water, 5 μl precipitating salts (3 M sodium
   acetate, pH 5), 1.5 μl 1 mg/ml glycogen and 150 μl cold ethanol. Precipitate, wash in 70% ethanol, briefly vacuum dry, and take up in several μl
   of TE (eg. 10–20 μl). The non-labelled DNA to be heteroduplexed is
   Genecleaned without the low melting agarose step. Our PCR reactions
   are normally 50 to 100 μl. The Genecleaned product is eluted in a 30 μl
   volume.
2. Making the heteroduplex: The actual reactions can be set up in a number
   of ways: (i) by labelling wildtype DNA and heteroduplexing to cold
   mutant DNA; (ii) by doing the reverse, (iii) by mixing the two DNAs 1:1,
   labelling the mix and heteroduplexing to the mix. We normally do (i) and
   (ii). 100–300 ng cold DNA is mixed with 10–30 ng ^{32}P-labelled DNA, 1 μl
   of $10 \times$ salts (3 M NaCl, 1 M Tris, pH 8) and distilled water to a volume
   of 10 μl in a 0.5 ml Eppendorf tube. This is overlayered with paraffin oil,
   heated to 100 °C for 5 min, then placed at 65 °C overnight.
       After heteroduplexing, the majority of oil is pipetted off and the
   product precipitated. The precipitation steps that follow appear several
   times in the protocol (essentially 200 μl of $1 \times$ precipitating salts, added
   carrier, or not, as stated, cold ethanol precipitation, 70% cold ethanol
   wash, brief vacuum drying and resuspension). To the heteroduplex is
   added 200 μl of 0.3 M sodium acetate, 0.1 mM EDTA and 50 μl chloroform. After vortexing and centrifuging, the top layer is removed to a
   1.5 ml Eppendorf tube and 1.5 μl 10 mg/ml mussel glycogen (carrier)
   added before 750 μl cold ethanol. The sample is placed in a dry ice/
   methanol bath for 5–15 min and centrifuged 5–10 min. The pellet is

washed twice in cold 70% ethanol dried briefly (1 min in vacuum) and taken up in 13 μl distilled water. The sample is divided into two 1.5 ml Eppendorf tubes, 7 μl for HA modification, and 6 μl for OsO_4 modification.

3. HA (hydroxylamine) modification: Make a fresh solution each time (0.94 g HA + 3.36 ml water. Vortex and add 1.02 ml diethylamine). Add 20 μl of this to the 7 μl of sample, mix and leave at 37 °C for 1 h.

4. OsO_4 modification: To the 6 μl of sample add 2.5 μl of 10 × OsO_4 buffer (100 mM Tris pH 7.7, 10 mM EDTA, 15% (v/v) pyridine). Add 15 μl freshly diluted OsO_4 (stock solution is 4%; use a 1/5 dilution to 0.8%). Mix, but do not centrifuge. Incubate at 37 °C for 5 min. After chemical modification the samples are precipitated as previously described, with 5 μl dextran being added with the 1 × precipitating salts.

5. Piperidine cleavage: Use freshly diluted piperidine (1/10 dilution to make 1 M solution). Add 50 μl to each tube. Vortex 1 min. Quickly spin and incubate 90 °C for 30 min. At the end of 30 min, precipitate in the usual way, adding no further carrier. Resuspend the pellet in 5 μl water and 2 μl formamide denaturing dye. (It is possible to do the initial PCR reaction with biotinylated primers and use Dynabead recovery to replace the multitude of precipitation steps).

6. Before loading the samples, heat to 90 °C for 2 min. Load on a 6% sequencing gel and run with 1 × TBE buffer until Bromophenol Blue reaches the bottom (about 1.5 to 2 h at 30–40 watts). The gel is fixed, dried and autoradiographed in the usual way. When long fragments are analysed, double loading of the sample is necessary, usually a 6 h and 2 h run. A BioRad sequencing apparatus is used.

Reagents

Geneclean and Mermaid kits from Stratech Scientific Ltd, 61 Dudley St, Luton LU2 0NP, England. From Aldrich: hydroxylamine Cat. No. 25,558.0; OsO_4, Cat. No. 25,175.0; Piperidine, Cat. No. 10,409.4; Pyridine, Cat. No. 27,040.7. OsO_4 is also supplied by Johnson Matthey, Material Technology, Orchard Road, Royston, Herts., SG8 5HE, England and is marginally preferable. The HA is weighed into 10 ml screw-capped conical tubes and stored at 4 °C until needed. 10 × OsO_4 buffer is aliquoted and stored frozen until use.

Results

The conditions of reaction favour the modification of mismatched Cs and Ts. Nevertheless, modification to a lesser degree occurs to correctly matched Cs and Ts. It is therefore necessary to run a normal track (labelled wildtype heteroduplexed to cold wildtype) so that bands due to mutations are detected against a normal, non-mutant pattern. For correct sizing, any sequencing reaction of known sequence can be run at the same time.

242 J. Prosser and A. Wright

13.8.3 Single-strand conformation analysis (SSCP)

SSCP is certainly the easiest scanning technique, but unless used under two or three different conditions, possibly the least efficient. The technique depends on the variation in electrophoretic mobility in non-denaturing acrylamide gels of short single-stranded DNA or RNA molecules whose sequences differ. The secondary structure of single-stranded nucleic acids varies according to sequence, and a single base change can result in an altered structure and altered mobility in the gel (Figure 13.6). SSCP is most effective on short fragments (around 150–200 bp in size) and should be used with more than one set of electrophoresing conditions to reach greater than 90% efficiency. There is certainly an upper size limit of fragments amenable to this analysis (at 400 bp, efficiency is reduced to 60–70%, at 600 bp to 3%!) but a lower size limit has also been mentioned (70–80% at around 100 bp) (Sheffield et al., 1993). SSCP is useful for a quick scan of any DNA sequence. If a mutation is found, it is very pleasing, if not, different SSCP conditions or a more efficient, usually more cumbersome, method can be tried. Little mention is made in the literature of what proportion of total sample must be mutant to be detectable

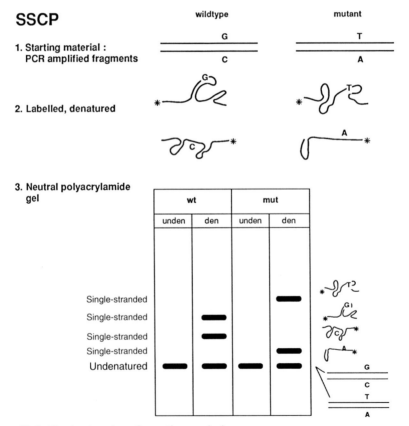

Figure 13.6. Single-strand conformation analysis.

by SSCP. Our experience with a multicopy gene of at least 30 copies permitted unambiguous detection of a single mutant allele (around 5% of total, J. Prosser, unpublished).

Heteroduplex analysis is very similar to SSCP and can be run in conjunction with it. Non-denaturing acrylamide gels separate heteroduplex molecules on the basis of molecular shape. Particular mismatches and their location in the double-stranded fragment distort the usual shape and alter the mobility of the heteroduplex. Mutant and wildtype DNAs can be PCRed together, or hetero-duplexed subsequently. It is possible that running both SSCP and heteroduplex analysis would be equivalent in efficiency to running two SSCP conditions.

13.8.3.1 Protocol (SSCP, Glavac and Dean, 1993)

This SSCP technique uses radioactive labelling. Samples are run in 8% acrylamide with 2% cross-linking at room temperature in $1 \times$ TBE. For each reaction, mix 1 μl $10 \times$ PCR reaction buffer, 0.2 μl oligo1 at 100 ng/μl, 0.2 μl oligo2 at 100 ng/μl, 1 μl DNA at 100 ng/μl, 1 μl dNTPs at 2 mM, 0.1 μl ^{32}P-dCTP (10 Ci/ml, Amersham), 0.2 μl Taq polymerase (Perkin Elmer Ampli-Taq at 5 U/μl) and 6.3 μl water. After a standard PCR reaction of 25 or 30 cycles, mix 2 μl of reaction with 10 μl SSCP denaturing dye (95% (v/v) formamide, 5 mM NaOH, 0.1% (w/v) Bromophenol Blue, 0.1% (w/v) xylene cyanol) and load 2 μl per track on the non-denaturing acrylamide gel described. Apparatus is Bio-Rad sequencing apparatus. Run at 10 watts, room temperature (about 5 h) until Bromophenol Blue reaches the bottom of gel.

13.8.3.2 Protocol (SSCP, Ravnik-Glavac et al., 1994)

The technique uses radioactive labelling. Samples are run in 10% acrylamide with 1.3% cross-linking and 10% glycerol at room temperature. The general protocol is as in previous SSCP method.

13.8.3.3 Protocol (SSCP with silver staining)

Detection by silver staining after running in 15% acrylamide with 3.3% cross-linking. From a normal 50 μl PCR reaction, a 10 μl sample is added to 6 μl formamide dye and run in 15% acrylamide, 3.3% cross-linking, at 10 watts for about 18 h. By the end of the running time, both dyes are long gone. The apparatus is a vertical Hoeffer rig using 9 cm $\times$ 16 cm plates. Silver staining is as follows: The acrylamide gel is placed in a 10% solution of ethanol for 5 min. This is removed by aspiration and replaced with a 1% solution of nitric acid for 3 min. This is again removed by aspiration and the gel rinsed with distilled water. It is then submerged in a solution of 0.012 M silver nitrate (BDH,

AnalR) and left for 20 min on a gentle shaker. This is removed, the gel rinsed
and placed in about 100 ml of a solution of anhydrous sodium carbonate (BDH
AnalR, 30 g/l with 0.5 ml formalin). The solution will be dark brown. This
should be discarded and a further ~ 400 ml added. Once the desired stain
intensity is reached, the reaction is stopped by a solution of 10% glacial acetic
acid (2–3 min) and by placing the gel hereafter in distilled water. A photograph can be taken, or the gel vacuum dried on Whatman 3 MM filter paper
for a permanent record.

13.8.4 Protein truncation test (PTT)

PTT is a recently developed technique with limited but useful application. In
genes, particularly large genes, where mutation frequently leads to premature
truncation of the protein, PTT permits analysis of large stretches of coding
sequence. Essentially, RNA is reverse transcribed and amplified by PCR.
Subsequently, nested PCR is performed with a modified primer containing a
T7 promoter and a eukaryotic translation initiation sequence to permit *in vitro*
transcription/translation of the PCR products. On SDS-PAGE analysis, the
appearance of shortened protein products permits fairly precise localization of

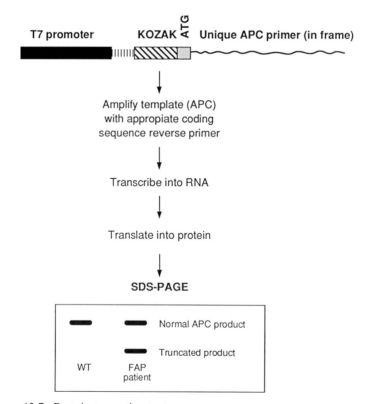

Figure 13.7. Protein truncation test.

the truncating mutation (Figure 13.7). The upper size limit of coding regions that can be examined remains to be determined but in practice, overlapping fragments of around 2–2.5 kb have been shown to work. For large genes in which terminating mutations are frequent (for instance, *DMD*, *APC*, *NF1*) this would be the first choice of mutation detection method. Subsequently, for point mutations not leading to terminating mutations in these large genes, CCM on RT-PCR would be most efficient.

13.8.4.1 Protocol (PTT, modified from Powell et al., 1993)

The method uses Promega TNT™ T7 coupled Reticulocyte Lysate System. It requires 0.2–2.0 μg DNA template from a PCR reaction amplified with a forward primer containing T7 promoter consensus and Kozak sequence (either a first amplification or nested second amplification). For each PTT reaction, mix 6.25 μl TNT rabbit reticulocyte lysate, 0.5 μl TNT reaction buffer, 0.25 μl TNT T7 RNA polymerase, 0.25 μl 1 mM amino acid mix (minus methionine), 0.5 μl (5 μCi) ^{35}S-methionine (Amersham) 10 mCi/ml and 2 μl TE. To this mix, add ~ 2 μg PCR product in 2.5 μl volume to a total of 12.5 μl.

Set up a positive control from the kit (luciferase, a 61 kDal monomeric protein) by using 0.5 μl luciferase control DNA at 0.5 mg/ml in 2.5 μl volume. Mix each of the reactions and incubate at 30 °C 1–2 h. Store at −20 °C or on ice if electrophoresing straight away. Gel conditions: To 9 μl SDS sample buffer (2 ml glycerol, 2 ml 10% SDS, 0.25 mg Bromophenol Blue, 2.5 ml 4 × buffer*, 2.75 ml water and, immediately before loading, 0.5 ml β-mer-captoethanol) add 3 μl TNT product. Place at 100 °C for 3 min, then load on 8%, 10% and 12% SDS-PAGE gels in Mini-Protean^{R}II Electrophoresis cells. Gels are electrophoresed at 200 V until Bromophenol Blue reaches the bottom (~ 0.5 h). Gels are fixed in 10:20:70 acetic acid:methanol:water for 30 min, dried on 3 MM Whatman paper and autoradiographed at room temperature overnight.

(*Buffer: 6.06 g Tris base, 4 ml 10% (w/v) SDS, pH 6.8, final volume 100 ml)

13.8.5 Sequencing protocols

Finally, sequencing in confirmation of the results of scanning techniques, to determine the exact base change and to eliminate simple polymorphisms, is easily done. With both SSCP and DGGE, the mutant and normal alleles can be individually excised from the gel for amplification and sequencing. This is useful when the mutant constitutes less than 50% of the total product and direct sequencing is not effective. Two protocols are given, both using the Sequenase kit (USB Sequenase version 2), although any other kit or procedure can be substituted. One method is for sequencing a double-stranded PCR product, in which primer annealing is done in the presence of DMSO at

$-70\,°C$. The other involves using one biotinylated primer in the PCR amplification in order to prepare single-stranded template using Dynabeads.

13.8.5.1 *Protocol for sequencing double-stranded PCR products (modified from Winship, 1989)*

Before beginning: Turn on a $100\,°C$ block and a $37\,°C$ waterbath. Label reaction tubes. Dilute Sequenase $1 + 7$ with Sequenase dilution buffer. Dilute the labelling mix 1:15 with water. Dispense termination mixes containing 10% (v/v) DMSO into labelled tubes. Then set up the annealing reaction: $3\,\mu l$ DNA (Genecleaned product excised from low melting agarose as described for CCM technique), $3\,\mu l$ water, $1\,\mu l$ DMSO, $1\,\mu l$ primer, $2\,\mu l$ Sequenase buffer. Place at $100\,°C$ for 2 min, then immediately into a dry ice/methanol bath for the annealing step. Estimates of the amount of template and primer are approximately 50 ng and 200–250 ng respectively.

Add to the annealing reaction (frozen) $1\,\mu l$ 0.1 M DTT, $0.5\,\mu l$ ^{35}S-dATP, $2\,\mu l$ diluted labelling mix, $1.5\,\mu l$ diluted Sequenase. Mix and leave at room temperature for 5 min. Add $3.5\,\mu l$ of this sequencing reaction to each of four tubes containing the $2\,\mu l$ termination mixes. Place at $37\,°C$ for 5–10 min. Stop with $4\,\mu l$ of formamide dye/stop solution.

13.8.5.2 *Protocol for single-stranded PCR products*

PCR the template for sequencing with one biotinylated primer and follow the Dynabead separation protocol. Sequence the single-stranded template with Sequenase kit. (Dynabeads and protocol from Dynal.)

13.8.6 *Concluding remarks regarding mutation detection*

All the mutation detection protocols offered have been used successfully by various researchers in the Edinburgh MRC Human Genetics Unit. Methods for automation of techniques are appearing in the literature, but these have not been discussed. Partly this is because the technology is expensive and not readily accessible and partly it is because the methods are in early stages of development. HOT has been adapted for the ABI automated sequencer by at least two laboratories (Ellis *et al.*, 1994; Lishanski *et al.*, 1994), and automation may in fact facilitate the analysis of large fragments as well as permitting limited multiplexing with the use of several fluorescent dyes. SSCP has been automated (Ellison, Dean and Goldman, 1993) and offers limited advantages, such as the precise measurement of mobility differences using internal standards and quantitative measurement of the products. The same sort of advantages might apply to running DGGE on an automated sequencer. Automated sequencing of PCR amplified products is a routine procedure, but there are problems in the analysis of gel ambiguities and heterozygous mutations which prevent the technique from having superceded all others. In situations where less than 50% of the starting material is mutant, direct sequencing of uncloned DNA is not useful.

13.8.6.1 Which detection technique is the best?

To screen for mutations in a limited number of patients in a gene of modest size and where the mutations are expected to be scattered throughout the gene, we would recommend SSCP, preferably in 150–250 bp sizes, and using more than one protocol. The technique is easy, requires no special apparatus, and if the mutations are there, a reasonable proportion of them will be detected. Where a mutation is not found, another technique can be tried.

To screen a large number of patients, or to routinely screen a particular gene or genes, we would recommend setting up DGGE. The time required to set the conditions for each of the fragments is well spent since the subsequent routine analysis is easier than SSCP. Only one set of conditions is required for nearly 100% efficiency of detection.

To screen very large genes, several considerations can be taken into account. If the gene is transcribed, either legitimately or illegitimately, in lymphocytes, then the screening is best done on RT-PCR products or transcripts from this readily available source. Straight RT-PCR is, in any case, sufficient to detect a proportion of mutations due to splicing errors. If mutations leading to protein truncations are prevalent, including either stop codons or splicing mutations leading to stop codons, then the PTT assay permits scanning of around 2.5 kb of coding sequence at a time. Although the limits of this technique are not known (the maximum size of fragment, the proportion of total transcript at which a mutation is detectable, the efficiency), it has been usefully applied to a number of very large genes. For non-protein truncating mutations in these large genes I would recommend CCM in overlapping 1 kb or 1.5 kb fragments. If no transcript is available then an SSCP slog might be unavoidable.

To look for a confirming mutation in a new candidate locus, I would not abandon the search until CCM has been used. If the mutation were not found with CCM, its presence could be seriously doubted. So, in such a situation, I personally would start and finish with CCM. But there is no reason not to start with SSCP.

Acknowledgements

The authors would like to thank several researchers at the MRC Human Genetics Unit and Molecular Genetics Service, Human Genetics Unit, University of Edinburgh for very generously sharing their protocols and experience: Stewart Morris for DGGE, Malcolm Dunlop and Morag Wright for PTT and microsatellite instability, Jon Warner and Lisa Strain for trinucleotide expansions, Richard Axton for SSCP with silver staining and Harris Morrison and Paul Perry for CGH.

References

Aaltonen, L. A., Peltomaki, P., Leach, F. S. *et al.* (1993) Clues to the pathogenesis of familial colon cancer. *Science* **260**, 812–6.

Aaltonen, L. A., Peltomaki, P., Mecklin, J.-P. *et al.* (1994) Replication errors in benign and malignant tumours from hereditary non-polyposis colorectal cancer patients. *Cancer Res.* **54**, 1645–8.

Barnes, W. M. (1994) PCR amplification of up to 35-kb DNA with high fidelity and high yield from λ bacteriophage templates. *Proc. Natl Acad. Sci. USA* **91**, 2216–20.

Barron, L. H., Rae, A., Holloway, S. *et al.* (1994) A single allele from the polymorphic CCG rich sequence immediately 3′ to the unstable CAG trinucleotide in the IT15 cDNA shows almost complete disequilibrium with Huntington's disease chromosomes in the Scottish population. *Hum. Mol. Genet.* **3**, 173–5.

Bodmer, W., Bishop, T. and Karran, P. (1994) Genetic steps in colorectal cancer. *Nature Genet.* **6**, 217–19.

Buetow, K. H., Weber, J. L., Ludwigsen, S. *et al.* (1994) Integrated human genome-wide maps constructed using the CEPH reference panel. *Nature Genet.* **6**, 391–3.

Condie, A., Eeles, R., Borresen, A.-L. *et al.* (1993) Detection of point mutations in the p53 gene: comparison of single-strand conformation polymorphism, constant denaturant gel electrophoresis, and hydroxylamine and osmium tetroxide techniques. *Hum. Mutation* **2**, 58–66.

Cooper, D. N., Smith, B. A., Cooke, H. J. *et al.* (1985) An estimate of unique DNA sequence heterozygosity in the human genome. *Hum. Genet.* **69**, 201–5.

Cotton, R. G. H. (1993) Current methods of mutation detection. *Mutation Res.* **285**, 125–44.

Cotton, R. G. H., Rodrigues, N. R. and Campbell, R. D. (1988) Reactivity of cytosine and thymine in single-base-pair mismatches with hydroxylamine and osmium tetroxide and its application to the study of mutations. *Proc. Natl Acad. Sci. USA* **85**, 4397–401.

Duncan, B. K. and Miller, J. H. (1980) Mutagenic deamination of cytosine residues in DNA. *Nature* **287**, 560–1.

Ellis, L. A., Taylor, G. R., Banks, R. and Baumberg, S. (1994) MutS binding protects heteroduplex DNA from exonuclease digestion *in vitro*: a simple method for detecting mutations. *Nucl. Acids Res.* **22**, 2710–11.

Ellison, J., Dean, M. and Goldman, D. (1993) Efficacy of fluorescence-based PCR-SSCP for detection of point mutations. *Biotechniques* **15**, 684–91.

Feinberg, A. P. and Vogelstein, B. (1983) A technique for radiolabelling DNA restriction endonuclease fragments to high specific activity. *Anal. Biochem.* **132**, 6–13.

Fodde, R. and Losekoot, M. (1994) Mutation detection by denaturing gradient gel electrophoresis (DGGE). *Hum. Mutation*, **3**, 83–94.

Fu, Y.-H., Kuhl, D. P. A., Pizzuti, A. *et al.* (1991) Variation of the CGG repeat at the fragile X site results in genetic instability: resolution of the Sherman paradox. *Cell* **67**, 1047–58.

Furrer, B., Candrian, U., Wieland, P. and Luthy, J. (1990) Improving PCR efficiency. *Nature* **346**, 324.

Ganguly, A. and Prockop, D. J. (1990) Detection of single-base mutations by reaction of DNA heteroduplexes with a water-soluble carbodiimide followed by primer extension: application to products from the polymerase chain reaction. *Nuc. Acids Res.* **18**, 3933–9.

Glavac, D. and Dean, M. (1993) Optimization of the single-strand conformation polymorphism (SSCP) technique for detection of point mutations. *Hum. Mutation*

2, 404–14.

Grompe, M. (1993) The rapid detection of unknown mutations in nucleic acids. *Nature Genet.* **5**, 111–17.

Gyapay, G., Morissette, J., Vignal, A. *et al.* (1994) The 1993–94 Genethon human genetic linkage map. *Nature Genet.* **7**, 246–339.

Hovig, E., Smith-Sorenson, B., Uitterlinden, A. G. and Borresen, A.-L. (1992) Detection of DNA variation in cancer. *Pharmacogenetics*, **2**, 317–28.

Hughes, A. E. (1993) Optimization of microsatellite analysis for genetic mapping. *Genomics* **15**, 433–4.

Ionov, Y., Peinado, M. A., Malkhosyan, S. *et al.* (1993) Ubiquitous somatic mutations in simple repeated sequences reveal a new mechanism for colonic carcinogenesis. *Nature* **363**, 558–61.

Jeffreys, A. J., MacLeod, A., Tamaki, K. *et al.* (1991) Minisatellite repeat coding as a digital approach to DNA typing. *Nature* **354**, 204–9.

Jeffreys, A. J., Wilson, V., Neumann, R. and Keyte, J. (1988) Amplification of human minisatellites by the polymerase chain reaction: towards DNA fingerprinting of single cells. *Nucl. Acids Res.* **16**, 10953–71.

Jeffreys, A. J., Wilson, V. and Thein, S. L. (1985) Hypervariable 'minisatellite' regions in human DNA. *Nature* **314**, 67–73.

Jen, J., Kim, H. G., Piantadosi, S. *et al.* (1994) Allelic loss of chromosome 18q and prognosis in colorectal cancer. *N. Engl. J. Med.* **331**, 213–21.

Kallioniemi, O.-P., Kallioniemi, A., Piper, J. *et al.* (1994) Optimizing comparative genomic hybridization for analysis of DNA sequence copy number changes in solid tumors. *Genes, Chromosomes & Cancer* **10**, 231–43.

Kallioniemi, A., Kallioniemi, O.-P., Sudar, D. *et al.* (1992) Comparative genomic hybridization for molecular cytogenetic analysis of solid tumors. *Science* **258**, 818–21.

Keen, J., Lester, D., Inglehearn, C. *et al.* (1991) Rapid detection of single base mismatches as heteroduplexes on Hydrolink gels. *Trens Genet.* **7**, 5.

Knudson, A. G. (1971) Mutation and cancer: statistical study of retinoblastoma. *Proc. Natl Acad. Sci. USA* **68**, 820–3.

Kwok, S. and Higuchi, R. (1989) Avoiding false positives with PCR. *Nature* **339**, 237–8.

Lishanski, A., Ostrander, E. A. and Rine, J. (1994) Mutation detection by mismatch binding protein, MutS, in amplified DNA: application to the cystic fibrosis gene. *Proc. Natl Acad. Sci. USA* **91**, 2674–8.

Mansfield, D. C., Brown, A. F., Green, D. K. *et al.* (1995) Automation of genetic linkage analysis using fluorescent microsatellite markers. *Genomics* **24**, 199–210.

Mashal, R. D., Koontz, J. and Sklar, J. (1995) Detection of mutations by cleavage of DNA heteroduplexes with bacteriophage resolvases. *Nature Genet.* **9**, 177.

Modrich, P. (1994) Mismatch repair, genetic stability, and cancer. *Science* **266**, 1959–60.

Modrich, P. (1995) Mismatch repair, genetic stability and tumour avoidance. *Phil. Trans. R. Soc. Lond. B* **347**, 89–95.

Monckton, D. G., Wong, L.-J. C., Ashizawa, T. and Caskey, C. T. (1995) Somatic mosaicism, germline expansions, germline reversions and intergenerational reductions in myotonic dystrophy males: small pool PCR analyses. *Hum. Mol. Genet.* **4**, 1–8.

Montandon, A. J., Green, P. M., Giannelli, F. and Bentley, D. R. (1989) Direct detection of point mutations by mismatch analysis: application to haemophilia B. *Nucleic Acids Res.* **17**, 3347–58.

Murray, V., Monchawin, C. and England, P. R. (1993) The determination of the sequences present in the shadow bands of a dinucleotide repeat PCR. *Nucleic Acids Res.* **21**, 2395–8.

Myers, R. M., Larin, Z. and Maniatis, T. (1985*a*) Detection of single base substitutions by ribonuclease cleavage at mismatches in RNA:DNA duplexes. *Science* **230**, 1242–46.

Myers, R. M. Lumelski, N., Lerman, L. S. and Maniatis, T. (1985*b*) Detection of single base substitutions in total genomic DNA. *Nature (London)* **313**, 495–8.

Nakamura, Y., Leppert, M., O'Connell, P. *et al.* (1987) Variable number of tandem repeat (VNTR) markers for human gene mapping. *Science* **235**, 1616–22.

Nancarrow, J. K., Kremer, E., Holman, K. *et al.* (1994) Implications of FRA16A structure for the mechanism of chromosomal fragile site genesis. *Science* **264**, 1938–41.

Orita, M., Iwahana, H., Kanazawa, H. *et al.* (1989) Detection of polymorphisms of human DNA by gel electrophoresis as single-strand conformation polymorphisms. *Proc. Natl Acad. Sci. USA* **86**, 2766–70.

Orr, H. T., Chung, M., Banfi, S. *et al.* (1993) Expansion of an unstable trinucleotide CAG repeat in spinocerebellar ataxia type 1. *Nature Genet.* **4**, 221–6.

Parrish, J. E., Oostra, B. A., Verkerk, A. J. M. H. *et al.* (1994) Isolation of a GCC repeat showing expansion in FRAXF, a fragile site distal to FRAXA and FRAXE. *Nature Genet.* **8**, 229–35.

Pinkel, D. (1994) Visualizing tumour amplification. *Nature Genet.* **8**, 107–8.

Powell, S. M., Petersen, G. M., Krush, A. J. *et al.* (1993) Molecular diagnosis of familial adenomatous polyposis. *New Engl. J. Med.* **329**, 1982–7.

Prosser, J. (1993) Detecting single base mutations. *Tibtech.* **11**, 238–46.

Prosser, J., Condie, A., Wright, M. *et al.* (1994) APC mutation analysis by chemical cleavage of mismatch and a protein truncation assay in familial adenomatous polyposis. *Br. J. Cancer* **70**, 841–6.

Rabbits, T. H. (1994) Chromosomal translocations in human cancer. *Nature* **372**, 143–93.

Ravnik-Glavac, M., Glavac, D. and Dean, M. (1994) Sensitivity of single-strand conformation polymorphism and heteroduplex method for mutation detection in the cystic fibrosis gene. *Hum. Mol. Genet.* **3**, 801–7.

Read, P. W., Davies, J. L., Copeman, J. B. *et al.* (1994) Chromosome-specific microsatellite sets for fluorescence-based, semi-automated genome mapping. *Nature Genet.* **7**, 390–5.

Roest, P. A. M., Roberts, R. G., Sugino, S. *et al.* (1993) Protein truncation test (PTT) for rapid detection of translation-terminating mutations. *Hum. Mol. Genet.* **2**, 1719–21.

Royle, N. J., Clarkson, R. E., Wong, Z. and Jeffreys, A. J. (1988) Clustering of hypervariable minisatellites in the proterminal regions of human autosomes. *Genomics*, **3**, 352–60.

Sambrook, J., Fritsch, E. F. and Maniatis, T. (1989) *Molecular cloning. A laboratory manual*, 2nd edn. Cold Spring Harbor Press, Cold Spring Harbor.

Sheffield, V. C., Beck, J. S., Kwitek, A. E. *et al.* (1993) The sensitivity of single-strand conformation polymorphism analysis for the detection of single base substitutions. *Genomics* **16**, 325–32.

Southern, E. M. (1975) Detection of specific sequences among DNA fragments separated by gel electrophoresis. *J. Mol. Biol.* **98**, 503–17.

Thibodeau, S. N., Bren, G. and Schaid, D. (1993) Microsatellite instability in cancer of the proximal colon. *Science* **260**, 816–19.

Verkerk, A. J. M. H., Pieretti, M., Sutcliffe, J. S. *et al.* (1991) Identification of a gene (FMR-1) containing a CGG repeat coincident with a breakpoint cluster region exhibiting length variation in fragile X syndrome. *Cell* **65**, 905–14.

Vogelstein, B., Fearon, E. R., Kearn, S. E. *et al.* (1989) Allotype of colorectal carcinoma. *Science* **244**, 207–11.

Weber, J. L. (1990) Informativeness of human $(dC-dA)_n(dG-dT)_n$ polymorphisms. *Genomics* **7**, 524–30.

Weber, J. L. and May, P. E. (1989) Abundant class of human DNA polymorphisms which can be typed using the polymerase chain reaction. *Am. J. Hum. Genet.* **44**, 388–96.

Weissenbach, J., Gyapay, G., Dib, C. *et al.* (1992) A second-generation linkage map of the human genome. *Nature* **359**, 794–801.

Wells, R. A., Green, P. and Reeders, S. T. (1989) Simultaneous genetic mapping of multiple human minisatellite sequences using DNA fingerprinting. *Genomics* **5**, 761–72.

White, M. B., Carvalho, M., Derse, D. *et al.* (1992) Detecting single base substitutions as heteroduplex polymorphisms. *Genomics* **12**, 301–6.

Willems, P. J. (1994) Dynamic mutations hit double figures. *Nature Genet.* **8**, 213–15.

Winship, P. R. (1989) An improved method for directly sequencing PCR amplified material using dimethyl sulphoxide. *Nucl. Acids Res.* **17**, 1266.

Youil, R., Kemper, B. W. and Cotton, R. G. H. (1995) Screening for mutations by enzyme cleavage with T4 endonuclease VII. *Proc. Natl. Acad. Sci. USA* **92**, 87.

Gene manipulation: molecular level

14

Basic cloning techniques

STEVE LEGON

14.1 Introduction

This chapter describes the fundamental procedures by which the DNA to be cloned is joined to the DNA of the vector and introduced into bacterial cells. This includes the purification of plasmids and phage, ligations and the transfection or transformation of bacteria. The preparation of the DNA to be cloned and the screening procedures by which the desired recombinants are selected are covered elsewhere (see quantitative and qualitative assessment of nucleic acids and proteins). The techniques described are in no sense novel and they are the basic building blocks from which the most sophisticated experimental protocols are constructed. Underlying these techniques, however, is an assumption that the reader knows how to handle DNA and will treat enzymes with the respect they deserve. As no such assumptions are made, this chapter also includes short sections on these elementary matters.

The reader of this chapter will undoubtedly be familiar with the major journals and will have noticed that there are many advertisements offering a variety of kits. They claim either to give improved efficiencies or to save time and, in general, these claims are true. Why then do we need to learn how to do these procedures for ourselves? There are several reasons that could be advanced ranging from a quasi-moral justification that you should do things the hard way to understand the subject properly to the view that cloning kits are for those with more money than sense. Neither position is entirely tenable. Twenty years ago many big laboratories made their own restriction enzymes rather than buy them. Nowadays, with the economies of scale achieved by the big companies, no one would seriously consider making the common enzymes. Grant-giving agencies are aware of these expenses and generally make allowances for these costs when funding work in this area. However, they are not generally sympathetic to the idea of funding projects for inexperienced workers more generously so that they can use cloning kits for all their work. Sadly, you will probably find that funds are limited and you will have to work long hours performing these basic procedures. In this chapter, I will try to steer a middle course indicating where outside help from commercial kits might be useful and where these simple techniques are perfectly adequate and essentially 'idiot proof'. In this respect, I am indebted to many members of my laboratory for all their efforts over many years to search out the weak points in these methods.

14.2 General scheme for making recombinants: an overview

Despite many variations in the technology, the overall scheme for producing clones of cells containing a particular DNA sequence is very simple in theory (Figure 14.1). It is equally applicable to cloning in bacterial or mammalian cells (See Chapter 4). A preparation of DNA containing the sequence of interest (e.g. genomic DNA, a cDNA copy of mRNA or a PCR product) is joined to the DNA of a cloning vector. The cloning vector is a DNA that has the capacity to replicate in cells (e.g. a virus or a plasmid) and retains this ability when foreign DNA is joined to it. The joining is achieved by a simple enzymic reaction using a DNA ligase. The DNA is then introduced into the cells in which it can replicate and colonies of cells (or plaques of virus) are then grown. Each of these represents a clonal expansion of the recombinant DNA that transformed the progenitor cell. The number of these 'clones' that are grown up will reflect the complexity of the original DNA preparation. If one is searching for a unique gene sequence in the human genome then one will typically need to grow more than a million clones to have a reasonable chance of isolating the desired sequence. If one is cloning a sequence from a poly-

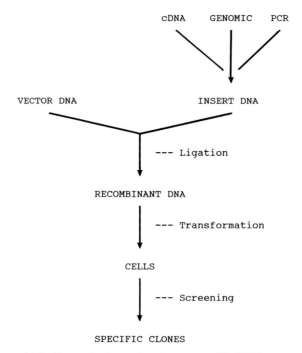

Figure 14.1. General scheme for cloning specific DNA sequences.

merase chain reaction (PCR) then half a dozen clones are usually enough to ensure one has the correct sequence. The final step in the procedure is to select the correct clone from the library of clones that have been generated. The options here range from detecting changes in the morphology of the cells, detecting the DNA sequence by hybridization, detecting the protein product expressed from the cloned sequence or simply determining DNA sequences blindly until the correct clone is found. I will now describe each of these steps in detail starting with the elementary skills involved in the handling of DNA and enzymes.

14.3 Handling DNA molecules

14.3.1 Shearing DNA

Many people are aware that double-stranded DNA molecules (dsDNA) are inflexible and may break if subjected to the shear forces produced by vortexing or pipetting. This is a problem with genomic DNA but is not a serious problem with vector DNAs. They may be pipetted normally though it is sensible not to vortex phage lambda DNA. It is better to risk a little damage than to be too gentle in the organic extraction procedures.

14.3.2 Organic extractions

Nucleic acid solutions are deproteinized by mixing with either phenol or chloroform or an equal mixture of the two. The phenol should be water saturated, neutralized and equilibrated to a suitable buffer such as TE (p. 261). Except where explicitly stated to the contrary, the 'chloroform' should contain 1% (v/v) isoamyl alcohol. Extractions are carried out by adding an equal volume of the organic solvent to the nucleic acid solution and mixing vigorously for about 30 s. The resulting emulsion is then centrifuged for a couple of minutes to separate the phases and the (upper) aqueous phase is removed carefully leaving behind any material at the interface. If a high yield is important then the organic phase can be re-extracted with an equal volume of TE buffer. The resultant two aqueous phases are then combined. If desired, traces of phenol may be removed from the aqueous phase. Add the salt necessary for ethanol precipitation (see below) then add an equal volume of ether. Mix vigorously then leave to stand. The phases should separate easily and the (top) ether layer can then be removed. The DNA is then recovered from the aqueous phase by ethanol precipitation.

14.3.3 Ethanol precipitation

DNA and RNA molecules are insoluble in 60–80% ethanol solutions in the presence of a little salt. Typically, to precipitate DNA a solution is first made

250 mM in sodium acetate, pH 5.2, and then 2–3 volumes of ethanol are added. If there is a lot of DNA then the precipitate will form immediately. Otherwise, the mixture is left at 0 °C or −20 °C for a period of time dependent on the original DNA concentration – an hour is sufficient for a 1 μg/ml solution of plasmid DNA (see Chapter 4). The mixture is then centrifuged at a speed that is dependent on the size of the expected precipitate and the supernatant is carefully removed. Often precipitation is performed in a microfuge tube. Since the supernatant will not pour off it is normally taken off by aspiration. A Pasteur pipette drawn out to a capillary point and attached to a water pump is ideal for this. Take off all the small droplets as these will contain the precipitating salt and possibly other substances. If the supernatant contains a substance like phenol then it is usual to rinse the pellet with ethanol to remove the last traces of phenol. Since DNA is actually soluble in 100% ethanol this must be done quickly. Carefully add a small volume of ethanol to the pellet making sure the pellet is not disturbed then immediately remove the ethanol. Finally, dry the pellet *in vacuo* for no more than a couple of minutes as very dry DNA pellets are difficult to dissolve.

14.3.4 *Estimating concentrations*

It is useful to be able to estimate DNA concentrations roughly where there is insufficient material for an absorbance measurement (< 50 ng/μl). There are specialized instruments for this purpose but a cheap alternative is to spot the DNA on an ethidium agarose gel. Make up 100 ml of agarose gel as for electrophoresis of DNA (i.e. 1% (w/v) agarose, TBE buffer and 250 ng/ml ethidium bromide). Pour about 30 ml into each of three 90 mm Petri dishes and allow to set. DNA concentrations are estimated by spotting 1 μl of the unknown and comparing the fluorescence with that given by 5–20 ng/μl standards. The range of this assay is limited but it is reasonably accurate.

14.4 Handling enzymes

14.4.1 *Storage*

Enzymes are generally stored at −20 °C in 50% glycerol. Avoid frost-free freezers, which indicates that the temperature periodically rises to melt the ice. For long-term storage most enzymes are more stable at −70 °C, but check the enzyme data sheets as some do not survive freezing.

14.4.2 *Buffers*

All enzyme suppliers provide the buffer with the enzyme, generally at ten times the working concentration. This will normally be stored frozen along with the enzyme. Always melt completely and mix thoroughly as most of the solute is in the first few drops to melt.

14.4.3 Reactions

Set up the reactions on ice and add the enzyme last so it is always in an appropriate buffer. As the enzyme will be in 50% glycerol, do not allow the total enzyme additions to exceed 10% of the reaction volume. Many enzymes are either inhibited or show abnormal specificities when in more than 5% glycerol. Pipette carefully and wipe the outside of the tip against the side of the enzyme tube as there may be more enzyme on the outside of the tip than within it. After adding the enzyme mix carefully but thoroughly. The glycerol will sink to the bottom of the tube and is reluctant to mix with the rest of the reaction. Do not vortex as proteins can be denatured in froth. Hold the tube firmly in one hand and flick the base of the tube with the other. This will splash liquid up the sides of the tubes, so spin the tube for a few seconds before incubating. Do not add a gross excess of enzyme (i.e. more than tenfold) hoping that this may make the reaction work. Concentrate on providing better substrates for the enzyme to act on.

14.5 Plasmid 'mini-preps'

14.5.1 Introduction

A mini-prep is a preparation of plasmid DNA on a 1–2 ml scale and is usually a first step in characterizing a clone. There are many kits on the market that will give high quality DNA in less than 30 min. The method described below takes about an hour and gives you plasmid DNA contaminated by all the *E. coli* cellular RNA. However, the cost is almost zero and the RNA contamination does not interfere with restriction analysis, sequencing or PCR, which are the most frequent reasons for making plasmid DNA on a small scale.

The procedure is known as a 'boiling mini-prep' and was described by Del Sal, Monfioletti and Schneider (1988). The advantage of this method over its rivals is that you only need one tube per preparation – a considerable benefit if you consider the tedium and error rate involved in labelling tubes for 100 mini-preps. Although possible, I do not find it convenient to scale up this method for large volumes and prefer to use the next method for preparative scale work.

14.5.2 Materials

STET buffer	8% (w/v) sucrose
	0.1% (v/v) Triton X-100
	50 mM EDTA
	50 mM Tris/HCl, pH 8.0
Lysozyme	50 mg/ml in STET buffer
CTAB solution (cetyl trimethyl ammonium bromide)	2% (w/v) in water
NaCl solution	1.0 M

14.5.3 Method

1. Spin down 1.5 ml of an overnight culture of plasmid-containing bacteria in a microfuge tube.
2. Resuspend (using a pipettor) in 400 μl of STET buffer.
3. Add 8 μl of lysozyme and incubate 10 min at room temperature.
4. Boil for 1 min then spin for 10 min.
5. Pick out the pellet with a toothpick and discard.
6. Add 40 μl of 2% CTAB then spin for 5 min at room temperature.
7. Remove supernatant and resuspend pellet in 300 μl of 1.0 M NaCl, vortexing vigorously – this may take some time.
8. Reprecipitate with 750 μl ethanol and spin for 10 min.
9. Resuspend in 180 μl water, add sodium acetate pH 5.2 to 250 mM and precipitate with 500 μl ethanol.

14.5.4 Comments

If the final pellet is resuspended in 20 μl then 2 μl is sufficient to visualize on an agarose gel and the remainder is more than enough to sequence a clone on both strands.

The DNA still has a lot of RNA in it but this doesn't seem to interfere with sequencing. However, it does make visualizing restriction digests difficult unless boiled RNAase (see p. 261) is included at 25 μg/ml in the restriction digest.

Sometimes the pellet after boiling is very gelatinous, leaving very little supernatant. If this happens it helps to spin for a further 10 min.

CTAB comes out of solution in a cold environment ($< 15\,°C$) and the precipitate is not easy to see. Check carefully before using if this might be a problem and redissolve by warming.

14.6 Plasmid 'maxi-preps'

14.6.1 Introduction

Kits for maxi-preps ($> 20\,\mu$g plasmid DNA) are widely available and give good results but at a price. The following method gives DNA pure enough for routine use but cleaner DNA is required for mammalian transfection. Although double banding on CsCl gradients gives the required purity, the best of the kits (e.g. Qiagen) achieve this in a fraction of the time.

The method given below is the traditional way of preparing plasmid DNA, first described by Birnboim and Doly (1979). There are a number of minor variations on this theme and the version given here includes two steps (the initial wash and the lysozyme treatment) that many would omit. However, it is extremely reliable and may be applied equally to preparations of plasmids or of the double-stranded form of M13 phage.

14.6.2 Materials

TSE washing solution	100 mM NaCl
	10 mM Tris/HCl, pH 7.5
	100 μM EDTA, pH 8.0
Glucose solution (GTE)	1% (w/v) glucose
	25 mM Tris, pH 8.0
	10 mM EDTA
Alkaline SDS (make fresh!)	200 mM NaOH
	1% (w/v) SDS
5 M Potassium acetate	3.0 M potassium acetate
	2.0 M acetic acid

   (For 1 l make 600 ml of 5 M potassium acetate and add 115 ml of glacial acetic and 285 ml of water).

Tris/EDTA (TE)	10 mM Tris/HCl, pH 7.5
	1 mM EDTA

PEG solution: 13% (w/v) polyethylene glycol 6000 in water
Lysozyme solution: dissolve at 5 mg/ml in GTE buffer immediately before use.
Heat treated RNAase A: dissolve crude RNAase A at 10 mg/ml in 10 mM Tris/HCl, pH 7.5, 15 mM NaCl and heat for 60 min at 80 °C.

14.6.3 Method

This protocol is given for a 400 ml overnight culture of *E. coli* but can be scaled up or down as appropriate.

1. Grow cells overnight in rich broth (e.g. LBM).
2. Spin down the cells, 5 min at 5000 r.p.m. (4000 *g*).
3. Resuspend in 100 ml TSE solution using a pipette.
4. Spin down the cells again.
5. Resuspend in 40 ml GTE.
6. Add 4 ml lysozyme solution.
7. Leave 10 min at room temperature.
8. Add 80 ml alkaline SDS, mix gently till clear.
9. Leave 10 min on ice.
10. Add 60 ml cold 5 M potassium acetate, mix by inversion.
11. Leave at least 15 min on ice (overnight is OK).
12. Spin 5 min at 8000 r.p.m. (10 000 *g*), cold.

13. Strain the supernatant through gauze into a centrifuge pot. Add 0.6 volume (about 150 ml) of isopropanol, mix and leave for 5 min at room temperature.
14. Spin down the DNA/RNA for 10 min at 9000 r.p.m. (13 000 g).
15. Drain carefully, dry and resuspend in 5 ml TE. (Although at this stage the DNA is restrictable and can be used for many purposes it is likely that a large-scale plasmid preparation is intended as a stock reagent and as such would benefit from removal of the cellular RNA and any proteins that may have been copurified.)
16. Transfer to a disposable tube, add RNAase A to 25 μg/ml and incubate at 37 °C for 20 min. Make 0.2% in SDS and add proteinase K to give final concentration of 100 μg/ml and incubate at 45 °C for 30 min. Extract with 3 ml of phenol/chloroform and then precipitate with sodium acetate/ethanol.
17. Dry and dissolve pelleted nucleic acids completely in 1.7 ml water. Add 0.3 ml 5 M NaCl and mix. Add 2.0 ml PEG solution, mix and leave on ice for 60 min.
18. Centrifuge at 10 000 r.p.m. for 10 min (cold), remove supernatant and resuspend in 2 ml water.
19. Add sodium acetate (to 250 mM) and 5 ml ethanol and reprecipitate.
20. Dissolve the DNA in an appropriate volume of GDW or TE and read the absorbance at 260 nm.

14.6.4 Comments

This procedure digests the bacterial cell wall to create protoplasts and then lyses the cells with alkaline SDS. The host DNA is then removed by coprecipitation of the chromatin with potassium dodecyl sulphate. It is important to avoid shearing the host DNA as low molecular weight DNAs do not precipitate. The mixing steps (8 and 10) therefore have to be gentle but complete.

The RNAase is heated to destroy DNAase activity – take extreme care not to contaminate solutions or glassware as RNAase is almost indestructible and definitely bad news if you ever intend to make RNA.

Isopropanol precipitation is needed rather than ethanol to make the volume suitable for standard centrifuge pots. Isopropanol is however much less volatile – after centrifugation leave the pots inverted for at least 5 min to drain and suck off any remaining droplets before drying *in vacuo*.

Yields should be 2–5 mg/l of culture for modern plasmids. Older plasmids based directly on pBR322 or pAT153 have a lower copy number and benefit from chloramphenicol amplification. Grow the culture to an A_{600} of 1.0 then add chloramphenicol to a concentration of 1.5 mg/ml. Allow to grow overnight – in theory the plasmid carries on replicating while the cells stop growing.

With M13 double stranded DNA preparations, grow cells to an A_{600} of 0.25 then infect with half an eluted plaque. Grow for 5 h then continue as for a plasmid preparation. M13 generates defectives (truncated DNA molecules) if infected at too high a multiplicity.

14.7 Lambda DNA preparations

14.7.1 Overview

Phage lambda is an excellent vector as the packaging system means that cloning efficiencies are very high. It is often the vector of choice for the primary cloning of rare sequences. Unfortunately it is difficult to prepare restrictable DNA from lambda clones and success cannot be guaranteed for any method unless one purifies phage particles on glycerol or caesium gradients. In my hands the method below has been as reliable as any of the kits I have tried but it is slower. My preference is to avoid this problem if possible. 'Phagemids' such as lambda-ZAP give all the advantages of lambda cloning without the problems of lambda DNA preparations. The phage DNA contains a plasmid sequence and the cloning site is in this plasmid. After screening, the plasmid is induced to replicate and the cloned sequence is analysed in the plasmid. Alternatively, it is often possible to analyse clones by using PCR to amplify the region of interest from boiled lambda phage.

14.7.2 Introduction

In a lytic lambda infection, the growth rate of the cells is critical in determining virus yield. Growing in liquid with good aeration, yields drop disastrously when the cells get too dense but it is very difficult to ensure that lysis occurs at the correct cell density (see comments section). For small scale preparations it is easier to carry out the infections on a plate as virus titre is much less critical and plate lysates as described below are adequate for most purposes.

14.7.3 Materials

14.7.3.1 Media

As lambda attaches to cells through their maltose receptor, media for use with lambda are supplemented with 0.2% maltose. Although L broth will give adequate yields, a twofold increase is achieved with NZYM medium. (Note: NZ amine is marketed by Dibco under the brand name of Peptone 140.)

> NZYM   10 g NZ amine
>   5 g yeast extract
>   5 g NaCl
>   2 g $MgSO_4$.

Dissolve in 1 litre and adjust to pH 7.5

14.7.3.2 Cells

Different strains of lambda require different cell types as there are various strategies for suppressing the growth of vector only phage. One should always try to obtain appropriate cell stocks with the phage.

TE buffer	see p. 261
Lambda diluent	10 mM Tris/HCl, pH 7.5
	10 mM MgCl$_2$

14.7.4 Method

1. On day 1, titre the virus to give plaques that are touching but not merging.
2. On day 2, plate out the virus at this density on 132 mm plates using agarose not agar for the 0.75% layer. This should be done early in the morning.
3. After 5–6 h the plaques will have developed. Leave at 4 °C for an hour to make the soft agarose set firmly then add 10 ml of diluent per plate and leave overnight in the cold, rocking gently.
4. Suck off the SM and pellet any cells by centrifuging for 5 min at 5000 r.p.m. Take off the supernatant into a clean tube.
5. Digest for 30 min at 37 °C with 1 μg/ml each of pancreatic DNAase and RNAase.
6. Add 10 ml of 2 M NaCl/20% PEG 6000 in SM. Leave 1 h at 0 °C.
7. Spin down the phage for 10 min, 12 000 r.p.m. (16 000 g), cold, and carefully remove all the supernatant.
9. Resuspend in 1 ml of diluent, and vortex for 30 s with an equal volume of chloroform (**note**: pure chloroform without isoamyl alcohol for this step only). Centrifuge for 5 min.
10. Take off the aqueous phase and add EDTA (to 10 mM) and SDS (to 0.5%) then proteinase K (to 100 μg/ml). Mix and incubate for 30 min at 37 °C.
11. Extract once with an equal volume of phenol. Mix by shaking the tube vigorously, not vortexing, as the lambda DNA has now been released from the phage particles. Then extract once each with phenol/chloroform and chloroform before ethanol precipitating.
12. Resuspend in 100 μl of TE and digest for 30 min at 37 °C with 50 μg/ml of heat-treated RNAase A (See p. 261). Extract with phenol/chloroform then ether and ethanol precipitate.
13. Resuspend in 50 μl of TE – 2 μl should be visible on a gel and 5 μl is OK for restriction digests.

14.7.5 Comments

The yield is about 2–5 μg/plate using 'cDNA vectors' such as gt10, rather less with genomic vectors such as EMBL4. Typically five plates will give enough

DNA to allow identification of appropriate restriction fragments for subcloning into a plasmid. If the DNA will not restrict, I have not found any treatment that can be relied upon to clean up the DNA.

For a large-scale preparation, it will be necessary to grow the phage in liquid culture. As with most virus preps, making lambda requires that you get the ratio of cells to virus right. With a plate lysate as above this is achieved by getting the plaque density right. With a liquid culture the ideal is to grow the cells to an A_{600} of, say, 0.25 and then add just the right amount of phage to get lysis of the whole culture when the A_{600} reaches 1.0–1.5. If it gets higher than this the yield will be derisory. One foolproof way of achieving this without even knowing the titre of the virus is to make up 5 l of medium for a 500 ml preparation. Grow the cells and infect with a little virus at an absorbance of 0.25 then continue monitoring the A_{600} until it reaches 1.5. Meanwhile warm 250 ml of fresh medium and when the A_{600} reaches 1.5, add half the culture to the fresh medium and discard the rest. Continue to grow the diluted culture and dilute as necessary to keep the A_{600} between 0.75 and 1.5. Eventually there will be lysis and the preparation is then treated as for the plate eluate above. The method seems extravagant but it does work and the medium is cheap.

To prepare lambda DNA for use as a cloning vector, you will need a more extensive purification of the phage particles before extraction. This can be done by banding the phage on step gradients of cesium chloride but it is advisable for a novice to buy the vector DNA, restricted and ready to ligate rather than try to make it.

14.8 Joining insert DNA to the vector DNA

14.8.1 Introduction

This section encompasses preparing the DNAs by restriction and phosphatase treatment and then joining them using T4 DNA ligase. These are straightforward enzyme reactions and appropriate buffers are supplied with the enzymes. Provided the DNA preparations are good then these steps should give few problems.

14.8.2 Method

14.8.2.1 Vector DNA

The vector DNA will be a double-stranded circular or linear molecule that must be cleaved by a restriction enzyme cutting uniquely within its cloning site so that the insert DNA can be joined to it. The choice of enzyme will be dictated by the nature of the insert DNA. A flush-ended insert will require a flush-cut vector (e.g. Sma 1) whereas inserts produced by restriction enzymes producing 'sticky ends' will require a vector cut by a compatible enzyme.

In most situations, vector molecules will be capable of ligating back to themselves without need for an insert and these vector-only molecules will be the major product of the ligation reaction. To improve the yield of recombinants it is common to treat the restricted vector with shrimp phosphatase to remove its 5′ terminal phosphates. These are required for ligation so their removal eliminates the vector background from the ligation. Note, however, that if the insert DNA also lacks 5′ terminal phosphates (e.g. PCR with non-kinased primers) then there will be no recombinants either. Phosphatase treatment is simply a matter of adding the enzyme at the end of the restriction digest with no change of buffer. The enzyme is then inactivated by heat treatment (15 min at 65 °C) and the cut and phosphatased vector is then purified by phenol/chloroform extraction and ethanol precipitation.

14.8.2.2 Insert DNA

One cannot generalize about insert preparation beyond saying that the DNA should be clean and of known concentration with ends that are compatible with the intended vector. Ideally the insert should have 'sticky ends' as these ligate more efficiently than flush ends. If the insert has been generated by PCR and restriction sites have not been incorporated into the primers then ligation can be a problem. DNA polymerases tend to add a nucleotide onto the 3′ ends of completed chains in a non-template dependent manner preventing ligation to flush-cut vectors. There are a number of solutions to this problem but perhaps the most effective is to trim away these unwanted residues using T4 DNA polymerase to leave perfectly flush ends for ligation (Wang *et al.*, 1994).

14.8.3 Ligation

In a ligation reaction it is important to get both the vector concentration and the molar ratio of vector to insert approximately right. The following conditions give good results and, if the vector is not phosphatased, then a good insert preparation will give about 25% recombinants when the ligation is plated out. In a 20 μl reaction:

> 10 ng of a 3 kb vector
> 3 ng of a 300 bp insert
> 1 unit of T4 DNA ligase
> + water and buffer for 20 μl
> Incubate overnight at 15 °C

The amount of vector and insert should be varied in proportion to their sizes, e.g. 20 ng of a 6 kb vector or 1.5 ng of a 150 bp insert. The aim is to get a 3:1 molar excess of insert over vector – some workers try a range of insert concentrations (e.g. 1:1, 3:1, 9:1) but 3:1 is usually optimal.

Ligation reactions can be accelerated by adding substances like polyethylene glycol. This can exacerbate the problem of self-ligation and my personal feeling is that it is no bad thing to have an overnight step in a protocol occasionally. A number of commercial kits offer quick ligation at a price.

14.9 Transforming *E. coli* with recombinant DNA

14.9.1 Introduction

Plasmids and M13 are introduced into *E. coli* as naked DNAs whilst lambda clones are first packaged into phage particles, which are used to infect cells. This section deals only with simple transformation with naked DNA. When lambda is used it is generally used for primary cloning when efficiency is of paramount importance. Home-made packaging mixes are unlikely to approach the efficiency of bought mixes and they are also quite difficult to make. Commercially prepared lambda packaging mixes are recommended.

The procedure for transformation with DNA consists of making *E. coli* competent to take up DNA and then introducing the DNA into these competent cells. Competent cells can be bought and if the highest efficiencies are needed then this is recommended as these levels are difficult to achieve in the laboratory. However, in general one does not need high efficiencies when cloning in plasmids or M13 phage and the method described below should give competent cells capable of about 10^6–10^7 colonies/μg of plasmid DNA.

14.9.2 Materials

100 mM $CaCl_2$.
100 mM $CaCl_2$/15% (v/v) glycerol.
BCIG (5-bromo-4-chloro-3-indolyl-beta-D-galactoside) (BCIG is often sold as X-gal). Make a 20 mg/ml solution in dimethyl formamide.
IPTG (Iso-propyl-beta-D-thio-galactopyranoside). Make 20 mg/ml in water.
Lawn is an exponential cell culture made 15% (v/v) in glycerol and frozen in 1 ml portions at $-70\,^\circ$C at an A_{600} of about 0.5–1.0.

14.9.3 Method

14.9.3.1 Preparing competent cells

1. Start from a single colony of cells appropriate for your vector. Grow overnight in 20 ml of L broth.
2. Subculture 1:50 into a litre of L broth and grow to an A_{600} of about 0.4.
3. Chill 5 min on ice and spin 5 min at 5000 r.p.m. (4000 g), 5 °C.

4. Resuspend cells in 500 ml of 100 mM $CaCl_2$ sucking up and down with a pipette.
5. Leave cells on ice for 30 min then spin down as before.
6. Resuspend cells in 40 ml of $CaCl_2$/glycerol. Leave overnight at $0\,°C$. Store at $-70\,°C$ in aliquots of 200 μl.

14.9.3.2 *Plasmid transformation*

1. Keep cells at room temperature until just thawed and then leave on ice for 10 min. Add DNA (up to 2/5 volume of cells; use no more than 100 ng per 200 μl cells). For ligations, 40 μl of competent cells with 10 μl of ligation using a phosphatase treated vector will probably give a suitable number of colonies for a 90 mm plate. Without phosphatase treatment use no more than 2 μl of ligation as the vector background will give extra colonies.
2. Leave on ice 15–45 min.
3. Heat shock at 42 °C, 90 s, then return to ice for 1–2 min.
4. Add 2–3 volumes of broth (at room temperature). Incubate at 37 °C for 45 min with gentle shaking. This is to allow the plasmid's antibiotic resistance gene to be expressed prior to plating.
5. Plate on suitable antibiotic containing plates and incubate overnight.

14.9.4 *M13 transformation*

1. Add 10 ml of broth to 1 ml of frozen lawn cells and grow with shaking for 1 h.
2. Melt a suitable amount of 0.7% nutrient agar to give at least 3 ml per plate and leave at 45 °C. Dry a nutrient agar plate and leave it and a sterile 5 ml vial to warm up to 37 °C.
3. Proceed with transformation exactly as for a plasmid up to the heat shock, which should be for 1 min at 45°.
4. In quick succession, add to the warm vial 3 ml of soft agar, 200 μl of lawn cells, 20 μl each of BCIG and IPTG and the transformation mixture. Mix and pour onto the warm plate.
5. Leave to set for 5 min and incubate inverted at 37 °C overnight. Plaques should begin to show in 5 h but the colour selection given by the BCIG + IPTG takes several hours longer to develop.

14.9.5 *Comments*

These procedures are very reliable and if there are no recombinant clones it is probably a problem with the ligation, which in turn is probably a problem with the DNAs being ligated. Always include a control plate with no insert DNA when carrying out transformations.

With M13, white plaques indicate recombinants and the blue plaques are vector only. If the plaques are difficult to see this is probably caused by premature setting of the soft agar. The protocol given should eliminate this problem though dispensing the soft agar can cause difficulties – it is easiest to use a disposable sterile graduated plastic Pasteur pipette. The 45 °C for 1 min heat shock for the M13 is used because a 45 °C bath is needed for the soft agar – the standard 42 °C heat shock is a little better if two baths are available.

Acknowledgements

It is difficult to provide adequate references to describe some of these techniques as they are constantly subject to detailed revision and generally the improvements are not published but spread by word of mouth. I am indebted to these workers and also to members of my own laboratory who have made many subtle changes to the published methods. In particular I thank Ms Z.-Q. Han, Ms R. Prathalingam and Ms M. Ahmed and Dr C. G. Nicholl for help in preparing this manuscript.

References

Birnboim, H. C. and Doly, J. (1979) A rapid alkaline extraction procedure for screening recombinant plasmid. *Nucl. Acids Res.* **7**, 1513–23.

Del Sal, G., Manfioletti, G. and Schneider, C. (1988) A one tube plasmid DNA mini-preparation suitable for sequencing. *Nucl. Acids Res.* **16**, 9878.

Wang, K., Koop, B. F. and Hoo, L. (1994) A simple method using T4 DNA polymerase to clone polymerase chain reaction products. *Biotechniques*, **17**, 236–8.

15

A practical approach to cloning strategies and screening of cDNA libraries

JAMES R. GOLDENRING and IRVIN M. MODLIN

15.1 Introduction

The structure of proteins is encoded in the genetic sequences transcribed in messenger RNAs (mRNAs). Thus, the identification, isolation and resolution of these sequences is often critical for detailed analysis of protein function. Over the past twenty years, an evolution in technology has led to approaches that now make molecular cloning a normal and essential part of the armamentarium of any cellular biologist. In this chapter we review briefly the classical methods of molecular cloning, which have been refined over the preceding decades. In particular, we concentrate the bulk of our discussion on practical approaches to cloning now available with the advent of the polymerase chain reaction (PCR). This analysis is meant to serve as a practical introduction to the process of molecular cloning.

15.2 Requirements for cloning: what information is needed?

To embark on the exercise of molecular cloning, one must first have a clear goal. Most often the impetus for cloning is the resolution of a specific protein. In the case of cloning particular proteins, information is required to identify, with some certainty, the protein target of interest. In most cases, this information represents either a resolved amino acid sequence or antibodies that specifically recognize the polypeptide. At the present time, resolved amino acid sequence is a reasonable prospect in most strategies. As little as 100 pmols of purified protein is required for microamino acid sequence determination for proteins resolved onto PVDF membranes. For even low abundance proteins, preparative two-dimensional gel electrophoresis and high resolution chromatographic methods have made isolations possible. Nevertheless, since the great majority of eukaryotic proteins are blocked on their amino termini, it should be kept in mind that significantly more protein will be required if peptide fragments are required. Furthermore, multiple peptide fragments may need to be resolved to obtain sufficient amino acid sequence for cloning strategies. For example, peptides with long repeats of aliphatic residues may exhibit excessive homology to other proteins to serve as an adequate template for primer synthesis.

Specific antibodies raised against proteins have been used as important tools in cloning, originally in the precipitation of polysomes involved in translation of the protein of interest. The most important prerequisites for antibodies to be used in cloning are their specificity and availability. Thus, specific and high titre antibodies have the most utility in expression cloning. A sufficient supply of antibody must also be available, since large volumes of antibodies are often required for expression cloning strategies.

A third and important piece of information, that is often used in cloning strategies, is the presence of specific motifs, either amino acid or nucleotide sequences. This strategy has been utilized to isolate and clone members of gene families that share motifs such as GTP-binding sites (Chavrier, Simons and Zerial, 1992; Goldenring *et al.*, 1993) or calmodulin-binding sites (Picciotto, Czernik and Nairn, 1993). Similarly, known nucleotide sequences can be utilized in strategies to investigate truncated forms of RNA splice variant products or for determining a specific sequence for a protein previously cloned from a different species (Graves, Tomer and Davies, 1992).

15.3 Sources of sequence: Total versus poly(A)-mRNA (Options and troubleshooting)

The general starting point for cloning is messenger RNA (mRNA). The mRNA, reverse transcribed into complementary DNA (cDNA), contains the required sequence information. The choice of total RNA or poly(A)-mRNA (selected from total RNA using chromatography on oligo-deoxythymidine, oligo(dT), resin) depends on the strategy to be used for cloning and the method to be used in the priming of cDNA synthesis. In general, poly(A)-mRNA has only one major disadvantage: the possibility of either the loss of particular messages that may be extremely long or the loss of messages without polyadenylated tails (especially in the nervous system). Concerns over the loss or incomplete transcription of long messages (greater than 5 kb (kilobases)) have been somewhat abrogated by the recent introduction of reverse transcriptases with increased processivity and lacking RNAase H activity (e.g. Superscript-RT, Life Technologies), that allow more efficient transcription of long mRNAs. These newer reverse transcriptase enzymes should be used whenever message length is greater that 4 kb. In addition, one should note that the size of the protein coding sequence does not always correlate with the length of the corresponding mRNA. Thus, small proteins may have very long messages due to extensive 5′ and 3′-untranslated regions. For example, both transforming growth factor-alpha (TGFα) and bcl-2 have messages that are four to five times longer that their coding sequences (Derynck *et al.*, 1984). Furthermore, if the

3'-untranslated region is extensive, as in the case of bcl-2, standard reverse transcriptase protocols may not yield cDNA containing coding regions, or may at least give only incomplete coding region sequence (Cleary, Smith and Sklar, 1986). Finally, the investigator should be aware that adequate cDNA library construction generally requires poly(A) selected mRNA.

The major disadvantage of total RNA is its high content of non-mRNA, ribosomal RNA species. Nevertheless, ideally all possible mRNA species should be present. Still, if extremely low abundance messages are of interest, total RNA samples may not contain a sufficient number of copies of the sequence to allow successful cloning. the choice of starting material is often dictated by sources of RNA and the strategy employed. If very little material is available, then use of total RNA may be required, However, if cDNA production requires the use of random promers, then poly(A)-selected mRNA is required.

As a final point, it should be noted that genomic DNA can often be used to obtain required sequence elements or even to clone entire coding sequences from intron-less genes. The major problem with using genomic DNA lies in the unpredictability of the position of introns that may break up coding regions. Nevertheless, genomic DNA can often be an effective template of last resort for resolving short sequences, especially through polymerase chain reaction (PCR)-based strategies. These short sequences can then often be used in subsequent cDNA cloning strategies.

15.4 Classical cloning methods: Options and troubleshooting

While this chapter is focused on alternative PCR-based cloning strategies, it is important to realize that established methods of cDNA library screening still represent the best method for rapid resolution of *full-length* cDNAs. Several manuals discuss the technical methods involved in creation of cDNA libraries (see Chapter 16 and Sambrook, Fritsch and Maniatis, 1989), but a few important points are worthy of elaboration. First, perhaps the most important factor for the successful creation of a cDNA library is the quality of the mRNA. This is the step over which the investigator has the most control and input. If the poly(A)-selected RNA is pure and without significant degradation, then the resulting library is more likely to contain adequate representation of rare messages. Second, and perhaps as important, the investigator must consider carefully the method to be used in creation of the cDNA from the RNA. The techniques employed in the reverse transcription will vary depending on the intended use of the library. Thus, if the investigator is targeting the resolution of sequences from large proteins or receptors, then methods should be employed to ensure that the library contains long messages. Several techniques are available to accomplish these goals. The standard method for reverse transcription is to prime the reaction with oligo-deoxythymidine, oligo(dT). This strategy results in cDNAs that usually average 1.5 to 2 kb in length. However, recently novel reverse transcriptases have been introduced that possess higher efficency and yield reliably longer cDNAs (e.g. Superscript RT, Life Technologies). In addition, for those interested in cloning large proteins, cDNAs can be size-selected to create libraries enriched for long messages (e.g. greater that 1.5 kb). Creation of size-selected libraries often requires significantly greater amounts of mRNA starting material. Another traditional method for preparing libraries that can be utilized in cloning larger proteins employs random hexanucleotides to prime the reverse transcriptase reaction. Clearly, the purity of the mRNA template will be critical when using random priming, since any contaminating ribosomal RNA will also be transcribed. Size selection can also be applied after random priming to enrich for longer cDNAs. Finally, for complete representation of cDNAs, the best method presently available may be to prime the reverse transcription with both oligo(dT) and random primers. Indeed, for most general-purpose library construction, where DNA sequence screening will be primarily performed, the method of choice should be double priming.

15.4.1 Choice of vector

Many vectors are available for cDNA cloning and the choice of vector reflects both personal preference and intended application. Libraries can generally be constructed in either plasmids or variants of the λ bacteriophage. For most investigators, libraries propagated in bacteriophage are simpler to screen in larger numbers. The investigator should choose a vector based on the anticipated screening strategy. If the library screening will be based on antibody recognition or protein binding, then it must be constructed in a vector that incorporates an inducible fusion protein production system (e.g. the variants of λgt11 or LambdaZAP, Stratagene). The original λgt11 and LambdaZAP phage systems were not directional. Thus, the cDNA sequence could be inserted in either the forward or reverse orientation and in any of three frames. In general, orientation does not matter for screening of DNA sequences. However, when fusion protein production is used, on average, only one out of six cDNA inserts will yield fusion protein from a bidirectional cloning system. More recently, unidirectional cloning systems have been introduced (e.g. UniZAP, Stratagene), which utilize a modified oligo(dT) primer to eliminate inverted insertion. This process should double the number of in-frame clones that produce fusion protein. It is important to note also that hybrid cloning systems are now becoming available. These facilitate the construction of unidirectional libraries that can be used for screening in either bacteria or eukaryotic cells (e.g. ZAP Express, Strategene). Such libraries may be especially useful for those interested in cloning receptors based on ligand interactions.

15.4.2 Choosing a library: Make, borrow or buy?

Given the foregoing discussion, it is clear that a knowledge of the anticipated
screening strategies should lead the investigator to the correct choice of
libraries. Often an investigator will be able either to obtain an appropriate
library from another investigator or purchase it from vendors such as Stratagene or Clontech. Several caveats are important. First, identify a library that
appears to contain an adequate representation of cDNAs. In general, such a
library would have at least one million primary recombinants and preferably
much more. Libraries with higher numbers of primary recombinants should be
sought for successful expression screening. Investigators cloning high abundance proteins can generally be less selective. Second, if possible, obtain a
library that contains cDNA derived from the same species as your 'screening
information' (amino acid sequence or antibody target) and preferably from the
same tissue or cell type. If you are crossing species with an antibody screening,
ensure that the antiserum recognizes the epitope in the species in question.
Third, in the case where the investigator hopes to resolve the sequence of a
protein from a minor cell of a particular tissue (e.g. from an endocrine cell type
in the gastrointestinal tract), a library prepared from the tissue as a whole may
not yield sufficient numbers of clones. In this case, preparation of a new library
from a population enriched for the particular cell type will be more likely to
yield success.

   If no appropriate library is available, then the investigator must decide
whether to invest the time and money in creation of a new library with the
appropriate characteristics. Several kits are available from commercial vendors
(Stratagene, Promega, Clontech, Invitrogen) for the construction of bacteriophage libraries. These kits usually contain the reagents for up to five libraries.
The realistic time for production and characterization of a new library is 30 to
40 days or more. For the investigator interested in creation of a single library,
it may be more cost- and time-efficient to contract for the library fabrication.
Custom library production services are available from a number of vendors
including Stratagene and Clontech, and usually have 30-day turn-around times
(present cost is approximately $3000). While some services offer mRNA
production from tissue, as noted above, mRNA production is best left in the
hands of the investigator. Most services request $50-100\ \mu g$ of poly(A)-selected
mRNA.

15.4.3 Screening of cDNA libraries

The mechanics of plating and screening cDNA libraries are well covered
elsewhere (Sambrook *et al.*, 1989). We will therefore confine our discussion to
screening strategies, especially as they compare with the methods described
later for PCR-based cloning. There are three basic methods for screeing cDNA
libraries: (1) by protein expression; (2) using previously cloned DNA sequences; or (3) using synthetic oligonucleotides (see Chapter 16). Screening for

fusion protein expression induced in suitable vectors can be accomplished using appropriate high-affinity antibodies or labelled binding ligands. These methods require high-affinity binding to nitrocellulose replicas of plated phage or bacteria. As noted above, the success of this method is dependent on the quality of the library as reflected in the number of possible in-frame fusion protein products.

In the second case, protein cloning will often be undertaken based on a previously cloned sequence. These types of approaches will be used for recloning a protein from a different species or for investigating the presence of homologous but different protein species. Since these scenarios both seek non-identical sequences, the investigator will employ labelled cDNA probes in low stringency hybridization to replicas of either plated phage or bacteria. Generally, probing of duplicate lifts from plated libraries will decrease the number of artifactual isolates.

The third strategy of oligonucleotide screening is perhaps the most important technique to compare with PCR-based strategies. Like the PCR-based techniques to be discussed subsequently, phage or bacterial libraries are screened with synthetic oligonucleotides, constructed based on resolved amino acid sequences or motifs. Much of the following discussion of PCR primer selection also applies to the choice of oligonucleotide construction for library screening (see below). However, several aspects of oligonucleotide cloning should be highlighted. Synthetic oligonucleotides of 20–30 bp (base pairs) should be constructed to at least two separate areas of the target protein. The short length of oligonucleotides usually requires lower stringency screening. This necessitates that replicate lifts of libraries be screened simultaneously with at least two specific oligonucleotides. Colonies or plaques are then chosen based on their hybridization to both oligonucleotides. In general, as will be discussed below, primers should be selected for maximum hybridization temperature without increasing degeneracy much above 256-fold. For high and moderate abundance proteins, when a good library is available, oligonucleotide screening usually represents a better alternative than PCR for rapid cloning of *full-length* cDNAs.

15.5 Polymerase chain reaction (PCR) methods for cloning

The advent of the polymerase chain reaction (Mullis and Faloona, 1987) has now facilitated a number of strategies for cloning not readily available through established library screening (Arnheim and Erlich, 1992). Most obviulsy, PCR enables cloning from small amounts of RNA and therefore permits resolution of sequences where tissue quantity is a limiting factor (see Chapter 16). Techniques of rapid amplification of cDNA ends (RACE) also allow amplification of either 3' or 5' ends of mRNAs and therefore may be used as primary cloning methods or in conjuction with library screening (Frohman, Dush and Martin, 1988; Frohman, 1992). One important consideration in PCR-based strategies, compared with library screening, is that most of the strategies provide only partial sequence by themselves. Full length sequence often requires either combinations of PCR strategies (e.g. both 3′ and 5′ RACE) or combinations of PCR and traditional library screening. In addition, the investigator should recognize that the fidelity of PCR cloned sequences is not as high as cDNA library sequences. Mistakes in incorporation by Taq polymerase are estimated at 0.1 to 0.3% (Gelfand, 1990). Therefore, multiple clones of isolated inserts should be sequenced to ensure fidelity, or overlapping strategies should be used to check proper sequence.

15.5.1 *Considerations for PCR-based strategies*

As in all strategies for cloning, the approach will depend on the amounts and types of information available. Antibody screening for protein expression is not generally appropriate for PCR-based cloning and is better approached through expression library-based strategies. However, if resolved micro-amino acid is available or DNA sequence motifs are known, then PCR-based methods can often provide the required sequences. There are basically three stages to PCR cloning: (1) primer fabrication; (2) choice of approach; and (3) optimization of reactions. Each of these issues will be discussed in turn.

15.5.2 *Creating the correct primers*

Primer design is the most important step in any PCR cloning strategy. Synthetic oligonucleotide construction is available through a number of commerical sources, as well as through in-house facilities at many universities. The investigator should approach this process with a great deal of thought. The basic principles for primer construction are that the primer should have the highest specificity possible with the lowest degeneracy and highest melting temperature. These principles are often difficult to achieve, since PCR cloning strategies usually require the use of degenerate primers. Degeneracy in synthetic oligonucleotide primers is often required because of sequence ambiguities, especially in the third position of many triplet amino acid codons. A great deal has been written concerning primer construction strategies (Goldenring *et al.*, 1994) (see Chapter 16, also Wadda *et al.*, 1991 and Nichols *et al.*, 1994).

15.5.3 *Choosing the right strategy*

The choice of exact strategy, as noted above, will depend on the amounts and types of information available. The following section describes the uses of three strategies that are readily available for most investigators.

15.5.3.1 3'-Rapid amplification of cDNA ends (3'-RACE)

The 3'-RACE strategy focuses on the amplification of the 3'-end of the coding region of a candidate protein along with the entire 3'-untranslated portion of the mRNA (Figure 15.1). In this procedure, a cDNA is constructed by reverse transcription with oligo(dT) linked on its 5'-end with a 17–20 bp sequence tag (3'-RACE amplimer). The resulting 3'RACE cDNA is 'tagged' at each 5' end with a known sequence. Thus, when this cDNA is used in a cloning strategy, the 3'-RACE amplimer sequence can be used as a known anti-sense primer anchor. Only a single degenerate sense primer is needed. Therefore, sequence from only a single peptide fragment sequence is necessary for construction of a primer and cloning. Many oligonucleotide sequences have been used for RACE strategies, and the exact RACE amplimer compostion can be altered to give a melting temperature which approximates the sense primer. In practice, we have used an established 17 bp sequence with a great deal of success (GACTCGAGTCGACATCG). Other general sequences have been used, and kits for 3'-RACE have recently become available (e.g. from Life Technologies

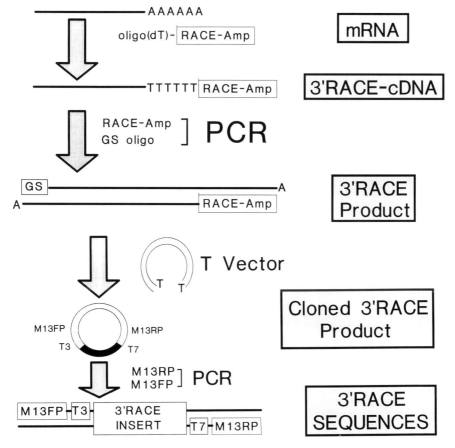

Figure 15.1. Scheme for 3'-RACE amplification with cloning into T-vector.

or Clontech). The primers in these kits possess longer sequences with higher T_ms. We have recently found that these longer primers may yield more efficient cloning when higher annealing temperatures can be used.

While 3'-RACE is an extremely versatile approach, it may often fail to yield adequate results. In all PCR-based strategies, one must realize that products with sized above 1500–2000 bp are not likely to be amplified as efficiently. While newer heat-stable polymerases with greater processivity may extend the bounds of clone size, for practical purposes the goals of most amplifications should be kept below 1500 bp. Thus, if a message includes an extremely long 3'-untranslated region, 3'-RACE may not be able to amplify adequate levels of product. Also, since the length of the resulting product may not be predictable, it is important that the investigator analyse specific amplification products across a range of sizes in comparison with amplifications performed with each of the sense and anti-sense primers individually. Bands resolved specifically only with the combination of both sense and anti-sense primers can then be isolated and tested for specificity by reamplification with the original primers.

15.5.3.2 5'-Rapid amplification of cDNA ends (5'-RACE)

The 5'-RACE strategy is the inverse of the 3'-RACE and is targeted at resolution of 5' coding region sequence along with the 5'-untranslated region (Figure 15.2). Compared with the 3'-RACE, the 5'-RACE strategy requires more sequence because two nested anti-sense gene-specific primers are required. The first gene-specific primer is utilized to create a specific cDNA by using the anti-sense oligonucleotide to prime reverse transcriptase. The resulting cDNA is then tailed using terminal transferase. This tailing step, in principle, could employ any deoxynucleotide and the original strategies used

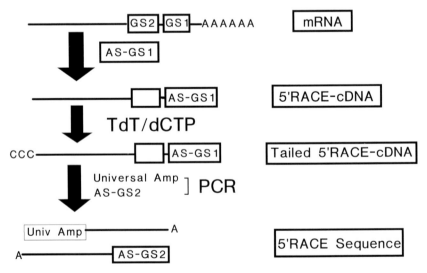

Figure 15.2. Scheme for 5'-RACE amplification.

deoxyadenosine. More recently we have found that a protocol of deoxycytosine tailing used a commercial kit from Life Technologies provides reliable tailing of relatively uniform length. Following tailing, second-strand synthesis is accomplished using a universal adapter primer (Life Technologies) containing a RACE amplimer sequence linked to an oligo-dC/dI 3' extension. PCR resolution then requires amplification using a combination of the universal adapter primer (or 5' amplimer sequence) with a second gene-specific anti-sense primer for a region nested towards the 5' side of the original primer used to make the cDNA. Once again, the RACE amplimer sequence in the original primer can be matched with the approximate T_ms for the gene-specific primers. In practice we have found that this has not been necessary.

As with the 3'-RACE protocol, the resolution of sequences is constrained by the ability of the PCR to resolve clearly sequences of less than 1500 bp. Thus the length of 5'-untranslated ends can also hinder resolution of the sequence. Also, like the 3'-RACE, it is not possible to predict a priori the exact length of the amplified product unless some knowledge of mRNA size is known from Northern blot. Since two anti-sense primers are needed, this strategy can often require two different degenerate primers. Certainly, in 5'-RACE more amino acid sequence is required for successful design of primers. The use of a single oligonucleotide to prime both reverse transcription and anti-sense PCR is not advised, and will usually not yield signficant product. Nevertheless, this method has been used as a primary method for the cloning of gene family members where two polypeptide motifs were known (Chavrier *et al.*, 1992). Clearly, this method works best when some preliminary sequence has been resolved, either through another PCR strategy (e.g. 3'-RACE) or through library screening. When some sequence is known definitively, the construction of two nested primers is simple and the yields are excellent (Goldenring *et al.*, 1993). This method can be especially helpful in completing sequences obtained from library screening, where there is concern as to whether the true 5' coding initiation site has been reached.

15.5.3.3 Resolution of defined sequence through PCR

A third method for at least partial cloning of genes utilizes a pair of sense and anti-sense primers derived from either amino acid sequences or motifs (Figure 15.3). As noted above, this type of strategy requires careful construction of primers with high T_ms and low degeneracy. The major advantage of this method is that it can be highly specific. In general, the critical factor in these targeted strategies is an ability to amplify with nested primers. This nesting can be performed in two ways: First, the reverse transcriptase reaction can be performed with an anti-sense primer followed by subsequent amplification with a sense primer and an anti-sense primer nested 5' to the original anti-sense primer (Figure 15.3). This procedure is similar to that used for 5'-RACE. Second, especially in the case of rare messages, amplification can be performed

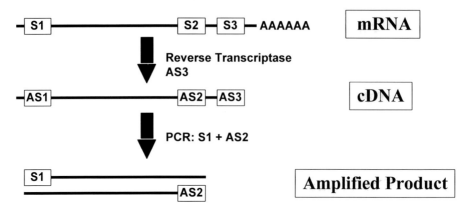

Figure 15.3. Scheme for nested amplification by gene-specific reverse transcription followed by nested PCR.

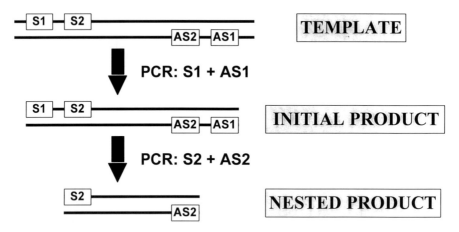

Figure 15.4. Scheme for amplification of rare sequences by initial PCR followed by nested primer reamplification.

initially with one set of sense and anti-sense primers, followed by reamplification with a second pair of primers nested inside the original set (Figure 15.4). This latter method can often resolve, with high specificity, extremely rare sequences.

These nested approaches have been highly successful in recloning sequences from other species when the sequence is known in one species (Goldenring *et al.*, 1994). Similarly, amplification of motifs such as calmodulin-binding sites has been used to obtain novel partial sequences corresponding to calmodulin-binding proteins (Picciotto *et al.*, 1993). Furthermore, when two or more peptides have been resolved from a protein under consideration, this strategy may provide the most specific sequence possible. If some knowledge is available of the positions of peptides or the positions of motifs in target proteins, predictions can be made concerning the anticipated size of amplification products. This knowledge aids the investigator immeasurably in differentiation of the specific product from artefact.

15.5.4. *Optimization of product amplification*

Since most PCR cloning strategies require the use of degenerate primers, the greatest impediment to success is amplification artefact. Differentiation of amplification artefact from correct product is often difficult, especially when the actual size of the products can not be predicted. Mispriming not only creates artefact bands, but also decreases the concentrations of key reagents (i.e., primers and nucleotides). As discussed above, much of the mispriming and primer dimer formation can be eliminated by thoughtful construction of primers. However, in the face of degeneracy, where exact sequences can not be predicted accurately, optimization of reaction conditions is often required. Several strategies are available to guide the investigator through the maze of PCR cloning artefact (see pages 191–197).

15.5.4.1 *Ensure that the reverse transcriptase reaction worked*

The success of cDNA production is critical for the subsequent amplification process. Therefore, if possible, a known PCR primer pair should always be included as a positive control. In the case of oligo(dT) or random primed cDNAs, along with 3'-RACE cDNAs, this can usually be a commercial primer pair for a constituitively expressed protein such as glyceraldehyde-3-phosphate dehydrogenase (GAPDH). In the case of 5'-RACE, the only positive control that is possible will be a parallel cDNA production using oligo(dT) and PCR amplification of a known sequence. In that case, at least the activity of the reverse transcriptase can be verified.

15.5.4.2 *Utilize 'hot start' procedures*

Protocols which utilize a 'hot start' paradigm expose the PCR amplification reactions to denaturing conditions prior to the addition of Taq polymerase. In this method, Taq polymerase is initially omitted from reactions, which are cycled to 95 °C to denature template and eliminate any primer associations. The temperature is then ramped down to and held at 80 °C. Taq is then added to this pre-denatured mixture and the cycling is initiated. The hot start protocol prevents false priming and extension during the initial ramping from ambient temperature to the 95 °C denaturation. Since the initial cycles are critical for accurate propagation of amplified materal, this method can significantly reduce artefactual priming.

An alternative to the traditional hot start employs antibodies to Taq polymerase to sequester Taq polymerase (see Chapter 11). These antibody preparations (e.g. TaqStart, Clontech) bind to Taq at ambient temperatures. However, when the reactions are ramped for the first time to 95 °C, the antibodies are denatured, releasing Taq without loss of activity. This antibody sequestration method appears to produce results similar to a regular hot start although the antibody reagent is an added expense. Since hot start strategies have no significant contraindications, they should be employed routinely in all PCR cloning strategies utilizing degenerate primers.

15.5.4.3 Optimize reaction mixture conditions

The yield and fidelity of priming is determined, as noted above, by the coordinated binding of sense and anti-sense primers and extension of product (see Chapter 11). All reaction components can impact on the reaction results:

(a) MgCl₂ concentration

The concentration of Mg^{2+} is the most critical component of the reaction mixture. Mg^{2+} ion is required for chelation of nucleotide triphosphates and the activity of the polymerase. The free Mg^{2+} ion concentration may be significantly altered by the concentration of nucleotides, template, and primers, all of which bind Mg^{2+}. Thus, optimization of the Mg^{2+} concentration should be the first goal. In practice, 2–3 mM $MgCl_2$ is usually sufficient to sustain most PCR mixtures. However, if large amounts of template or primers are present, then $MgCl_2$ should be optimized throughout a concentration range of 1.0 to 4.0 mM $MgCl_2$, in steps of 0.5 mM. This is rather easily accomplished since many Taq preparations are now furnished with magnesium-free buffer systems to which variable amounts of a standard $MgCl_2$ solution (25 mM) may be added separately.

(b) Reaction pH

Buffer pH may alter both the efficiency of primer binding to template as well as Mg^{2+} binding to nucleotides, template and primer. Thus optimization of reaction pH, often in conjuction with optimization of Mg^{2+}, can significantly alter artefactual priming. Several kits have recently been introduced (e.g. PCR Optimizer, InVitrogen) which allow simultaneous optimization of pH and $MgCl_2$.

(c) Single-stranded DNA-binding protein

E. coli ssDNA-binding protein is a heat-stable protein that binds to unpaired single-stranded regions of DNA. This binding can disrupt mispaired priming by destabilizing incomplete hybridization of primers with template. This effect decreases mispriming significantly and can therefore allow cycling to be performed at lower annealing temperature. Thus, successful priming by an oligonucleotide sequence with a low T_m can be accomplished with increased efficiency. Clearly this can be an advantage in degenerate PCR cloning strategies. Commercial sources of ssDNA-binding protein are now available (e.g. Bind-Aid, USB). These preparations are best utilized in combination with a hot-start protocol that blocks mispriming in the initial cycle. It should be noted that the use of these binding proteins may not be optimal for some strategies, especially in motif cloning where there may be some measure of uncertainty concerning the exact sequence.

(d) Taq polymerase

We have observed few differences in the results obtained from different Taq polymerase preparations. Nevertheless, we have observed that certain universal buffer system formulations (e.g. Promega) appear to be more tolerant of alterations in Mg^{2+} concentrations. It should also be noted that polymerases with 5′-exonuclease proofreading capacity (e.g. Pfu and Vent) should be considered when anticipated PCR product length is above 1.5 kb.

(e) Nucleotide concentrations

In general, nucelotide concentrations should be in excess for most reactions. However, since the nucleotide concentration alters Mg^{2+} concentration, in some cases examination of a range of nucleotide concentrations may yield better amplification results.

(f) Primer concentrations

In the case of degenerate primers used in PCR cloning, the actual concentration of oligonucleotide with the correct sequence among a group of degenerate sequences will be crucial for proper amplification. For example, if, under the parameters of annealing and extension, only a single primer species out of a 32-fold degenerate oligonucleotide is able to prime the reaction, then the concentration of that primer may limit the amount of product amplified. The problems can be compounded if several of the other species are involved in the production of spurious products. The optimization of primers is easier in anchored strategies such as 3′-RACE where only one degenerate primer is required. In the case of 3′-RACE, PCR reactions can be performed across a wide concentration range for the degenerate primer, while the RACE amplimer sequence is fixed. In the course of such a concentration dependent optimization, the investigator should note that, while non-priming oligonucleotide sequences may not be participating in the reaction, they will be binding Mg^{2+} ions. Thus, primer dependence should be performed in the presence of higher Mg^{2+} concentrations (usually 3.0 to 4.0 mM).

(g) Optimizing cycling temperatures

Cycling temperatures are clearly critical to the success of any amplification process. While much of the choice of cycling temperature is empirical, several points are worthy of consideration. First, one of the major causes of failed amplification is insufficient denaturation of template strands. For this reason, it is always advisable to begin cycling with a 3–5 min denaturation cycle at 95 °C. The half-life of Taq polymerase at 95 °C is 40 min (Gelfand, 1990). If extended high temperature melting is necessary over many cycles, it may be advisable either to increase the concentration of Taq in the reactions or to use one of the newer heat-stable polymerases such as Pfu (Stratagene) or Vent (New England Biolabs).

Second, when considering the correct annealing temperatures, evaluate the predicted annealing T_ms of the particular primers. In the case of degenerate primers, choose a minimum T_m from the possible combination of nucleotides. In general, a 'safe' annealing temperature of 5 °C below the T_m for the oligonucleotides. The efficiency of priming and extension is generally considered better at annealing temperatures of 55 °C or greater. However, it may not be possible to achieve this goal, especially when limiting degeneracy is a major concern. Nevertheless, if degeneracy can be minimized, it is often better to construct longer primers that may have higher T_ms (i.e. greater than 60 °C).

Third, it is always important to consider that the results of amplification are ultimately determined by the success of product formation in the first five cycles. In most cases, especially where degenerate primers are employed, the initial one to five cycles should be performed at a low stringency temperature (i.e. 37–42 °C). While low stringency temperatues might also lead to mispriming, the likelihood for the sequence having both sense and anti-sense primable sequences is relatively small. Thus, misprimes should generally only be amplified arithmetically, rather than exponentially, as in the case of correctly amplified material. One important exception to this is found in RACE protocols where the cDNAs are uniformly anchored with the amplimer sequences. Especially in 3'-RACE, mispriming from the sense primer can lead to significant amounts of spurious product. Despite these possible concerns, the importance of early product formation generally prevails, especially when rare message cDNAs are sought. In addition to the low stringency temperature in the initial cycles, a slow ramping speed between annealing and polymerization temperatures (e.g. 0.5 degrees or less per second) has often yielded improved initial product formation. Following the initial cycles, the annealing temperature can be stepped to an intermediate temperature for a further five cycles (e.g. 10 °C below the T_m followed by 20 to 30 cycles at 5° below the T_m). As noted above, addition of ssDNA binding protein can assist in reactions where uncertainty concerning T_ms of degenerate primer pairs may make the choice of annealing temperature difficult. In general, greater than 40 total cycles should be avoided since larger number of cycles are often complicated with plateau effects. This problem reflects a limiting amount of correct product in the face of increasing quantities of artefactual amplifications and even hybrid products. In addition, when greater than 40 cycles are employed, the loss of Taq activity may become a limiting factor.

(h) Dimethyl sulfoxide (DMSO)

DMSO is a potent solvent, which can improve the specificity and stability of hybridization reactions. The actual mechanism of its reaction is not clear, but it has been used as an important adjunct in PCR performed with the Klenow fragment of DNA polymerase (Scharf, Horn and Erlich, 1986). Unfortunately, the 10% DMSO concentraion that has been used with Klenow causes a 50% reduction in Taq activity. Still, addition of DMSO in concentrations of 1–10% may be used as a method of last resort when other methods have failed to elicit successful amplification.

15.5.5 Cloning of PCR products

Several methods are available for the expedited cloning of the products of PCR reactions. First, restriction sites can be engineered in the primers. When restriction sites are added on the 5'-ends of oligonucleotide primers, it is necessary to add three to four extra bases past the restriction recognition sequence to ensure effective cutting. A sequence of GGCC is a convenient tetrabase sequence for this purpose. The major advantage of the addition of restriction sites is their applicability for directional cloning into the plasmid vector of choice. This may aid in later construct formation and simplify subsequent sequencing. The major disadvantage of incorporation of restriction sites accrues during degenerate cloning strategies, where the extension may elicit additional spurious priming and increase background.

Second, a number of methods have been developed recently that allow direct cloning of PCR fragments into plasmid vectors. These strategies utilize 'T-vectors' constructed by the method of Marchuk, *et al.* (1991). This method utilizes the propensity of Taq polymerase to leave an extra adenosine moiety at the 3' end of a polymerized sequence. T-vectors are constructed with complementary single thymidine residues at the 3' free ends of the linearized vectors. The T to A pairing allows rapid ligation of PCR products into vector, usually without specific purification of PCR insert. We have found that the process in blue–white screenable vectors is fast and efficient. Several vendors now sell T-vector cloning systems (e.g. InVitrogen, Novagen), but in our experience the original method for pBluescript often yields the best results (Marchuk *et al.*, 1991).

Third, whatever the method used to clone the PCR-derived inserts, one should choose the particular plasmid vector based on the anticipated applications. Plasmid flanking regions can often furnish important 5' or 3' restriction sites that are necessary for subsequent steps leading to the production of recombinant proteins and expression vectors. For example, we have found that the pBluescript T-vector allows rapid production of full length constructs for transfer to pET expression plasmids (Novagen), since the cloning site (at the original EcoRV site) is flanked by *Bam*HI and *Xho*I restriction sites necessary for the 3′ end of the insert. Similarly, when the inserts will be used for RNAase protection assay development, one should consider vectors that have 3′ and 5′ RNA polymerase initiation sequences.

15.5.6 *When to reamplify and advanced screening techniques*

The question will often arise during the course of a cloning strategy as to either the fidelity or presence of amplification products. This is especially true when the target of amplification is a rare message, or when gene family or motif cloning produces a heterogeneous mixture of sequences. We have found that several approaches may help clarify these situations. First, as mentioned above, comparison of amplification results obtained from reactions containing both sense and anti-sense primers, with results when cycling is performed with each of the primers alone, can often clarify questions of specificity.

Second, if some knowlege of sequence characteristics is known, then Southern blotting of amplification products resolved on agarose gels can be invaluable. For example, when recloning a protein from another species, intermediate stringency hybridization of the known sequence to a Southern blot of the amplified sequences can identify the presence of authentic product. This information can then be used for gel isolation and subsequent reamplification. Alternatively, the identified region of products can be gel isolated, T-cloned and then the subsequent transformed bacteria can be screened for the proper sequence on colony replicas.

Third, often size selection can be used to eliminate artifactual bands and enrich for appropriate products. For example, if the product is expected to be between 600 and 1000 bp, then amplification products can be resolved on agarose gels and the products in the correct size range gel isolated. The resulting inserts can then either be cloned and screened or reamplified. Reamplification should be carried out at higher annealing temperatures than the original reactions amplifying cDNA. In general, copy number will not be limiting in these reamplifications. Also, reamplifications should generally be limited to 25–30 cycles, again using reactions with sense and anti-sense primers alone as controls.

Fourth, as mentioned above, if nested primers are available, gel isolation of size-selected products, followed by reamplification will often provide some of the most specific results for minor cDNAs (Figure 15.4).

Finally, for motif cloning or resolution of gene family members, it may be necessary to resolve sequences by direct cloning (Goldenring *et al.*, 1993). In this circumstance, we have found that size selection followed by immediate cloning in T-vector provides a large number of suitable clones that can be rapidly sequenced using flanking RNA polymerase initiation sequences (e.g. T3 or T7 in pBluescript). Much of this screening can be expedited by PCR amplification from plasmid micro preps or insert using flanking M13 sequences, followed by doubled-stranded sequencing using a PCR-based system (e.g. Fmol, Promega) with T3 and T7 sequences (which are nested inside the M13 sequences).

15.5.7 *Alternate and hybrid techniques*

Particularly in the case of low abundance proteins, the process of cDNA cloning requires both creativity and perserverance. The investigator should have full command of the complete armamentarium of cloning techniques, and be willing to move between them when necessary. Hybrid strategies are currently the norm. Thus, a short length of sequence may be discerned through PCR amplification of a particular motif and then used as a high stringency probe in cDNA library screening. Similarly, several PCR techniques have recently been employed to screen cDNA and genomic libraries using primers resolved from a short length of PCR-derived sequence. Through PCR screening of libraries subdivided into subaliquots, one can rapidly reduce the task of library screening from one million plaques to less than a thousand (Takumi and Lodish, 1994). This technique has been combined with robotic technology to allow resolution of genomic sequences in less than 10 days based on the presence of a known 150–200 bp segment of coding region (Genome Systems).

As a last method, the investigator should recognize that cDNA libraries themselves are often an excellent source of DNA sequences for PCR amplification. The most important aspect of cDNA libraries is that they often represent modalities for anchored PCR since the insert sites are flanked by known sequences for RNA polymerases (T3, T7 and SP6) and M13 recognition sites. Complementary oligonucleotides may therefore be used in anchored PCR strategies or even to amplify the entire spectrum of inserts in a library. In practical terms, due to the large amount of homologous phage or plasmid DNA, the primers require higher annealing temperatures than those employed in cDNA-based PCR strategies. Nevertheless, when an excellent library is available, this method may be a rapid way to resolve sequences.

Further innovations are likely in the coming years. New technologies will expand the range of sequence lengths that are amenable to amplification, and higher fidelity polymerases will improve the accuracy of cloned sequences. As these new modalities become available, each investigator should adopt whichever strategy optimally facilitates his or her individual goals.

References

Arnheim, N. and Erlich, H. (1992) Polymerase chain reaction strategy. *Meth. Enzymol.* **61**, 131–56.
Chavrier, P., Simons, K. and Zerial, M. (1992) The complexity of the rab and rho GTP-binding protein subfamilies revealed by a PCR cloning approach. *Gene* **112**, 261–264.

Cleary, M. L., Smith, S. D. and Sklar, J. (1986) Cloning and structural analysis of cDNAs for bcl-2 and a hybrid bcl-2/immunoglobulin transcript resulting from the t(14;18) translocation. *Cell* **44**, 19–28.

Derynck, R., Roberts, A. B., Winkler, M. E. *et al*. (1984) Human transforming growth factor-alpha: precursor structure and expression in *E. coli. Cell* **38**, 287–97.

Frohman, M. A. (1992) Rapid amplification of cDNA ends (RACE): user-friendly cDNA cloning. *Amplifications* **2**, 11–15.

Frohman, M. A., Dush, M. K. and Martin, G. R. (1988) Rapid production of full-length cDNAs from rare transcripts: amplification using a single gene-specific oligonucleotide primer. *Proc. Natl. Acad. Sci., USA* **85**, 8998–9002.

Gelfand, D. H. (1990) Thermostable DNA polymerases. In *PCR protocols: a Guide to Methods and Applications* (ed. M. A. Innis, D. H. Gelfand, J. J. Sninsky and T. J. White), pp. 129–41. Academic Press, San Diego.

Goldenring, J. R., Shen, K. R., Vaughan, H. D. and Modlin, I. M. (1993) Identification of a small GTP-binding protein, Rab25, expressed in the gastrointestinal mucosa, kidney and lung. *J. Biol. Chem.* **268**, 18419–22.

Goldenring, J. R., Soroka, C. J., Shen, K. R. *et al*. (1994) Enrichment of rab11, a small GTP-binding protein, in gastric parietal cells. *Am. J. Physiol.* **267**, G187–94.

Graves, P. N., Tomer, Y. and Davies, T. F. (1992) Cloning and sequencing of a 1.3 kb variant of human thyrotropin receptor mRNA lacking the transmembrane domain. *Biochem. Biophys. Res. Commun.* **187**, 1135–43.

Marchuk, D., Drumm, M., Saulino, A. and Collins, F. S. (1991) Construction of T-vectors, a rapid and general system for direct cloning of unmodified PCR products. *Nucl. Acids Res.* **19**, 1154.

Mullis, K. B. and Faloona, F. A. (1987) Specific synthesis of DNA in vitro via a polymerase-catalyzed chain reaction. *Meth Enzymol.* **155**, 335–50.

Nichols, R., Andres, P.C., Zhang, P. and Bergstrom, D. E. (1994) A universal nucleoside for use at ambiguous sites in DNA primers. *Nature* **369**, 492–3.

Picciotto, M. R., Czernik, A. J. and Nairn, A. C. (1993) Calcium/calmodulin-dependent protein kinase I: cDNA cloning and identification of autophosphorylation site. *J. Biol. Chem.* **268**, 26512–21.

Sambrook, J., Fritsch, E. F. and Maniatis, T. (1989) *Molecular cloning: a laboratory manual*. Cold Spring Harbor Laboratory Press, Cold Spring Harbor, New York.

Scharf, S. J., Horn, G. T. and Erlich, H. A. (1986) Direct cloning and sequencing analysis of enzymatically amplified genomic sequences. *Science* **233**, 1076–8.

Takumi, T. and Lodish, H. F. (1994) Rapid cDNA cloning by PCR screening. *Biotechniques* **17**, 443–4.

Wadda, K., Wada, Y., Doi, H. *et al*. (1991) Codon usage tabulated from the GenBank genetic sequence data. *Nucleic Acids Res.* **19**, 1981–5.

16

Identification and isolation of novel DNA and cDNA species

FIONA K. BEDFORD
and JANUSZ A. Z. JANKOWSKI

16.1 Introduction and background

Novel genes are defined as newly identified coding portions of the genome. They can be isolated by a number of different approaches including methods biased either to the function of the protein or more specifically to the DNA sequence. The methodologies described in this chapter encompass those based on isolating novel genes using specific characteristics in their DNA sequence. A simple overview of each approach is given, including the disadvantages and advantages of each technique.

Historically, novel genes have been isolated as a result of the unique cellular phenotype they encode, either as a wild-type or mutated gene. In humans the phenotype of mutated genes is most easily observed when they encode novel oncogenic potential, resulting in deregulated proliferation and aberrant differentiation, characteristic of either solid tumours or leukaemia (Cory, 1986). Mutation of specific genes, however, can also be associated with an alteration in normal psychology and behaviour, genetic birth defects and other physiological processes in the body. Whilst the techniques described below can still be employed to isolate novel mutant genes, several of these are also sensitive enough to isolate non-mutated genes without prior knowledge of the phenotype they encode.

The isolation of novel genes is important for several reasons. Firstly, the isolation of novel mutated genes associated with pathogenic diseases may help to reveal the origin of the disease and possibly lead to a treatment. An example of this is acute promyelocytic leukaemia (APL) in which one copy of a transcription factor, the retinoic acid receptor, is mutated by a chromosomal translocation, resulting in aberrant differentiation of the myelomonocytic haematopoietic lineage (Borrow *et al.*, 1990; De The *et al.*, 1990). This discovery gave a scientific rationale to the previously known successful treatment for this leukaemia, namely the induction of differentiation of the leukaemic cells with all-trans retinoic acid. Secondly, the identification of novel wild type genes that are closely related to known oncogenes may aid in the discovery of new potential oncogenes.

To date it is estimated that only 5% of the genes encoded within the human genome have been identified and therefore the map of the human genome is incomplete. Nevertheless, knowledge of the human genome will have enormous beneficial effects in the diagnosis and treatment of genetically based diseases. The rate at which novel genes are being identified has more recently increased as a direct result of identifying large gene families. In this regard, cloning techniques similar to low stringency hybridization and degenerate oligonucleotide primed PCR have been developed to isolate several genes simultaneously by exploiting their structural and sequence homologies within functional domains. This is particularly noteworthy of transcription factors, where gene families containing more than 500 members exist, which encode highly sequence conserved DNA binding motifs. Other functional domains such as protein–protein interaction domains and extracellular ligand-binding domains of membrane receptors are also highly conserved and suitable for this kind of experimental manipulation.

Although the topic of this chapter is the isolation of novel genes, the techniques used primarily identify differences in gene expression between different RNA samples and have been modified for novel gene identification. Therefore, it is noteworthy to remember that differences in gene expression will not always lead to the isolation of novel genes. Another factor to be cautious of is the possibility of isolating false positive or false negative clones, in essence, the isolation of cDNA clones that do not adhere to the criteria for selection but are merely a by-product of the cloning technique. In this regard, it is essential that extensive analysis of the isolated clones be performed after their isolation, to confirm that they are novel and do adhere to the criteria for selection.

Novel gene identification is clearly an experimental approach that is or should become an integral part of most clinical laboratories. Moreover, these techniques have several applications to medical science: Novel gene identification will aid in diagnosis (i.e. identification of genetic lesions), prognosis (understanding the molecular origin and progression of the disease) and therapy (identifying appropriate treatment, i.e. specific growth factors or steroids).

16.2 Methodological overview

16.2.1 Subtractive hybridization

The technique of subtractive hybridization is particularly useful in identifying specific genes that are uniquely expressed in one tissue or cell type. This involves comparing RNA (cDNA) species between two different tissues or cell extracts. For example, this technique can be used to identify which genes are specifically induced in cell lines after culture with inducing agents such as growth factors. A specific example where this technique has been used successfully is in the cloning of a novel GTP-binding protein (Schenker *et al.*, 1994).

Subtractive hybridization most commonly involves preparing a cDNA library from the RNA of interest and isolating cDNA clones that do not hybridize to labelled cDNA (RNA) prepared from the control sample, i.e. by subtraction. This method has been well documented (Yancopoulos and Alt, 1990). The advantages of this technique are that it requires no prior knowledge of the cDNA species to be isolated and that several non-related cDNA species can be isolated at the one time. The disadvantages are primarily that the technique relies heavily on hybridization kinetics, requiring a huge excess of RNA to cDNA otherwise false positives can be isolated and false negatives missed.

An alternative approach has been described (Ausubel *et al.*, 1987), which may increase the selectivity of the method as well as being a non-radioactive approach. In this method the cDNA of interest is designated [+] while the cDNA to be used for the subtraction is designated [−]. Briefly, [+] cDNA with *Eco*RI ends and [−] cDNA with blunt ends are prepared either from pre-made cDNA libraries or *de novo* from poly(A)$^+$ RNA. The [−] cDNA is digested with *Rsa*I and *Alu*I to generate small blunt-ended fragments and then mixed at a 50-fold excess with the [+] cDNA. The mixed cDNAs are heated to melt the dsDNA and then complementary strands are allowed to anneal. After the annealing step, double-stranded *Eco*RI fragments are ligated into *Eco*RI-cut cloning vectors. These fragments are unique to the [+] cDNA source, as [+] cDNA will only generate clonable *Eco*RI fragments if there are no complementary fragments present in the [−] cDNA. For example, a [+] sequence with a complementary [−] sequence will hybridize to one of the *Alu*I/*Rsa*I [−] fragments and become a non-clonable, partially single-stranded, partially double-stranded molecule.

16.2.2 Low stringency hybridization

Low stringency hybridization is the traditional approach used to isolate related cDNA species. For example, members of the *PBX* (Monica *et al.*, 1991) and *MOX* (Candia *et al.*, 1992) families of homeobox transcription factors have been isolated using this technique. This approach can also be used to isolate equivalent genes from phylogenetically diverse organisms, in which studying the function of the gene may be simpler, e.g. *Caenorhaebiditis elegans*.

The basic tools required for using this technique, therefore, are the availability of a cDNA or genomic DNA library and a suitable probe, also of either DNA or cDNA origin. If a suitable probe is not available, one can be generated simply by PCR (see Chapter 11), as long as the desired sequence to be amplified is available. Degenerate oligonucleotides (described further in the next section) have also been used successfully as alternative probes for low stringency hybridization screening (Deguchi *et al.*, 1991). The protocol below is a suitable screening procedure that can be used to screen bacteriophage, plasmid or cosmid libraries. It is important when hybridizing at low stringency to keep the background to a minimum (signal:noise ratio), as this will reduce the number of false positives. Probing Southern and Northern blots under various conditions prior to screening will enable suitable hybridization conditions to be optimized. Hybridization to Northern blots may also indicate which cDNA libraries will be most useful to screen. The advantages of using low stringency hybridization are that whole cDNAs for several members of a single gene family can be isolated at one time with few false positives. The disadvantage of this approach is that it is limited to isolating novel genes that are related to previously isolated genes.

16.2.2.1 Basic protocol

1. Plate out a cDNA or genomic library on appropriate agar plates and make two sets of duplicate filters using nylon membranes (Sambrook, Fritsch and Maniatis, 1989) in order to reduce false positives. Orientation marks should be made with the agar plate on the filter by using a needle and India ink.

2. Place the filters (colony side up) on 3MM Whatman paper soaked in 10% (w/v) SDS for 5 min (omit this stage when bacteriophage screening), then transfer to paper soaked in denaturing solution (1 M NaCl; 0.5 M NaOH) for 5 min. Transfer to paper soaked in neutralization solution (1.5 M NaCl; 0.5 M Tris-HCl, pH 7.2) for 5 min and finally to paper soaked in $2 \times$ SSC for 5 min. Allow the filters to air-dry for 30 min then bake them at 80 °C for 2 h under vacuum to fix the DNA to the nylon membrane. Alternatively, crosslink the DNA to the membrane using UV.

3. Prewash the filters in solution W ($5 \times$ SSC; $0.5\% \times$ SDS; 1 mM EDTA, pH 8.0) at 42 °C for 30 min and rinse briefly in $2 \times$ SSC, to remove bacterial debris and therefore reduce non-specific background.

4. Prehybridize the filters in prehybridization solution (0.72 M NaCl; 40 mM NaH_2PO_4, pH 7.6; 4 mM EDTA; 0.2 mg/ml denatured salmon sperm DNA; 2 g/L PVP; 2 g/L Ficoll; 0.1% SDS) for 1 h at 65 °C.

5. Remove the prehybridization solution and replace with hybridization solution (same as prehybridization solution with the addition of 90 g/L dextran sulphate, sodium salt), and kept at 65 °C (minimum of 1 h) until the probe (^{32}P labelled) is added. Hybridization is carried out overnight at 65 °C.

6. Following hybridization, wash the filters at low stringency. Suggested wash conditions are: two washes in $5 \times$ SSC; 0.5% SDS at 65 °C for 20 min and one wash in $2 \times$ SSC; 0.1% SDS at 65 °C for 20 min. The filters should then be placed in a cassette with a film and left to expose for several hours, so as to monitor the level of background signal.

7. After autoradiography, higher stringency washes might be required if the background signal is still high ($0.1 \times$ SSC, 0.1% SDS, 20 min at 65 °C),

so care should be taken to avoid letting the filters dry out during auto-radiography.

8. The position of positive clones should be identified and a plug of agar containing the region removed using the more open end of a Pasteur pipette. The plug should be transferred into 500 μl of either bacterio-phage storage buffer containing a few drops of chloroform or L broth containing the appropriate antibiotic, depending on what type of library was screened. This pool of partially purified clones should then be plated at a lower concentration and rescreened (steps 1–7). This process should be repeated until a single purified clone is obtained.

9. Depending upon the type of library used, the clone may be ready to analyse and sequence (plasmid library) or require further subcloning (bacteriophage library) into a suitable bacterial plasmid, such as pBlue-script II (Stratagene).

10. The clones should be sequenced by standard techniques and compared to a suitable database as discussed in the next section.

16.2.3 Degenerate oligonucleotides

Degenerate oligonucleotide primed PCR (DOP-PCR) is an adaptation of the standard PCR protocol (see Chapter 11) and it is used specifically to amplify closely related sequences. DOP-PCR is most useful in identifying novel genes, from a specific tissue or cell of interest, which encode a motif that is highly conserved among a family of functionally important proteins. This simple technique has been used successfully, for example, to identify *HEX*, a novel homeodomain-related gene expressed during haematopoiesis (see Figure 16.1)

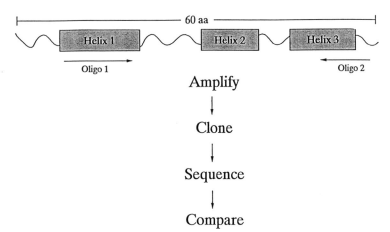

Amplify

Clone

Sequence

Compare

Figure 16.1. Strategy used to identify related homeobox genes by degenerate PCR. A cartoon representation of the 60 amino acid (aa) homeodomain showing the 3α-helices is presented at the top. The positions of the two degenerate PCR primers, oligo 1 and oligo 2, relative to the α-helices are shown along with the basic strategy used to amplify homeodomain transcripts underneath.

(Bedford et al., 1992). It has also been employed to identify E-cadherin homologues, which have an important role in colorectal cancer (Bedford and Jankowski, 1995).

DOP-PCR can be used successfully to amplify gene sequences from both DNA and RNA (cDNA), with only a minimum requirement for available tissue (compared to several other techniques described in this chapter). For example, a minimum of 100 ng of DNA or 10 ng cDNA synthesized from ≤ 1 μg total RNA is usually sufficient to amplify the majority of desired gene sequences. This means that several different DOP-PCR analyses can be performed on a limited amount of tissue, as is often the case with partially degraded or highly purified pathological samples.

The most important criterion to be considered when choosing this technique, however, is the choice of gene sequence to be amplified and therefore the design of the PCR primers. The particular sequence to be amplified should encompass a conserved functional domain, such as the DNA-binding domain of transcription factors, ligand-binding domain of cellular receptors or any other functionally conserved domain. Genomic DNA may be preferable over cDNA, as no prior knowledge of expression patterns is required, but attention should be paid to restrict amplification to exons and avoid amplifying across intronic regions. In designing the PCR primers, one should first focus on the regions of the conserved functional domain that are most *highly* conserved between the gene families and across species. These are often determined by aligning related peptide sequences within the conserved domain, as is shown in Figure 16.2 for the DNA-binding homeodomains of closely related homeobox genes. In addition, if two regions within a protein are highly conserved, then several sets of PCR primers can be constructed. This may enable an internal ('nested') set of primers to be used, which can increase the specificity of the PCR greatly. The following criteria are critical in choosing the desired peptide sequence and designing the degenerate oligonucleotide primers:

1. Peptide sequences that dictate minimal degeneracy should be chosen. For example, methionine and tryptophan residues are preferable as they are specified by unique codons, in contrast to leucine, arginine and serine residues, which should be avoided if at all possible, as they will result in a large increase in degeneracy of the primer. Successful priming has been reported with primers with 128-fold degeneracy but for routine use 32–64-fold degeneracy is optimal. Primers should generally be 20–25 bp in length, derived from 7–9 amino acids of contiguous sequence. In special circumstances for motif cloning, one may not be able to avoid degenerate amino acid codons even at the 3′ end.

2. An approximate amplified product of 200–500 bp is optimal, as fewer PCR artifactual bands are generated when amplifying small products.

3. Degeneracy should be avoided at the 3′ end of the PCR primer as these residues are the most important in conferring specificity on the PCR.

```
                    Helix 1              Helix 2           Helix 3
                 _____          _____        _____

HEX         KRKGGQVRFSNDQTIELEKKFETQKYLSPPERKRLAKMLQLSERQVKTWFQNRRAKWRRL
Hox11L2     ---KPRTS--RV-IC----R-HR----ASA--AA---S-KMTDA----------T----Q      58%
Hox11L1     ---KPRTS--RS-IC---RR-LR----ASA--AA---A-RMTDA----------T----Q      56%
Hlx         --SWSRAV---L-RKG---R--I---VTK-D--Q--A--G-TDA---V------M---HS      56%
Bsh         R-RKARTV--DP-LSG---R--G-R-------VE--TA-G---T----------M-HKKQ      56%
Hox11       -K-KPRTS-TRL-IC----R-HR----ASA--AA---A-KMTDA----------T----Q      56%
GBX1        -SRRRRTA-TSE-LL----E-HCK----LT--SQI-HA-K---V---I---------K-I      53%
ceh-9       X--KARTT--GK-VF----Q--AK----SSD-SE---R-DVT-T---I------T--KKI      53%
OM(1d)      -QRKARTA-TDH-LQT---S--R-----VQ--QE--HK-D--DC-----Y----T--M-Q      53%
NKCH4       -KRKSRTA-T-Q-IF----R-LY------AD-DEI-GG-G--NA--I---------LK-D      53%
BarH1       -QRKARTA-TDH-LQT---S--R-----VQ--QE--HK-D--DC-----Y----T--K-Q      53%
Smox-5      --RKTRTT---C-LN---NN-NR-R--T-TD-D-I--H-G-TNT--I---------LK-E      53%


CONSENSUS  K        FS  Q   LE  F   KYLS  ER  LA  L  LS  QVK  WFQNRR  K  RR
```

Figure 16.2. Homeodomain peptide sequence alignment. The aligned peptide sequences represent homeodomains that share 53% or more identity with the murine *Hex* homeodomain and were identified from a search of the OWL database (Bleasby and Wooton, 1990) using the BLASTP computer homology program (Altschul *et al.*, 1990). Amino acids identical with the murine *Hex* homeodomain are denoted by a dash and the percentage is shown on the right. The relative positions of the three homeodomain α-helices are shown at the top and the peptide consensus sequence is shown at the bottom with underlined amino acids representing conserved amino acid changes.

4. Degeneracy in the primer can be reduced by eliminating CpG dinucleotides, which are under-represented in vertebrate genomes. For example, a histidine followed by glutamic acid would normally be CA(T,C) GA(A,G), however CAC is unlikely to be followed by a G.

5. Codon preference (codon usage) tables are also available (Lathe, 1985), which can be used to further reduce degeneracy at other positions. These preference tables may indicate that certain codons occur only rarely and are especially useful for decreasing degeneracies due to sixfold coding amino acids such as arginine. In addition, inosine can be included at positions of degeneracy (Wadda *et al.*, 1991; Rossolini *et al.*, 1994). Inosine will bind to all four natural bases and can replace all fourfold degenerate codons. Unfortunately, inosine lowers the melting temperature further making inappropriate hybridization of the primer more likely (Nichols *et al.*, 1994).

6. If the primer has a hairpin at the 3′ end or self-complementarizes (primer-dimers) PCR will probably be highly inefficient. It is therefore best to design primers with computer programs such as Oligo 5.0.

Figures 16.3 and 16.4 show examples of degenerate PCR primers that have been used successfully to isolate novel sequences (Bedford *et al.*, 1992; Bedford and Jankowski, 1995).

	Gln	Thr	Phe/Leu	Glu	Leu	Glu	Lys	Glu	Phe/Leu

DP1:5'-CA(AG) ACN (TC)TN GA(AG) (TC)TN GA(AG) AA(AG) GA(AG) TT-3'

	Lys	Ile	Trp	Phe	Gln	Asn	Arg/Ser	Arg/Ser	Met	Lys/Asn

DP2:3'-TT(TC) TA(TGA) ACC AA(GA) GT(TC) TT(AG) (GT)CN (GT)CN TAC TT-5'

Figure 16.3. Homeodomain degenerate PCR primers. The primers relate to oligos 1 (DP1) and 2 (DP2) in Figure 16.1 and were used successfully in the isolation of a number of homeodomain transcripts from human and murine haematopoietic tissue including the novel homeobox gene *HEX* (Bedford *et al.*, 1992). The possible codon usages are also shown and N = T, G, C or A.

forward primer 5' end

5'- GGD GGR GAR GAR GAY CAR GA-3'

2390bp GGC GGA GAA GAG GAC CAG GAC...//...GAC CCC ACA GCC CCG CCT TAT 2590bp

5'-CTR GGN TGN CGN GGD GGD AT-3'

reverse primer 3' end

Figure 16.4. E-Cadherin degenerate PCR primers. These degenerate primers were used to amplify homologues of E-cadherin from colonic tissue and are based on sequences from exons 13 and 14 (cytoplasmic tail) of E-cadherin (Bedford and Jankowski 1995). D = T, G or A; R = G or A; Y = T or C; N = T, G, C or A.

16.2.3.1 Sample protocol

The following protocol utilizes standard PCR conditions and is sufficient for most degenerate primers, however attention to the optimization parameters below might be required for certain sets of degenerate primers.

1. Each separate PCR reaction is performed in a 0.5 ml Eppendorf tube using sterile pipette tips. If several different samples are to be analysed a single working solution should be used to reduce tube-to-tube variation. To each PCR tube add:
   - 10 × PCR buffer (150 mM Tris-HCl, pH 8.8; 600 mM KCl; 22.5 mM $MgCl_2$)   2.5 μl
   - 10 mM dNTPs (Pharmacia or Boehringer Mannheim)   1 μl
   - of each primer (10 pmol/μl)   0.5 μl
   - Taq polymerase* (Perkin Elmer Cetus, 5 units/μl)   0.5 μl
   - Double distilled water   to 24 μl

   *Other available polymerases such as Vent (Stratagene) and Pfu (NE Biolabs) can be used instead of Taq as they have 'proofreading' actions and therefore incorporate fewer errors into the sequence. However, it is important that longer extension periods are used for these alternative polymerases.

2. Add the template (genomic DNA 50–500 ng, cDNA 10–500 ng) in a 1 μl aliquot (preferably). The buffer containing the template should not

contain excessive Mg^{2+} ions or EDTA as these could change the buffering of the reaction.

3. The samples should then be overlaid with 150 μl light mineral oil.

4. The PCR is performed in an automatic thermal cycler and a minimum of 30 cycles of denaturation, annealing and extension should be performed. An appropriate PCR cycle using degenerate primers is: 94 °C for 1.5 min, 55 °C for 2 min and 72 °C for 2 min. After the 30 cycles, an extended extension incubation of 5 min at 72 °C should be included to ensure that all amplified molecules are full length.

5. To visualize the products, 20 μl should be removed from beneath the oil layer, mixed with 2 μl loading buffer (0.25% (w/v) Bromophenol Blue, 0.25% (w/v) xylene cyanol, 30% (v/v) glycerol) and loaded on to a 4% (w/v) agarose gel (3% NuSieve molecular biology grade agarose (FMC)/1% regular agarose).

6. Subcloning PCR products: If a single PCR species is amplified then it can be sequenced directly (as described by Ashworth, 1991), however several gene products are more usual with degenerate primers and therefore the product needs to be subcloned to be analysed further. Several techniques are available to subclone PCR products such as blunt-ended cloning or TA cloning. In our experience the method of TA cloning as supplied in kit form by Invitrogen is far superior (10–100 × greater efficiency) and should be chosen if at all possible. The basis of TA cloning is to take advantage of the A residue which is most frequently added to the 3′ end of the PCR product and subclone the product into prepared T-tailed vectors.

7. Once sub-cloned the products can be sequenced by standard techniques and analysed. Novel sequences should be compared to a sequence database to ensure that they are indeed novel. A suitable database is the OWL database, which is a non-redundant database comprising the sequences from both the SWISS and EMBL databases (Bleasby and Wooton, 1990). Screening of the database should be done at the peptide level and a homology program such as BLASTP can be used (Altschul *et al.*, 1990).

16.2.3.2 *Optimizing parameters*

It is often necessary to optimize the efficiency of the PCR when using highly degenerate primers and the following are some of the parameters that can be varied.

1. Number of cycles: In certain cases a greater number of cycles than 30 might be required in order to amplify sufficient PCR product to visualize on agarose gels. However, artifactual amplification can occur if excessive numbers of cycles are performed and therefore a range of 30–50 cycles should be maintained.

2. Magnesium concentration: The concentration of magnesium used in the PCR buffer can be varied widely to optimize the specificity of PCR reactions and this is well documented (Ashworth, 1991).

3. Annealing temperature: The temperature of annealing can be varied in the range of 37–70 °C and this can have large effects on the specificity of the PCR reaction (see Figure 16.5), depending upon the degeneracy and nucleotide content of the primers. When first using the degenerate primers several individual reactions should be set up with different annealing reactions. The method of 'touchdown PCR' has also proven useful when trying to optimize DOP-PCR, as well as reducing artifactual amplification and gives a selective advantage to highly matched sequences early in the PCR. An initial annealing temperature above the expected annealing temperature is set and reduced by 1–2 °C every 1–2 cycles until the touchdown annealing temperature is achieved, at which 10–20 cycles are performed.

The advantages of DOP-PCR are that it requires little starting material and is highly versatile. The main disadvantage is that sequences not corresponding to the genes of interest are produced. These by-products of the DOP-PCR are usually a different size and can easily be detected on agarose gels. Furthermore, base pairs from the degenerate primer sequence can also become incorporated into the products of interest and therefore care should be taken to obtain the accurate sequence before publication. In addition, amplification of one particular gene can occur preferentially over other sequences using certain degenerate primers, so that novel genes may not be detected using one set of primers compared to another. If amplification from vertebrate genomic DNA is desired one should be aware that a large number of pseudogenes are present in the vertebrate genome.

16.2.4 Differential display

This technique was first described by Liang and Pardee in 1992, as a method of identifying and isolating genes that are differentially expressed in various cells. Novel genes can thus be identified by this technique, by association either to a specific cell or tissue type. This technique is similar to subtractive hybridization in theory, but does not involve the laborious generation and subtraction of

---

Figure 16.5. The affects of annealing temperature on the specificity of PCR amplification using degenerate oligonucleotides. (a) The optimal temperature for the degenerate primers is 57 °C at which a single amplified product is visualized. Lower temperatures (54 °C–56 °C) yield numerous additional artifactual bands. The DNA ladder is the 1 kb ladder (Gibco BRL). (b) Very low annealing temperature (37 °C–42 °C) can produce different sized amplified products (B–H) using degenerate primers in comparison to the control annealing temperature of 57 °C (A), where a single 200 bp band of the correct size is amplified. The DNA ladder is the 1 kb ladder (Gibco BRL).

D OP PCR

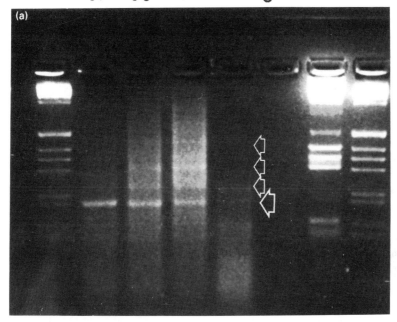

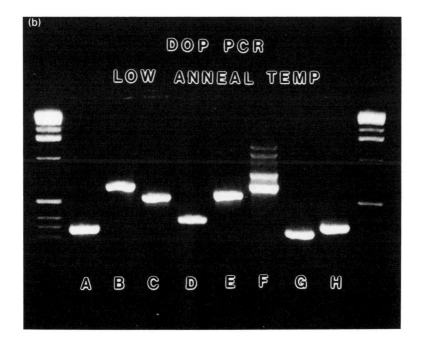

cDNA libraries and is not limited to the comparison of only two samples. This method has been used widely by several groups to isolate novel genes, for example it has been used to identify the candidate tumour suppressor gene α 6 integrin (Sager et al., 1993). It has also been used as a means of RNA fingerprinting specific cell-types involved in spermatogenesis (Miller et al., 1994).

Differential display is a PCR-based technique that utilizes a set of oligonucleotide primers that will preferentially amplify a subset of mRNA molecules. The first primer is designed so that it will anneal to the polyadenylated tail of a subpopulation of mRNAs, while the second primer is short and non-specific, designed so that it will anneal at a variety of positions upstream (5′) relative to the first primer. The PCR products are analysed using polyacrylamide gel electrophoresis, which separates the PCR products according to size. The different amplifed band patterns on the gel represent cell-specific cDNA (RNA) fingerprints and novel bands appearing in one lane relative to another may represent either expression of a new gene or loss of gene expression in the other sample. Novel bands in the sample of interest can then be gel isolated and analysed further. The advantages of this technique are that it is very sensitive, rapid and versatile and that several cell types can be analysed simultaneously. The disadvantages of the technique are similar to any PCR-based technique, such as contamination and problems with optimization of primers (see above).

16.2.4.1 Sample protocol

The experimental procedure requires a five-stage process: (1) tissue processing (mRNA purification); (2) selection of suitable pairs of primer oligonucleotides; (3) creation and amplification of cDNA; (4) gel fractionation and recovery of cDNA; (5) cDNA cloning and sequencing.

1. Messenger RNA can be isolated using a variety of commercially available kits (e.g. Fastrack mRNA kit, Invitrogen), which are simple to use and give a high yield of poly(A)$^+$ mRNA. The integrity of the mRNA can be checked by using Northern blot analyses to known mRNAs.

2. Pairs of oligonucleotide primers should be chosen so as to give no more than 50–100 amplified cDNA products from the purified mRNAs. This is the optimal number of cDNA products that can be visualized on a vertical sequencing gel plate. The 3′ primer should contain a chain of ten thymidine nucleotides, followed by two other dissimilar nucleotides allowing only one-tenth of the mRNA populations to be hybridized. The 5′ primer should contain < 10 unique base pairs, allowing only a fraction of the cDNA containing a complementary sequence to this primer within 2–3 kb of the poly(A) tail to be efficiently amplified. Primer pairs can be selected by pilot reactions and their amplified products analysed on 1.5% agarose gels.

3. cDNA is synthesized from mRNA by reverse transcriptase (Invitrogen, USA) and amplified by standard polymerase chain reaction protocols (see Quantitative gene analysis chapter) for 40 cycles and also incorporating ^{35}S-labelled ATP.

4. The resultant cDNA products should be loaded onto a standard polyacrylamide sequencing gel. An autoradiograph of the sequencing gel will then reveal bands that are identical between the samples and ones that are non-identical. New bands present in the sample lanes compared with the corresponding normal 'control' may represent novel genes that have been transcribed. The DNA in these bands can then be recovered by electroelution with Hoefer's gel eluter and ethanol precipitation and reamplified with the same set of primers. After ensuring that an adequate yield of 'novel DNA' has been purified (by running a sample on an agarose gel) the DNA sequence can be further amplified with primers that have unique restriction enzyme sites on the 5' and 3' flanking regions, so that the DNA can be inserted into a general M13 cloning vector.

5. The unique DNA band can then be sequenced and the new sequence(s) cross-checked against known sequences in genetic sequence libraries (see above).

References

Altschul, S. F., Gish, W., Miller, W. *et al*. (1990) Basic local alignment search tool. *J. Mol. Biol*. **215**, 4043–410.

Ashworth, A. (1991) Cloning Transcription Factors by homology. In *Transcription factors a practical approach* (ed. D. S. Latchman), pp. 125–42. Oxford University Press, Oxford.

Ausubel, F. M., Brent, R., Kingston, R. E. *et al*. (1987) *Current protocols in molecular biology*. Wiley Interscience, New York.

Bedford, F. K., Ashworth, A., Enver, T. and Wiedemann, L. M. (1992) *HEX*: A novel homeobox gene expressed during haematopoiesis and conserved between mouse and human. *Nucl. Acids Res*. **21**, 1245–9.

Bedford, F. K. and Jankowski, J. (1995) Gene analysis and manipulation in clinical tissue. *J. Clin. Pathol*. (Molecular Clinical Pathology supplement) **1**, 3–9.

Bleasby, B. J. and Wooton, J. C. (1990) Construction of validated, non-redundant composite protein sequence databases. *Prot. Engin*. **3**, 153–9.

Borrow, J., Goddard, A. D., Sheer, D. and Solomon, E. (1990) Molecular analysis of acute promyelocytic leukaemia breakpoint cluster region on chromosome 17. *Science* **249**, 1577–80.

Candia, A. F., Hu, J., Crosby, J., *et al*. (1992) *Mox-1* and *Mox-2* define a novel homeobox gene subfamily and are differentially expressed during early mesodermal patterning in mouse embryos. *Development* **116**, 1123–36.

Cory, S. (1986) Activation of cellular oncogenes in haematopoietic cells by chromosome translocations. *Adv. Cancer Res*. **47**, 189–234.

Deguchi, Y., Moroney, J. F., Wilson, G. L. *et al*. (1991) Cloning of a human homeobox gene that resembles a diverged *Drosophila* homeobox gene and is expressed in activated lymphocytes. *New Biol*. **3**, 353–63.

De The, H., Chomienne, C., Lanotte, M. *et al*. (1990) The t(15;17) translocation of acute promyelocytic leukaemia fuses the retinoic acid α gene to a novel transcribed locus. *Nature* **347**, 558–61.

Lathe, R. (1985) Synthetic oligonucleotide probes deduced from amino acid sequence data. Theoretical and practical considerations. *J. Mol. Biol.* **183**, 1–12.

Liang, P. and Pardee, A. B. (1992) Differential display of eukaryotic messenger RNA by means of the polymerase chain reaction. *Science* **257**, 967–70.

Miller, D., Tang, P.-T., Skinner, C. and Lilford, R. (1994) Differential RNA finger-printing as a tool in the analysis of spermatozoal gene expression. *Hum. Reprod.* **9**, 864–9.

Monica, K., Galili, N., Nourse, J. *et al*. (1991) *PBX2* and *PBX3*, new homeobox genes with extensive homology to the human proto-oncogene *PBX1*. *Mol. Cell Biol.* **11**, 6149–57.

Nichols, R., Andrews, P. C., Zhang, P. and Bergstrom, D. E. (1994) A universal nucleoside for use at ambiguous sites in DNA primers. *Nature* **369**, 492–3.

Rossolini, G. M., Cresti, S., Ingianni, A. *et al*. (1994) Use of deoxyinosine-containing primers vs. degenerate primers for polymerase chain reaction based on ambiguous sequence information. *Mol. Cell. Probes* **8**, 91–8.

Sager, R., Anisowicz, A., Neveu, M. *et al*. (1993) Identification by differential display of alpha 6 integrin as a candidate tumor suppressor gene. *Faseb J.* **7**, 964–70.

Sambrook, J., Fritsch, E. F. and Maniatis, T. (1989) *Molecular cloning; a laboratory manual*. 2nd edn Cold Spring Harbor Lab Press, Boston.

Schenker, T., Lach, C., Kessler, B. *et al*. (1994) A novel GTP-binding protein which is selectively repressed in SV40 transformed fibroblasts. *J. Biol. Chem.* **269**, 25447–53.

Wadda, K., Wada, Y., Doi, H. *et al*. (1991) Codon usage tabulated from the GenBank genetic sequence data. *Nucl. Acids Res.* **19**, 1981–85.

Yancopoulos, G. D. and Alt, F. W. (1990) Subtractive hybridization techniques. In *Methods for cloning and analysis of eukaryotic genes*, pp. 115–138. Boston, Jones and Bartlett Publishers, Boston.

17

Manual and automated DNA sequencing

MICHAEL D. JONES

17.1 Introduction

The ultimate characterization of a DNA molecule is the determination of its nucleotide sequence, and there is now no reason for scientists to work with unsequenced DNA clones. The current methods of sequence analysis are both rapid and reliable. DNA sequence analysis falls into two major areas; firstly, the determination of the sequence of new unknown 'virgin' clones, and secondly, the analysis of known genes in the search for variations. This second area has arisen due to the powerful methodology of the polymerase chain reaction (PCR) for studying gene sequences (Saiki *et al.*, 1988). The method that is predominantly used to sequence DNA is the dideoxy chain termination or Sanger procedure (Sanger, Nicklen and Coulson, 1977). At present there are several automated sequencing instruments commercially available, and the one most widely used is the ABI 373A. These instruments eliminate the post-electrophoresis manipulations involved in sequencing, such as processing of the polyacrylamide gel, developing of the X-ray autoradiograph and entering the sequence data into a computer. Sequencing reactions still have to be performed, polyacrylamide gels poured and samples loaded on gels, just as with manual sequencing. However, the advantage of automated instruments is their sample capacity (the ABI 373A can analyse 36 samples at once), a detection system that does not involve radioactivity but fluorescence, and the fact that once the gel has been loaded and the electrophoresis started, there is nothing left to do but watch the computer screen (well almost nothing). Regardless of the technology employed, manual radioactive or automated fluorescence, determination of DNA sequence relies upon denaturing polyacrylamide gel electrophoresis. The amount of sequence that can be read from such gels is in the order of 500–800 bases. In practice it is more usual to read 300–400 bases for manual systems, and 350–550 for automated systems. Obtaining sequence beyond 700–800 bases is usually only obtained with 'good' templates and a very efficient sequence reaction, and is not to be expected on a routine basis. It should be noted that in general you will get better data, both in terms of quality and quantity, using single stranded (ss) M13 DNA templates, compared with the double stranded (ds) plasmid DNA (see Chapter 14).

17.2 M13 shotgun cloning

For the characterization of a new 'virgin' clone, the best procedure is to shotgun clone random DNA fragments into a suitable M13 bacteriophage vector (Messing and Vieira, 1982; see Figure 17.1) and let a computer piece together the various gel readings into the complete sequence of the clone (Sanger *et al.*, 1980; Bankier, Weston and Barrell, 1987; Smith *et al.*, 1993). To generate an accurate DNA sequence of an unknown piece of DNA, both strands have to be determined completely. DNA fragments under 600 base pairs (bp) in length can be analysed without further fragmentation by sequencing from both ends. Fragments greater than ~1000 bp in size will require fragmentation into smaller sizes for sequence analysis. Although restriction endonucleases can be used to digest the target DNA, they are not ideal. Restriction sites are not randomly distributed along DNA, so sequencing very small fragments (< 200 bp) wastes sequence gel space and large fragments (> 1000 bp) have to be further fragmented. Also, to join the data into a

M13 VECTORS

Cloning Sites in M13 Vectors:

```
                                    EcoRI
M13mp2:                5' ATGACCATGATTACGAATTCACTGGCCGTCGTTTTAC 3'
                          M  T  M  I  T  N  S  L  A  V  V  L

M13mp7:                5' GAATTCCCCGGATCCGTCGACCTGCAGGTCGACGGATCCGGGGAATTC 3'
                          EcoRI     BamHI SalI  PstI   SalI   BamHI    EcoRI

M13mp8:                5' GAATTCCCGGGGATCCGTCGACCTGCAGCCAAGCTTGG 3'
                          EcoRISmaI BamHI SalI   PstI      HindIII

M13mp9:                5' CCAAGCTTGGCTGCAGGTCGACGGATCCCCGGGAATTC 3'
                          HindIII PstI  SalI   BamHISmaI EcoRI

M13mp18:               5' GAATTCGAGCTCGGTACCCGGGGATCCTCTAGAGTCGACCTGCAGGCATGCAAGCTT 3'
                          EcoRI SacI   KpnISmaI BamHI XbaI   SalI   PstI  SphI    HindIII

M13mp19:               5' CCAAGCTTGCATGCCTGCAGGTCGACTCTAGAGGATCCCCGGGTACCGAGCTCGAATTC 3'
                          HindIII    PstI  SalI   XbaI   BamHISmaIKpnI  SacI  EcoRI
                          SphI
```

Synthetic Oligodeoxynucleotide Primer Sites in M13 vectors

```
CAGGAAACAGCTATGAC LMB3
    AACAGCTATGACCATGATT RevP
                                              (+ strand)
CAGGAAACAGCTATGACCATGATTACG-insert-ACTGGCCGTCGTTTTACAACGTCGTGACTGGGA
          M  T  M  I  T        L  A
                        TGACCGGCAGCAAAATG LMB2
                                TGCAGCACTGACCCT MPP
```

Figure 17.1. The sequence of the Messing series of M13 mp vectors.

contiguous sequence a second library of M13 clones generated with a second restriction enzyme is required. Sequencing by primer walking, i.e. sequencing about 300–400 bases of a large fragment, then designing a primer to extend another 400 bases, etc., is not recommended. You can only sequence one step at a time and for a large fragment the cost of custom primers becomes very expensive.

M13 vectors are (usually) ideal for cloning fragments of DNA under 2000 base pairs, and as such are perfect for sequence determination. However, depending upon the actual sequence you are trying to clone, fragments may only clone in one orientation or not at all. The reason for these problems in M13 cloning, is most likely because the polylinker cloning site in the Messing M13mp vectors is within the origin of replication of filamentous phages (Messing and Vieira, 1982). Thus, DNA sequences that can interfere with the replication of M13 will in general be difficult to clone. It is in fact amazing that you can clone into the origin region and obtain stable clones!

By far the best methods to fragment DNA randomly are either sonication or DNAase I digestion. Both methods essentially cleave DNA into fragments with no sequence specificity at the site of breakage, thus generating an assortment of overlapping DNA fragments. A general rule of thumb in a shotgun strategy is that you need to obtain sequence data to ~6 times the length of the target fragment. Thus for a 4000 bp fragment, you need to generate approximately 24 000 bases of data. Assuming an average gel reading of 400 bases, this means you need to analyse about 60 random M13 clones, which can be sequenced within two to three days using an ABI 373A sequencer. At this point the data is assessed for completeness, and then a more directed approach using a few specific primers can be pursued to finish the sequence.

The target fragment can be purified away from the original cloning vector before fragmentation. To eliminate any end effects of the random digestion, the target fragment is self-ligated into high molecular weight concatamers or circles. Thus, fragmentation will take place at any point in the target sequence, as it effectively has no ends. If you do not do this, then degrading a linear fragment may lead to an overrepresentation of clones derived from the ends of the target DNA (Deininger, 1983). However, the laborious procedure of fragment isolation and self-ligation can be avoided by performing the fragmentation on the intact recombinant clone. The random M13 clones are then screened for vector only clones before sequence analysis. One benefit is that any vector clones that are inadvertently sequenced will provide an accuracy check on the sequence data obtained. For sequence analysis of entire cosmid clones (30–40 kbp inserts in a vector ~5 kbp in size) it is easiest to sequence everything, rather than waste time screening out the relative small proportion of vector containing clones.

To obtain enough DNA fragments after random degradation for cloning into M13, you will need to start with at least 10–20 μg of DNA depending upon the size of the target.

17.2.1 *Fragment isolation and self-ligation*

Digest the clone DNA with a suitable restriction enzyme to release the insert. Depending upon the volume of the digest, and the complexity of the fragments produced, it may be advisable to extract with buffered phenol/chloroform and precipitate with ethanol, before running the DNA on an agarose gel (see also Chapter 11).

The agarose gels are typically 0.4–1.0% (w/v) agarose in 0.5–1 × TBE buffer, containing ethidium bromide at 0.5–1 μg/ml, and are run at a constant current of between 30–50 mA. The gel is inspected on a long UV wave light box to check for separation of the vector and insert DNA bands. The two methods described below are useful ones for extracting DNA out of the agarose gel. All methods for extracting DNA from agarose gels are a compromise between the complexity of the procedure, and the quality and yield of the recovered DNA.

For trough elution of DNA fragments, visualize the gel under UV light and cut a narrow trough in front of the DNA band to be eluted. Fill the trough with 2 × TBE buffer and run the gel on top of the UV light box at 50–70 mA. The high buffer concentration slows down the migration of the DNA, thus trapping it within the trough. The electrophoresis has to be carefully monitored, or else the DNA will pass through the trough and into the rest of the agarose gel. When the DNA has migrated into the trough, switch off the power, and remove the buffer, containing the DNA. Extract with buffered phenol/chloroform, and to the upper aqueous phase add 3 M sodium acetate to 0.3 M, and precipitate the DNA with ethanol (3 volumes) either overnight at −20 °C or at −70 °C for 1 h. If the volume is large and the amount of DNA is small, add $MgCl_2$ to 10–20 mM to aid precipitation. Centrifuge for 10 min in a microfuge and decant off the supernatant. Wash the DNA pellet with 500 μl of cold 100% ethanol and respin for 30 s. Decant off the ethanol and leave the pellet to air dry for 10 min. Add TE8 buffer and wait 10 min, then briefly vortex to resuspend the DNA.

For NA45 elution (Dretzen *et al.*, 1981), run the gel as above and then cut a thin slit (a little wider than the band) in the gel, in front of the DNA band to be eluted. Insert a piece of NA45 paper and electrophorese the DNA onto the paper, at 70 mA. To prevent any larger unwanted DNA bands contaminating the required target DNA, insert a piece of NA45 behind the target to trap larger fragments. Remove the paper and wash briefly with water. You should see the ethidium-stained DNA on the paper. Trim away excess unstained paper, then cut the paper into small pieces and place in a tube with enough NA45 elution buffer to cover the paper (50–100 μl) and incubate at 70 °C for 20–30 min. Remove the buffer to a clean tube and add another aliquot of elution buffer to the paper pieces, and heat at 70 °C for a further 20–30 min. Check the recovery of DNA, by viewing the NA45 paper under UV light; it should have lost most (if not all) of its ethidium staining. Pool the eluted DNA and extract with an equal volume of buffered phenol/choroform, and then

precipitate by adding 2.5 volumes of ethanol. There is no need to add salt.
MgCl$_2$ may be added to 10 mM if the concentration of DNA is low. Depending
upon the size of the DNA fragments being eluted, the recovery of DNA is
usually 50–80%.

NA45 elution buffer is 1.5 M NaCl in TE8 buffer. NA45 paper is pretreated
by cutting it into suitable sized strips, and washing in 2.5 M NaCl (in TE8
buffer) for 1 h at room temperature. The NA45 strips are then rinsed and
stored in 1 mM EDTA at 4 °C. Prior to use the strips are rinsed in water
and then soaked in electrophoresis buffer for 1–2 min.

Set up a self-ligation in a 1.5 ml tube as below. You need to use at least
5–10 μg of target DNA. Incubate at 15 °C for 2–3 h, or overnight at 4 °C.

> 20 μl (5–10 μg) DNA in TE8 buffer
> 3 μl 10 × ligase buffer
> 3 μl 10 mM ATP
> 2 μl water
> 2 μl T4 DNA ligase (20 units)

17.2.2 Sonication

Use a suitable sonicator and monitor the extent of fragmentation by agarose
gel electrophoresis (Deininger, 1983). To calibrate the sonicator, sonicate some
intact λ DNA or a circular plasmid (5 μg DNA in 200 μl TE8 buffer). Once
suitable conditions have been found, repeat exactly with your target DNA.
Sonication generates a smear of DNA fragments, and the aim is to obtain
conditions that generate most ethidium staining in the region of 1–2 kbp, and
then to isolate fragments in the 800–1500 bp size range.

If you are using a direct probe sonicator, clean the probe first by boiling in
0.5% SDS, wash thoroughly with distilled water, and then sonicate in sterile
water at full power for 60 s. Cool the tube of DNA in ice, insert the probe and
sonicate for x bursts of 5 s (4 or 5 bursts may be needed, depending on sample
and conditions, to achieve the required size range), with cooling for 40 s
between bursts. The tip of the sonicator should be inserted just under the
surface of the DNA and must not touch the sides of the tube. Keep the tube
surrounded by ice all the time. After sonication, to 200 μl DNA solution add
20 μl 3 M sodium acetate, pH 7.5; 1 μl 2 M MgCl$_2$ and 600 μl ethanol. Mix well
and leave overnight at −20 °C or at −70 °C for 60 min to precipitate the DNA.
Resuspend the sonicated DNA in 30 μl of TE8 buffer.

17.2.3 DNAase I digestion

For DNAase I treatment the DNA has to be dissolved in water or TE8 buffer
(Anderson, 1981), because the digestion buffer contains manganese and must
not contain magnesium (which is present in ligase buffers). Do a trial digestion

to ascertain the best conditions that will generate fragments in the required size range (800–1500 bp). Take 5 μg of test DNA and digest at 15 °C or 20 °C in 40 μl of 1 × DNAase buffer with 0.001 unit of DNAase I. Remove 5–10 μl samples at 5 min intervals, and stop the digestion by adding 1 μl 250 mM EDTA, pH 8. Analyse the digestion time points by agarose gel electrophoresis. Repeat using the same conditions exactly, processing 10–20 μg of target DNA. After DNAase I digestion, extract once with an equal volume of buffered phenol/chloroform. The DNA fragments are recovered by precipitation with sodium acetate and ethanol as described above.

17.2.4 End repair and size fractionation

Both of the above methods of fragmentation yield DNA molecules with ragged overhangs, 5' and 3', which have to be repaired to generate blunt-end fragments suitable for ligation into M13 vectors. This end repair is best carried out by a mixture of T4 DNA polymerase and Klenow DNA polymerase enzymes, as described below.

30 μl random DNA fragments in TE8 buffer
4 μl 10 × Hin buffer
4 μl dNTP Mix (0.25 mM each dNTP)
10 units T4 DNA polymerase
10 units Klenow polymerase

Incubate the above at room temperature for 30 min, and then heat inactivate the enzymes by heating at 68 °C for 15 min. Add 5 μl of sucrose dye mix and load all the sample onto a 50 ml 1.0–1.5% agarose/TBE minigel and run at 30–60 mA. Elute the required size range (800–1500 bp) either by trough elution or onto NA45 paper (see above). The eluted DNA is extracted with phenol/chloroform and precipitated with ethanol. The DNA fragments are redissolved in 30 μl TE8 buffer. It is advisable to monitor recovery by analysing 3 μl on an agarose gel. It is recommended that you leave a space between the size-markers and your DNA fragments, to eliminate the possibility of accidentally isolating and cloning size-marker DNA.

It should be noted that sonicated and repaired DNA fragments will clone inefficiently compared to DNA generated by restriction enzyme digestion. This reflects the damage caused to the DNA by sonication and the inefficient polishing of the ragged ends of the DNA fragments. Povinelli and Gibbs (1993) have used mung bean nuclease to digest the single-stranded DNA overhangs, before treatment with T4 DNA polymerase and Klenow polymerase enzymes to generate the blunt ends.

10 μg sonicated DNA in water
5 μl 10 × Mung Bean buffer
10 units mung bean nuclease
water to a final volume of 50 μl

Incubate at 37 °C for 30 min, then extract with 50 μl buffered phenol/chloroform, and ethanol precipitate the DNA fragments as described above. The mung bean nuclease buffer contains zinc, so the DNA has to be precipitated before and after digestion to enable the correct buffers to be used.

17.2.5 M13 Ligation and transformation protocols

Prepare vector DNA by digestion with *Sma*I restriction enzyme to generate a blunt-end vector:

> 5 μl M13mp8 RF DNA (1 μg/μl)
> 5 μl 10 × Sma buffer
> 40 μl water
> 20 units *Sma*I enzyme

Incubate at 37 °C for 1–2 h. Commercial suppliers recommend incubation at 25 °C. Check a sample (5 μl) on a 0.8% agarose/TBE minigel to see that digestion is complete. Then heat inactivate the enzyme at 65 °C for 15 min, extract with buffered phenol/chloroform and then precipitate with ethanol (add 5 μl 3 M sodium acetate, pH 7.5 and 150 μl ethanol). Resuspend the DNA in TE8 buffer at 20 ng/μl. Store the vector DNA at −20 °C in 50 μl samples. The vector can be treated with phosphatase to eliminate vector self-ligations, but I tend to rely upon the blue/white colour selection. Alternatively, you can purchase cut and phosphatased vectors from various commercial suppliers.

Set up ligations at 15 °C or 4 °C overnight in 1.5 ml tubes as below:

> 1 μl *Sma*I cut vector DNA (20 ng/μl)
> 1–4 μl DNA fragments, (try a range of DNA, e.g. 1, 2 and 4 μl)
> 1 μl 10 × Ligase buffer
> 1 μl 10 mM ATP
> 0.5 μl T4 DNA ligase (100 units)
> water to a final volume of 10 μl

It is advisable to include controls to test the efficiency of the ligations and transformations, so include the following ligation controls:

Tube number	1	2	3
Vector	+	+	+
T4 DNA ligase	+	+	−
BE DNA	+	−	−

BE DNA fragments are any source of blunt ended DNA; a good source is λ DNA digested with *Alu*I. Use about 50–70 ng with 10–20 ng of *Sma*I cut M13mp8. You should expect to see around ~100 white plaques (Tube 1). Tube 3 should yield no plaques, or just a few blue ones, and Tube 2 only blue plaques, or none if a phosphated vector was used. An alternative to blunt-end cloning is to use synthetic adaptors (Andersson *et al.* 1994) (see Figure 17.2).

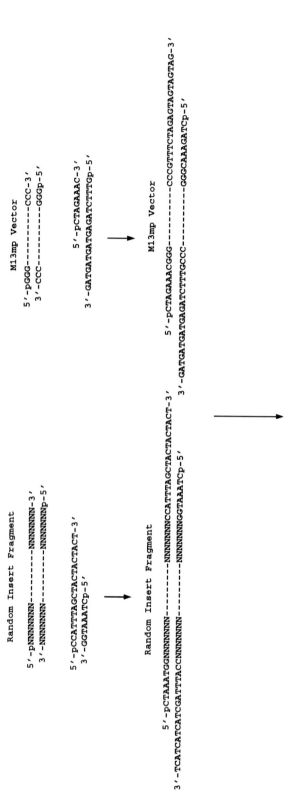

Figure 17.2. The adaptor strategy for cloning shotgun fragments into M13 vectors.

17.3 Template preparation

The quality of the DNA sequence is highly dependent upon the purity of the template DNA that is used for the sequencing reactions. The following methods have been tried and tested, and do yield excellent DNA for sequence analysis.

17.3.1 *M13 phenol protocol*

It is preferable to pick, grow and process M13 plaques as soon as possible on the day after plating, as this prevents diffusion of the phages through the agar. If this is not feasible, then pick plaques into 20 μl TE8 buffer in a 1.5 ml tube or 96-well microtitre plate and store sealed at 4 °C. For phage growth use sterile Falcon 2059 tubes, which have a capacity of 14 ml. To each tube add 2 ml of a 1 in 100 dilution of an overnight JM101 culture in 2 × TY broth, and toothpick into each tube a single white plaque. Grow at 37 °C with *vigorous* shaking, 350–400 r.p.m., for 5–5.5 hours *only*. Transfer the culture to a 1.5 ml tube and spin for 5 min at room temperature at 13 000 r.p.m. (11 000 *g*).

Decant the supernatant to a new tube (the supernatant can be kept in a sterile tube at 4 °C as a virus stock), and add 200 μl of 20% (w/v) PEG 8000/2.5 M NaCl. Do *not* transfer over any of the cell pellet. Mix well and leave at room temperature for 15 min, spin for 10 min in a microfuge at 13 000 r.p.m. (11 000 *g*) and aspirate off the supernatant. You should see a phage pellet. Take care not to touch the phage pellet when aspirating. Respin for ~1 min, and aspirate off the last traces of the supernatant. It is important to remove all traces of the PEG supernatant.

Add 100 μl TE8 buffer and leave for 10 min. Vortex vigorously for 30 s, and then add 50 μl freshly buffered phenol (See Reagents and Buffers section). Vortex for 60 s (at least) and then centrifuge for 3–5 min. Remove the upper aqueous phase to a clean tube (it is important not to inadvertently carry over any of the phenol and/or interface material) and add 10 μl 3 M sodium acetate, pH 7.5, and 300 μl ethanol, mix well and leave at −20 °C overnight or −70 °C for 60 min. Centrifuge for 10 min, decant off the supernatant and wash the DNA pellet with 500 μl of cold 100% ethanol. Respin briefly, decant off the supernatant, and air dry the DNA pellet. Resuspend the DNA in 30 μl TE8 buffer and store at −20 °C. Expect the yield of single stranded DNA to be ~5 μg (check 2–3 μl on 1% agarose/TBE minigel if desired). It is feasible to prepare 48 or more clones per day by this method.

17.3.2 *M13 thermo-extraction protocol*

Process the recombinant M13 clones as described above as far as PEG precipitation of the phage particles. Add 100 μl TTE buffer (10 mM Tris/HCl, pH 8.0; 0.1 mM EDTA, 0.25% Triton X-100) (125 μl Triton X-100 in 50 ml TE8 buffer) to the phage pellet and leave for 5–10 min at room temperature.

Then vortex vigorously for 1 min and respin to collect all the liquid at the bottom of the tube. Heat at 80 °C in a water-bath or an oven for 10 min. Centrifuge for 2 min to collect all the liquid and pellet any debris.

Remove ~90 μl to a new 1.5 ml tube, and add 10 μl 3 M sodium acetate and 300 μl ethanol. Mix well and leave at −20 °C for 60 min (preferably longer). Spin for 10–15 min, wash the DNA pellet with 500 μl cold ethanol, respin for 1–2 min, and decant off the ethanol. Air dry the DNA pellet for ~10 min at room temperature.

Resuspend the DNA pellet in 20–30 μl TE8 buffer. Transfer to a new 1.5 ml tube or 96-well microtitre plate for storage at −20 °C, particularly if any undissolved material is present after dissolution of the DNA.

This is essentially the method described by Chissoe *et al.* (1991). As it does not involve phenol extraction, it is possible to adapt the method for processing the M13 clones grown in Beckman 96-deep well microtitre plates and thus process 96 clones at once. As you need to balance the centrifuge it is no hardship to process two plates at once, and so prepare 192 clones.

> Pick M13 plaques into 800 μl of diluted overnight JM101 in each well of a Beckman 96-deep well microtitre plate. Place firmly in a container with wet paper towels on the bottom to provide moisture, seal the container and shake at 37 °C at 300 r.p.m. overnight.
> Centrifuge the plate at 3500 r.p.m. (1500 g) for 15 min.
> Carefully remove 600 μl of supernatant and add to the wells of a new Beckman 96-deep well microtitre plate.
> Add 120 μl PEG/NaCl solution to all wells. Seal with 96-plate cap and mix thoroughly.
> Incubate at room temperature for 15 min.
> Centrifuge for 15 min at 3500 r.p.m. (1500 g).
> Pour off the supernatant, and leave the plate upside down to drain for 2–3 min.
> Spin the plate inverted on paper towels for 1 min at 200 r.p.m., (500 g) to remove the last traces of the PEG/NaCl.
> Add 100 μl TTE buffer per well, wait 10 min, then cap and vortex for 1 min. Spin briefly to bring all the liquid to the bottom of the wells, and then heat at 80 °C for 10 min.
> Cool the plate and then respin at 3500 r.p.m. for 5 min, then remove ~90 μl to the wells of a third Beckman 96-deep well microtitre plate, and add 10 μl 3 M sodium acetate and 300 μl ethanol. Seal the plate, mix well and leave at −20 °C overnight.
> Spin for 15 min at 3500 r.p.m., decant off the supernatant, and wash the pellet with 500 μl cold 100% ethanol, respin and then drain upside down on tissues for 3 min.
> Air dry the pellet for 15 min, and then dissolve in 20 μl TE8 buffer.
> Transfer to a normal 96-well microtitre plate for storage at −20 °C.

Mardis (1994) has modified the above procedure to prepare M13 DNA, which is an excellent template for fluorescent Dye-primer cycle sequencing with the thermostable polymerase Sequitherm (Epicentre Technologies). The phage pellets are resuspended in 20 μl of TTE buffer, heated at 80 °C for 10 min. After a brief centrifugation, the DNA is transferred to a 96-well microtitre plate and diluted by adding 20–40 μl of water (depending upon the size of the phage pellet). This procedure keeps the manipulations to a minimum and allows for automation of M13 template DNA preparation by robotic work-stations.

17.3.3 Filamid ssDNA preparation

Filamid vectors are useful in the event of any cloning problems with M13 vectors. Filamids are plasmid vectors that contain the M13 filamentous phage origin of replication as well as the normal plasmid origin. As they contain both phage and plasmid origins of replication they are also known as phagemids and/or plages. They replicate like plasmids inside bacterial cells, and upon infection with a helper bacteriophage (to provide all the filamentous phage proteins), the ssDNA of the filamid is packaged into phage particles. Thus, it is extremely easy to isolate filamid ssDNA suitable for sequence analysis.

Colonies are grown on antibiotic-selection plates overnight and single colonies are picked and grown at 37 °C with shaking at 350 r.p.m. for 90 min, in 2 ml of 2 × TY broth (+ antibiotic). Do not place the plates at room temperature or at 4 °C prior to picking colonies, as this reduces the growth and yield of ssDNA. 5 μl of M13CO8 helper phage ($\sim 10^9$–10^{10} p.f.u./ml) is added and the culture is shaken for a further 30 min at 37 °C, and then chloramphenicol is added (to 12.5 μg/ml) and the culture shaken overnight at 37 °C (Davies and Hutchison, 1991). The culture can then be processed as described above for the preparation of M13 ssDNA. The helper phage, M13CO8, is propagated as infected *E. coli* DH5αF' cells on chloramphenicol plates (12.5 μg/ml), and grown overnight in 2 × TY media plus chloramphenicol. The bacterial cells are removed by centrifugation and the phage supernatant stored at 4 °C.

17.3.4 Plasmid dsDNA Preparation

The following methods have been used very successfully to prepare plasmid dsDNA for automated fluorescent sequencing (Pan *et al.*, 1994).

Grow overnight culture at 37 °C with shaking in 5 ml 2 × TY broth plus antibiotic in a 20 ml sterile disposable plastic universal. Pellet the cells by centrifugation at 4000 r.p.m. (3000 *g*). Aspirate off the supernatant and re-suspend the cell pellet in 200 μl 50 mM glucose, 25 mM Tris/HCl, pH 8.0, 10 mM EDTA, and transfer to a 1.5 ml tube. After 5 min, add 100 μl of 1% SDS, 0.2 M NaOH solution (made fresh each day), and mix gently by inversion. Place on ice for 5 min and then add 150–200 μl 5 M potassium acetate, pH 4.8, mix well and place at −20 °C for 15 min. Centrifuge for 15 min to pellet the cellular debris, etc. Transfer the supernatant to a fresh 1.5 ml tube, add 1 μl of RNAase A (20 μg/μl), mix well and incubate at 37 °C for 20 min.

For fluorescent dye-labelled primer cycle sequencing proceed as follows: extract with an equal volume of buffered phenol/chloroform and precipitate the plasmid DNA by addition of 1 ml of ethanol. Redissolve the DNA pellet in 50 μl of TE8 buffer and store the DNA at −20 °C.

For fluorescent dye-terminator cycle sequencing, add 1 ml of DNA binding matrix (See Reagents and Buffers section) and allow the DNA to bind for 5 min. Centrifuge for 2 min and carefully remove the supernatant, as the matrix is loose at this stage and would come off the walls of the tube. Wash the pellet twice with 1 ml of 50% ethanol, 10 mM Tris/HCl, pH 8.0, 1 mM EDTA, decanting off the wash solution after centrifugation. Wash a third time with 1 ml of acetone, decant off the acetone and dry the matrix briefly at 68 °C. Resuspend the matrix in 100 μl of 10 mM Tris/HCl, pH 8.0, 1 mM EDTA, and place in an oven at 68 °C for 5 min. Centrifuge the matrix for 5 min (13 000 r.p.m., 11 000 *g*) and carefully remove the supernatant. Precipitate the DNA as usual by addition of sodium acetate and ethanol, and resuspend the DNA pellet in 50 μl of TE8 buffer.

17.3.5 PCR products

The widespread use of PCR means that in many cases target DNA does not have to be cloned first before sequencing. Also amplification of inserts from bacteriophage λ or plasmids is a reliable way of obtaining DNA for sequence analysis. Growth of the recombinant is thus unnecessary as PCR amplification can be performed directly on a small amount of phage particles or bacterial cells (Rosenthal *et al.*, 1993). PCR is invaluable for finishing a sequence project along with custom primer walking. For example, if a gap exists between two contigs, and one M13 clone extends into the gap, PCR amplification and sequence analysis from the far end of the insert (M13 reverse sequencing) may close the gap and join the contigs. This is the advantage of cloning into M13 inserts of between 1–2 kbp. Although you can sequence only 500–600 bases with the universal primer, the possibility is there to extend the sequence by primer walking or PCR sequencing.

Polymerase chain reactions that yield a single DNA band product can easily be prepared for sequence analysis by a simple PEG (polyethylene glycol) precipitation step (Rosenthal, Coutelle and Craxton, 1993). To the PCR reaction mix add an equal volume of PEG/Mg/NaAc mix (26% (w/v) PEG 8000, 6.5 mM $MgCl_2$, 0.6 M sodium acetate, pH 6–7). Mix thoroughly and leave at room temperature for 10–15 min. Centrifuge for 10 min to pellet the DNA, and carefully remove the supernatant. Wash the pellet twice with 100% ethanol, allow to air dry and redissolve in 20 μl TE8 buffer. The PEG precipitation removes unincorporated dNTPs and excess primers. The PCR product can be sequenced reliably with one of the original primers used for the amplification.

17.3.6 Vector screening

Spot 1 μl of M13 clone DNA onto a nitrocellulose filter in an ordered array. The filter is then baked at 80 °C for 2 h and is then ready for probing. The M13 DNA is single stranded and so no denaturation is required. The filter can be probed with the insert DNA to detect clones containing insert sequences. If it is difficult to purify some insert to use as a probe, then duplicate filters can be probed with the complete recombinant clone (insert + vector) and the cloning vector alone. This will indicate which M13 clones to sequence.

17.4 Sequence reactions

There are now several DNA polymerases available for manual and automated DNA sequencing. The one described here is for *Thermus aquaticus* DNA polymerase sequencing (Innis *et al.*, 1988), as provided in the Taquence kit (USB/Amersham). This has been found to be a reliable enzyme for sequencing M13 ssDNA. Sequenase (USB/Amersham) enzyme, a modified T7 DNA polymerase, is excellent for sequencing plasmid dsDNA.

Anneal the primer to the ssDNA template by incubating at 60 °C for 30 min as below:

Universal primer	1 μl (0.5 pmol, ~2.5 ng)
M13 ssDNA	10 μl (~1 μg)
Taquence buffer	2 μl

Allow to cool to room temperature. Add to the annealed primer/template 2 μl labelling mix (1.5 μM dGTP, dCTP and dTTP), 1 μl [α-^{35}S]dATP (10 μCi, 1000 Ci/mmol) and 2 μl diluted *Taq* DNA polymerase (2 units), mix gently and incubate at 45 °C for 4–5 min. Then dispense 4 μl of this solution into each of four tubes containing 4 μl of the respective termination mixes. Mix gently by pipetting up and down a few times, taking care not to introduce bubbles. Place the tubes at 72 °C for 4–5 min, and then add 4 μl of formamide dye mix to each tube to stop the reactions.

For multiple samples (say 24 M13 clones), the annealings and labelling extensions are carried out in 1.5 ml tubes and the termination reactions in a flexible polycarbonate V-shaped 96-well microtitre plate (Hybaid or Costar). The termination mixes are added to the respective wells and the plate stored on ice until required. A master mix of diluted *Taq* polymerase, labelling mix and [α-^{35}S]dATP can be made and then 5 μl is added to each clone after the annealing step.

Sequenase sequencing follows a similar protocol, primer and template are annealed, then a labelling step is performed at room temperature (~20 °C) and the termination steps at 37 °C. If the target sequence is GC-rich, then termination mixes containing 7-deaza-dGTP or inosine triphosphate (dITP) can be used.

Linear amplification cycle sequencing is possible with *Taq* DNA polymerase (Craxton, 1991). In this method, the primer is end-labelled with $[\gamma\text{-}^{32}\text{P}]\text{ATP}$ and T4 polynucleotide kinase as below.

> 2 μl universal primer (5 pmol/μl)
> 15 μl $[\gamma\text{-}^{32}\text{P}]\text{ATP}$ (30 pmol, specific activity 4500 Ci/mmol)
> 5 μl 10 × Kinase buffer
> 2 μl T4 polynucleotide kinase (10 units/μl)
> 26 μl water

Incubate at 37 °C for 30 min and use directly for sequencing.

> 8 μl M13 ssDNA (~0.2 μg)
> 5 μl 5'-$[^{32}\text{P}]$-labelled primer (1 pmol)
> 2 μl *Taq* cycle sequence buffer
> 2 μl Diluted *Taq* DNA polymerase (2 unit)

Dispense 4 μl of the termination mixes into 4 tubes, and then add 4 μl of the above mix. Overlay with a drop of mineral oil, and place in a thermal cycler held at 94 °C. Cycle the reactions through the following programme: 95 °C, 30 s; 55 °C, 30 s; 72 °C, 2 min; 20–30 times. At the end of the reaction add 4 μl of formamide dye mix to each tube and briefly centrifuge to mix. Carefully remove the aqueous layer from under the oil to a clean tube.

For sequencing many clones the reactions can be performed in a flexible polycarbonate V-shaped 96-well microtitre plate. Many of the commercial thermal cyclers have heating plates that accept 96-well plates, for example the MJ Research PTC-100. A master mix can be prepared and added with a multiple dispensing pipette. For 24 clones, make up a master mix as follows: 100 μl 5'-$[^{32}\text{P}]$-labelled primer (25 pmol), 50 μl *Taq* cycle sequencing buffer and 50μl diluted *Taq* DNA polymerase (100 units). Dispense 2 μl of template DNA per four wells and add 4 μl of the respective termination mix, and then 2 μl primer/enzyme mix to all wells. Finally add a drop of mineral oil to all wells and place on the thermal cycler.

If the reactions are not being run on the same day as they are prepared, it is better not to add the formamide dye mix, but directly to freeze the reactions at −20 °C, and then add the formamide dye just before heating the samples for the gel electrophoresis.

17.5 Gel electrophoresis

The float glass plates used for sequencing are 20 cm × 50 cm × 0.5 cm in size. Wash the glass plates thoroughly in water and rinse, then clean with either acetone or ethanol and dry. The notched plate is then siliconized. Do this in a fume cupboard. Gel moulds are assembled using 0.35–0.4 mm thick Plastikard spacers. Combs for the wells should yield 30–50 wells per gel. I use a comb with well sizes 2 mm wide, 5 mm deep and with a spacer of 1 mm. This allows the loading of 8–10 clones (32–40 tracks) per gel.

For normal 6% PAGE gels (Sanger & Coulson, 1978) you need 50 ml of gel mix. This is made up from 23 g urea dissolved in 5 ml 10 × TBE buffer, 7.5 ml 40% acrylamide/bisacrylamide (A/B mix) and water to a final volume of 50 ml. Polymerization is initiated by adding 300 μl 10% ammonium persulphate (APS) and 50 μl TEMED (N, N, N', N'-tetramethylenediamine); mix well and pour immediately, insert the comb, and let the gel set for at least 60 min before use. These gels can easily be poured by clamping the sides of the gel, leaving the bottom open, and placing the gel in a horizontal position. The gel mix is taken up into a 50 ml syringe, and squeezed into the mould from the top end. The mix is taken up by capillary action and flows down to the bottom end of the plates. Once the gel mix has filled the plates, the comb is inserted, and the top and bottom are covered with Saran wrap, and the gel is left to set for ~1 h.

For buffer gradient 6% PAGE gels (Biggin, Gibson and Hong, 1983) two solutions are required, made up as below:

7 ml 5 × TBE gel mix +14 μl 25% APS +14 μl TEMED
45 ml 0.5 × TBE gel mix +90 μl 25% APS +90 μl TEMED

Use a 10 ml pipette and take up 6–8 ml of 0.5 × TBE gel mix solution and then all the 5 × TBE gel mix solution. In the process of taking up the acrylamide solutions into the pipette, they mix to form the crude gradient. Just watch the blue dye. Lift the plates to an angle of 45°, and pour carefully down the edge of the gel plates, lowering to the horizontal to stop the flow, and then top up the gel mould with the rest of the 0.5 × TBE gel mix solution. Try to pour as a broad band to keep the gradient at the bottom of the gel. These gels are not very difficult to pour, and they do yield 300–350 bases of sequence data with 50 cm plates.

Wedge gels produce a similar effect to buffer gradient gels. They are simpler to set up and use 1 × TBE/acrylamide mix as for normal 6% gels. The gel mould is set up by adding an extra thickness of Plastikard, ~3 cm in length, to the side spacers at the bottom of the plates. This increases the thickness of the gel at the bottom, and effectively slows down the DNA fragments as they migrate into the thicker acrylamide at the bottom of the gel. Unfortunately these gels take longer to dry down, and the thickness can reduce the sharpness of the DNA bands.

For gel electrophoresis, denature the samples at 80 °C for 15 min, uncovered to evaporate away the water, concentrate the solution and denature the DNA. Immediately before loading flush out the wells, which will contain urea leeched from the gel. Load 1–3 μl of sample/well and run the gel at a constant 28–30 mA for normal gels or a constant 37–37.5 watts for the gradient gels, for 2–3 h until the Bromophenol Blue dye has run off the end of the gel. Do not load within 2 cm of the edge of the gel, as this will help eliminate 'smiling' effects. Gels are run with $1 \times$ TBE buffer in the top and bottom reservoir troughs. To eliminate uneven heating effects across the gel, which give rise to 'smiling' or 'frowning', thin aluminium sheets can be clamped in front of the gel to even out the heating across the whole gel.

After electrophoresis, let the gels cool for 10 min before separating the glass plates and then acid-fix the gel by carefully submerging in a bath of 10% acetic acid for 15 min. Carefully remove the glass plate and gel, drain well and carefully blot dry with paper towels (Kim wipes), and then transfer to a sheet of Whatman 3MM paper. The trick here is to make sure you blot away most of the liquid from the gel. Spread the Whatman paper onto the gel, and firmly press down, and then lift the paper with the gel attached from the glass plate. Cover with Saran wrap and dry on a gel dryer for 30 min at 80 °C with a vacuum pump. Remove the Saran wrap and place the dried gel in an X-ray cassette with X-ray film (Fuji RX) and expose overnight at room temperature. Although the notched (or eared plate) is siliconized to prevent the gel sticking to it, it is our experience that the gel will stick to the siliconized plate 3–4 times out of 10.

17.6 Automated sequencing

Automated DNA sequencing instruments come with detailed protocols for their use, and it is not intended here to reproduce them, but to give a brief summary of the methods used. The sequencers cost around £60 000–100 000, and are best run by a trained operator, rather than as an instrument that anybody can come and use. The instruments work by detecting fluorescence as the DNA moves past a detector, and thus is real-time detection. Once the DNA fragments have passed the detector there is no going back. As a result of this each DNA fragment in the sequence ladder has to have enough fluorescent tag attached to allow detection as it passes the scanning laser. In the early days of automated sequencing this meant that large amounts of DNA had to be prepared for the sequence reactions. However, these initial problems in the amount of DNA required have been overcome by *Taq* cycle sequencing procedures.

Although Sequenase can be used to sequence DNA with fluorescence detection, it does require more DNA, which ideally should be single-stranded. *Taq* cycle sequencing requires much less DNA and can be applied to both single- and double-stranded DNA templates.

17.6.1 *Dye primer sequencing chemistry*

The following discussion applies to sequencing with an ABI 373A sequencer, although some of the points will apply to other fluorescent sequencers. One major advantage of the 373A is the fact that the sequencing chemistries utilize four different dyes, one specific for each base. This means that the sample is electrophoresed in one lane, and the scanning laser detects all four dyes (Smith *et al.*, 1986). The instrument is able to analyse 36 samples at once. All other automated instruments use just one label which then requires four lanes per sample, and thus reduces the number of samples that can be analysed per run.

Taq dye-primer cycle sequencing is set up in a similar manner to that described above for radioactive cycle sequencing. The sequencing reagents provided by ABI are already pre-mixed, PRISM^{TM} ready reaction kits, so only a sample of DNA has to be added. The PRISM^{TM} mixes contain all four dNTPs and a specific terminator and a corresponding 5'-fluorescently labelled M13 universal primer, buffer and *Taq* DNA polymerase. The reactions can be performed in individual 0.2 ml thin-walled PCR tubes or a flexible polycarbonate V-shaped 96-well microtitre plate.

For sequencing M13 ssDNA the concentration of the DNA has to be ~100 ng/μl, and for plasmid dsDNA ~200 ng/μl. Reaction mixes required are:

A reaction	C reaction	G reaction	T reaction
1 μl DNA	1 μl DNA	2 μl DNA	2 μl DNA
4 μl A mix	4 μl C mix	8 μl G mix	8 μl T mix

Cover each reaction with a drop of mineral oil.
Place in a thermal cycler already heated to 95 °C.
Thermal cycle as follows: 95 °C for 30 s; 55 °C for 30 s; 70 °C for 1 min for 15 cycles, then 95 °C for 30 s; 70 °C for 1 min for 15 cycles.

The first three steps correspond to an initial amplification phase, and the second two-step cycle to a chase reaction. To allow for the different spectral properties of the dyes, you use twice as much reagent for the G and T reactions.

To recover the samples from under the mineral oil, add 15 μl of water to each tube or well and centrifuge briefly to force the water below the oil. Remove the aqueous phase (it doesn't matter if you also carry some oil over) and add to a tube containing 1 ml chloroform. For each sample pool the A, C, G and T reactions. Remove the ~100 μl aqueous phase to a fresh tube and add 10 μl 3 M sodium acetate, pH 7.5 and 300 μl ethanol. Allow the DNA to precipitate at −20 °C for 30–60 min. Centrifuge for 10 min, decant off the supernatant, and wash the DNA pellet with 500 μl cold 100% ethanol. Let the pellet air dry for 10–15 min and then dissolve in 2–3 μl formamide mix (deionized formamide:50 mM EDTA, 5:1 by volume).

The use of a thermal cycler with a heated lid eliminates the need of mineral oil to prevent evaporation of the samples. It also does away with the chloroform extraction of the oil, so the four base reactions can be pooled and precipitated directly. However, because of the small volumes involved care has to be taken to make sure there is no evaporation during the linear cycling.

Fulton and Wilson (1994) have obtained excellent results by reducing the cycling parmaters to five cycles for the first three step amplification and to ten cycles for the second two step chase. This reduces the time taken to carry out the cycle sequencing reaction and did not affect the quality of the data.

17.6.2 Dye-terminator sequencing chemistry

One advantage of the use of dye-terminator chemistry is that it only requires one tube, as the reaction mix contains four different fluorescently-labelled dideoxynucleoside triphosphates, the four normal dNTPs, buffer and *Taq* DNA polymerase (Prober *et al.*, 1987). Also, you can use any oligonucleotide primer, which allows for the ability to sequence any cloned DNA, and in particular any PCR generated DNA fragment. The set up is extremely easy:

> 5 μl DNA
> 1 μl primer
> 5 μl water
> 9 μl ABI Prism^{TM} dye-terminator mix

Cover the reaction with a drop of mineral oil and place the tube in a cycler held at 96 °C. The cycling parameters are as follows: 96 °C for 30 s, 50 °C for 15 s and 60 °C for 4 min, for a total of 25 cycles. The lower extension temperature of 60 °C is required to allow the *Taq* DNA polymerase to incorporate the fluorescent terminators.

For M13 ssDNA you need 0.5 μg of template, for plasmid dsDNA 1 μg, and for PCR generated DNA ~0.5 pmol (Hyder *et al.*, 1994). For ssDNA you need 0.8 pmol of primer (~5 ng of a 20-mer) and for dsDNA 3.2 pmol(~20 ng).

After cycling it is imperative to remove excess dye-terminators, because they will mask the sequence. This is achieved simply by two phenol extractions: Add 80 μl of water to the reaction mix, centrifuge to mix, then remove the aqueous phase from under the mineral oil, and extract twice with 100 μl phenol/chloroform/water reagent (68:14:18 by volume). The DNA in the aqueous phase is then precipitated by addition of 10 μl 3 M sodium acetate, pH 7.5 and 300 μl ethanol, and treated as above for fluorescent dye-primer DNA.

Alternatively, the excess dye-terminators can be removed by gravity flow gel filtration using Sephadex G50 superfine, as described by Rosenthal and Charnock-Jones (1992). This eliminates the tedious and time-consuming phenol extractions, particularly if you have 36 reactions to process.

17.6.3 Gel electrophoresis

The composition of the gels is essentially the same as those for manual radioactive sequencing. However, great care has to be exercised to eliminate any background fluorescent material that will mask the actual sequence. This includes detergents and organic solvents used traditionally for cleaning the sequencing gel plates. The glass plates are special optically clear glass to eliminate background fluorescence, and they are also very expensive, around £200 for each plate. The plates are washed in water only, and finally rinsed with deionized water, then left to air dry at room temperature. The plates can be scanned by the laser to check for any high background fluorescence. Although it can be extremely frustrating at first, eventually it becomes very easy to wash the plates so they have minimal background fluorescence. We have found it very useful to use a black rubber car windscreen wiper to remove excess water from the plates.

The 6% acrylamide/7.5 M urea gel mix is deionized with a mixed bed resin, Amberlite MB-1 (Sigma), prior to addition of the $10 \times$ TBE buffer. The gel mould is set up without tape, by placing clips along the sides of the glass plates. Remember to place the clips over the spacer, and not any further onto the mould, as this will distort the plates and hence the gel. The gel plates are placed horizontally on a support and the gel mix squeezed from a 50 ml syringe into the mould. Any bubbles are removed by tapping on the glass plates, which causes the bubbles to move with the flow of acrylamide. When the mould is filled completely, the comb is inserted and the bottom and top covered with Saran wrap. We have found it best to use a square tooth comb, rather than a shark tooth's comb. Square tooth combs yield a clear space between successive sequence lanes making it easier for the analysis software to track the individual lanes.

The gel is pre-run for ∼30 min prior to loading of the DNA samples. The samples are heated at 80 °C for ∼5–10 min, and carefully loaded into the wells. Remember to flush the wells of urea before loading. The odd-numbered lanes are loaded first, and then electrophoresed for 5 min before loading the even numbered lanes. This helps in tracking the lanes at the analysis stage. Another initial frustration with automated sequencing is the fact that there is no dye in the formamide buffer, and so you are loading a colourless liquid onto the gel. However, the denser formamide solution can be observed sinking into the well if you look carefully enough. Once the electrophoresis run has been started, all you have to remember is to start the data collection on the computer. The gels are run overnight for 12–14 h. This long electrophoresis time is required because to obtain data out to 500 bases all the fragments have to pass the scanning laser.

17.7 Data analysis and troubleshooting

17.7.1 *Manual sequencing*

It is advisable to load the gels in the order AGCT or TCGA. This is because most problems occur with the G and C tracks, and it is best to place them in the middle next to each other. The *Taq* DNA polymerase does not utilise the dideoxynucleotides with equal efficiency at every position, and so the band intensity can be variable. However, this effect is useful, because it can help in reading a sequence in problem areas. The artefacts described below are the most frequent ones. Sequenase yields nearly equal intensity bands at all positions, and this feature has made it a very popular enzyme for DNA sequencing. This property of Sequenase allows for the detection of hetero-zygote sequences, particularly with PCR amplification of DNA from diploid cells.

For *Taq* DNA polymerase sequencing the C track can be the most difficult to read, because a single C band can be weak. With runs of more than one C, in general the first C is weak and the second C the strongest (particularly if the preceding base is a G). G residues may behave like C residues, in that the first G in a run of G bands may be faint compared to the others. This effect is found particularly when Gs are preceded by a T residue. Sometimes the first A band in a run of As is more intense than succeeding bands. This is sometimes found with T residues.

There are two major gel reading artefacts, compressions and pile-ups. Compressions are areas on the autoradiograph where bands suddenly appear very close together, usually with gaps or increased band spacing above the compression. They appear to be caused by GC base pairing forming a stable hairpin loop which persists even under the denaturing gel electrophoresis conditions. As this is a gel problem, it is possible to resolve it by running the gel at a higher temperature, i.e. increasing the power, but there is the danger of cracking the glass plates. Inclusion of formamide in the gel to 25% or more will aid denaturation of the DNA, but in general the formamide leads to higher background bands. To destabilize the GC base pairing, then analogues of dGTP can be used, such as dITP which forms a weaker base pair. Inspection of the area around a compression in general reveals a sequence capable of forming a GC-rich stem with a loop of 3–4 bases. The sequence can be read into the loop from both sides if opposite strands of the DNA are sequenced, thus resolving the compression.

Pile-ups are positions on the gel where a band appears in all four lanes at one or several positions, and can be accompanied by a reduced intensity of bands above the pile-up. They are caused by the DNA template forming a stable stem-loop structure, which slows down the DNA polymerase such that it pauses at that position. If the stem-loop is very stable, then polymerase extension may be completely blocked, resulting in no bands above the pile-up. This effect is stabilized by salt in the sequencing buffer. The problem can be alleviated by performing the reactions at a higher temperature. This type of problem can be seen when Sequenase is used as the enzyme, but is not seen in general with *Taq* DNA polymerase sequencing, because there is less salt in the *Taq* buffer and the reactions are carried out at a high temperature, conditions which destabilize DNA secondary structures.

In a shotgun sequencing project, it is better not to try and read out as far as possible the sequence from an individual autoradiograph. It is always difficult reading the top part of a gel because of the reduced band spacing. Although you can load more of the sample onto a second gel and electrophorese for 5–8 h to enable you to read further, in general the quality is not as good as that seen with the first 300 bases of data. It is better to sequence more shotgun clones and use gradient gels. Obviously, in the latter stages of a sequencing project running gels for 8 h (or more) may be required, but these should not be used on a regular basis. It should be pointed out that the best separation and hence the best and most reliable sequence that can be read from autoradiographs is the first 300 bases.

17.7.2 Automated sequencing

Figure 17.3 shows an example of the data obtained with dye-primers on a M13 ssDNA template. Data out to beyond 330 bases has been obtained, and the best advantage of all is that it is already in a computer readable format.

Most of the problems and artefacts associated with manual sequencing also apply to automated sequencing. Pile-ups and compressions will occur. Compressions are the most problematic for any form of sequencing, particularly if they are weak, because they can be missed. Strong compressions are clearly obvious! Figure 17.4 shows three examples of artefacts with computer interpretation of sequence data. In Figure 17.4(a), because of variation in peak spacing, the software has called a T base at position 91, where the sequence should read just three Cs. The same effect is seen in Figure 17.4(b), where at position 171–172 two As have been called, when it is obvious that there is only one A peak. Finally, in Figure 17.4(c) there is a compression around position 274 due to the GC-rich sequence. This results in increased spacing beyond, leading the analysis software to overcall two Ts and five As, when the sequence is clearly one T and four As. This overcalling of bases is probably the most common problem seen with automated sequencers. That is why it is always best to visually inspect all trace data.

Dye-terminator sequencing has its own peculiarities. One benefit is that the presence of the bulky fluorescent dye moiety at the 3′ end of the DNA tends to inhibit stem-loop formation, thus compression artefacts are not as common. Also, pile-ups tend not to be seen. In dye-primer or radioactive sequencing, the pile-up is a result of the polymerase pausing at a particular position, and a peak or band is seen because the label is at the 5′ end of the synthesized DNA. Dye-terminator chemistry has the label at the 3′ end, so only correct termination yields a peak. Any pausing which does not incorporate a terminator base is in effect invisible.

Figure 17.5 shows two examples of 'classic' artefacts seen with dye-terminator chemistry. In Figure 17.5(a) are three examples of very weak C peaks due to a preceding G base, at positions 187, 198, and 207. When there is a run of Ts followed by a G base, the T peaks become weaker and there is an increase in the intensity of the G peak, as shown in Figure 17.5(b).

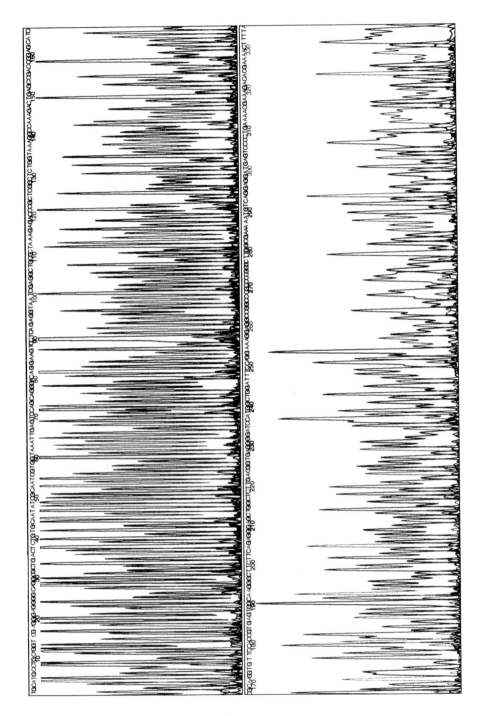

Figure 17.3. Sequence trace of a dye-primer sequence using M13 ssDNA and the universal M13 primer (ABI PRISM™ ready reaction kit). The analysis software generated sequence is shown, it has not been manually edited to remove potential errors in base calling.

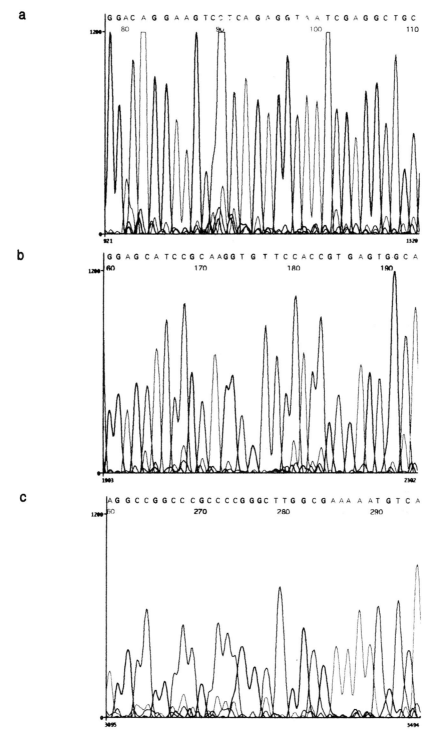

Figure 17.4. Enlargement of sections of the trace shown in Figure 17.3(a) Sequence trace showing incorrect software calling of a T base at position 91. (b) Sequence trace showing software overcalling of 2 As instead of a single A at position 171. (c) Sequence trace showing a compression, centred around base 274, and the overcalling of 2 Ts and 5 As.

a

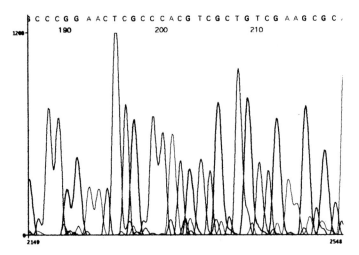

b

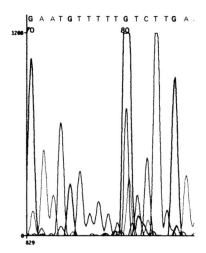

Figure 17.5. Enlargement of sequence traces showing dye-terminator artefacts. (a) Weak C peaks due to a preceding G base, at positions 187, 198 and 207. (b) Run of Ts followed by a G. The intensity of subsequent T peaks falls off, coupled with a very strong G peak (position 80).

For compilation of shotgun sequence data, the Staden computer package is recommended (Dear and Staden, 1991). For management of gel trace data a second computer is essential, as the previous night's data can be transferred off the data collection computer, freeing the instrument for another run. The Staden package, which requires a SUN computer or an X-windows emulator, contains a trace editor (ted), which allows you to inspect the sequence traces and carry out any editing prior to data assembly. Another excellent feature of the data assembly software (xbap) is that once data is assembled into contigs, any problem areas can be visualized directly by calling up the relevant region of the sequence trace. This is a major time-saving feature. With manual autoradiograph sequencing, checking out problem sequences means looking at the original X-ray film and reading the sequence to find the relevant problem area, and then making a decision on the uncertainty in gel reading.

I have included tips in the above protocols at the relevant places. An extensive trouble-shooting guide for DNA sequencing would be an entire volume in itself, but an excellent guide has been written by Ward and Howe (1989).

17.8 Reagents and buffers

TE8 buffer	10 mM Tris/HCl, pH 8
	0.1 mM EDTA Na$_2$, pH 8
10 × Ligase buffer	500 mM Tris/HCl, pH 7.5
	100 mM MgCl$_2$
	10 mM DTT (dithiothreitol)
10 × DNAase buffer	500 mM Tris/HCl, pH 7.5
	10 mM MnCl$_2$
	1 μg/μl bovine serum albumin
10 × Hin buffer	500 mM NaCl
	100 mM Tris/HCl, pH 7.5
	100 mM MgCl$_2$
	10 mM DTT
10 × Mung Bean buffer	300 mM sodium acetate, pH 4.6
	10 mM zinc acetate
	0.01% (v/v) Triton X-100
10 × Sma buffer	150 mM Tris/HCl, pH 8
	150 mM KCl
	50 mM MgCl$_2$
	10 mM DTT
2 × TY Broth	1.6% Bacto Tryptone
	1.0% Bacto Yeast Extract
	0.5% NaCl
H Top Agar	0.8% Bacto Agar
	1.0% Bacto Tryptone
	0.8% NaCl

H Plates	1.5% Bacto Agar
	1.0% Bacto Tryptone
	0.8% NaCl

MM Plates	Autoclave then add:	20 g Bacto Agar in 800 ml water
		200 ml sterile 5 × A Salts (see below)
		1 ml sterile 20% (2 M) $MgSO_4$
		10 ml sterile 20% glucose
		1 ml sterile 1 M Thiamine/HCl

5 × A Salts	5.25% K_2HPO_4
	2.25% KH_2PO_4
	0.5% $(NH_4)_2SO_4$
	0.25% sodium citrate.$2H_2O$
	Autoclave for 15 min at 15 p.s.i.

Sucrose dye mix	20% sucrose in 1 × TBE
	0.1% BPB (Bromophenol Blue)

Freshly buffered phenol — Phenol is distilled and stored at $-20\,°C$ in foil-wrapped glass universals in 5–10 ml aliquots. To buffer add 20 ml 100 mM Tris/HCl, pH 8; 10 mM EDTA and melt at 37 °C, and shake periodically to mix and buffer thoroughly. Use on the day for M13 preparations *only*.

Phenol/chloroform — Buffered phenol (buffered with 1 M Tris base, and then 1 M Tris/HCl, pH 8) to which is added chloroform and isoamyl alcohol in the following ratios by volume. Phenol:chloroform:isoamyl alcohol 24:24:1. 8-hydroxyquinoline is then added, 100 mg per 200 ml phenol/chloroform. This serves as an antioxidant and gives the phenol layer a yellow colour. The phenol/chloroform is stored in a foil wrapped bottle at 4 °C.

DNA binding matrix — Shake up 25 g of diatomaceous earth (Sigma D5384) in 500 ml of water and let it settle for 3 h. Aspirate off the supernatant containing the fine particles that still remain in suspension. Add water to 100 ml and shake to resuspend the matrix. Store in 2 × 50 ml Falcons at 4 °C. Add 1 ml of the resuspended (250 mg/ml) matrix to 25 ml of 6 M guanidine hydrochloride, 50 mM Tris/HCl, pH 7.0, 20 mM EDTA and store at room temperature in a foil-wrapped flask.

Taquence buffer	260 mM Tris/HCl, pH 8.8 26 mM MgCl$_2$
Taq cycle sequencing buffer	260 mM Tris/HCl, pH 9.5 65 mM MgCl$_2$
Taquence dilution buffer	25 mM Tris/HCl, pH 8.8 0.1 mM EDTA 0.5% Tween 20 0.5% NP 40 (omit for Taq cycle sequencing)
10 × Kinase buffer	500 mM Tris/HCl, pH 7.5 100 mM MgCl$_2$ 50 mM DTT
Universal primer	The 'universal' primer can be purchased from many sources. Dissolve to 5 pmol/μl (25 ng/μl) in TE8 and store at -20 °C. Use at 0.5 pmol/clone for Taquence sequencing. Sequence: 5′-GTAAAACGACGGCCAGT-3′

Termination mixes (USB/Amersham)

A Mix 15 μM each dGTP, dATP, dTTP, dCTP
       600 μM ddATP (300 μM for cycle sequencing)

G Mix 15 μM each dGTP, dATP, dTTP, dCTP
       45 μM ddGTP (15 μM for cycle sequencing)

C Mix 15 μM each dGTP, dATP, dTTP, dCTP
       450 μM ddCTP (150 μM for cycle sequencing)

T Mix 15 μM each dGTP, dATP, dTTP, dCTP
       1200 μM ddTTP (450 μM for cycle sequencing)

Formamide Dye mix	100 ml deionized formamide 0.1 g xylene cyanol FF 0.1 g Bromophenol Blue 2 ml 0.5 M EDTA Na$_2$, pH 8
10 × TBE buffer	108 g Tris base 55 g boric acid 9.3 g EDTA Na$_2$ Dissolve in 1000 ml water
40% A/B	380 g acrylamide 20 g bisacrylamide

Dissolve in 1 l water and deionize with Amberlite MB1, 20 g, for 20 min and then filter. Store at 4 °C

0.5 × TBE mix
150 ml 40% A/B mix
50 ml 10 × TBE
460 g urea
Dissolve in 1000 ml water, foil wrap. Store at 4 °C

5 × TBE mix
150 ml 40% A/B mix
500 ml 10 × TBE
460 g urea
50 mg BPB
Dissolve in 1000 ml water, foil wrap. Store at 4 °C

Acknowledgements

I would like to thank the following for their generosity over the years for advice and instructions in the arts of DNA sequence analysis: Alan Bankier, Bart Barrell, Carol Churcher, Bruce Roe, Chris Davies, Clyde Hutchison, Rick Wilson and Elaine Mardis.

References

Anderson, S. (1981) Shotgun DNA sequencing using cloned DNase I-generated fragments. *Nucl. Acids Res.* **9**, 3015–27.

Andersson, B., Povinelli, C. M., Wentland, M. A. *et al.* (1994) Adaptor-based uracil DNA glycosylase cloning simplifies shotgun library construction for large-scale sequencing. *Anal. Biochem.*, **218**, 300–8.

Bankier, A. T., Weston, K. M. and Barrell, B. G. (1987) Random cloning and sequencing by the M13/dideoxynucleotide chain termination method. *Methods Enzymol.* **155**, 51–93.

Biggin, M. D., Gibson, T. J. and Hong, G. F. (1983) Buffer gradient gels and ^{35}S label as an aid to rapid DNA sequence determination. *Proc. Natl Acad. Sci. USA* **80**, 3963–5.

Chissoe, S. L., Wang, Y.-F., Clifton, S. W. *et al.* (1991) Strategies for rapid and accurate DNA sequencing. *Methods: Comp. to Meth. Enzymol.* **3**, 55–65.

Craxton, M. (1991) Linear amplification sequencing, a powerful method for sequencing DNA. *Methods: Comp. to Meth. Enzymol.* **3**, 20–6.

Davies, C. J. and Hutchison III, C. A. (1991) A directed DNA sequencing strategy based upon Tn3 transposon mutagenesis: application to the ADE1 locus on *Saccharomyces cerevisiae* chomosome I. *Nucl. Acids Res.* **19**, 5731–8.

Dear, S. and Staden, R. (1991) A sequence assembly and editing program for efficient management of large projects. *Nucl. Acids Res.* **19**, 3907–11.

Deininger, P. L. (1983) Random subcloning of sonicated DNA: application to shotgun DNA sequence analysis. *Anal. Biochem.* **129**, 216–23.

Dretzen, G., Bellard, M., Sassone-Corsi, P. and Chambon, P. (1981) A reliable method for the recovery of DNA fragments from agarose and acrylamide gels. *Anal. Biochem.* **112**, 295–8.

Fulton, L. L. and Wilson, R. K. (1994) Variations on cycle sequencing. *BioTechniques*, **17**, 298–301.

Hyder, S. M., Hu, C., Needleman, D. S. *et al*. (1994) Improved accuracy in direct automated DNA sequencing of small PCR products by optimising the template concentration. *BioTechniques*, **17**, 478–82.

Innis, M. A., Myambo, K. B., Gelfand, D. H. and Brow, M. A. (1988) DNA sequencing with *Thermus aquaticus* DNA polymerase and direct sequencing of polymerase chain reaction-amplified DNA. *Proc. Natl Acad. Sci. USA*, **85**, 9436–40.

Mardis, E. R. (1994) High-throughput detergent extraction of M13 subclones for fluorescent DNA sequencing. *Nucl. Acids Res.*, **22**, 2173–5.

Messing, J. and Vieira, J. (1982) A new pair of M13 vectors for selecting either DNA strand of double-digest restriction fragments. *Gene*, **19**, 269–76.

Pan, H.-Q., Wang, Y.-P., Chissoe, S. L. *et al*. (1994) The complete nucleotide sequences of the SacBII Kan domain of the P1 pAD10-SacBII cloning vector and three cosmid cloning vectors: pTCF, svPHEP, and LAWRIST16. *Genetic Analysis, Techniques & Applications*, **11**, 181–6.

Povinelli, C. M. and Gibbs, R. A. (1993) Large-scale sequencing library production: an adaptor-based strategy. *Anal. Biochem.* **210**, 16–26.

Prober, J. M., Trainor, G. L., Dam, R. J. *et al*. (1987) A system for rapid DNA sequencing with fluorescent chain-terminating dideoxynucleotides. *Science*, **238**, 336–41.

Rosenthal, A. and Charnock-Jones, D. S. (1992) New protocols for DNA sequencing with dye terminators. *DNA Sequence*, **3**, 61–4.

Rosenthal, A., Coutelle, O. and Craxton, M. (1993) Large-scale production of DNA sequencing templates by microtitre format PCR. *Nucl. Acids Res.* **21**, 173–4.

Saiki, R. K., Gelfand, D. H., Stoffel, S. *et al*. (1988) Primer-directed enzymatic amplification of DNA with a thermostable DNA polymerase. *Science* **239**, 487–91.

Sanger, F. and Coulson, A. R. (1978) The use of thin acrylamide gels for DNA sequencing. *FEBS Lett.* **87**, 107–10.

Sanger, F., Coulson, A. R., Barrell, B. G. *et al*. (1980) Cloning in single-stranded bacteriophage as an aid to rapid DNA sequencing. *J. Mol. Biol.* **143**, 161–78.

Sanger, F., Nicklen, S. and Coulson, A. R. (1977) DNA sequencing with chain-terminating inhibitors. *Proc. Natl Acad. Sci. USA*, **74**, 5463–7.

Smith, L. M., Sanders, J. Z., Kaiser, R. J. *et al*. (1986) Fluorescence detection in automated DNA sequence analysis. *Nature* **321**, 674–9.

Smith, V., Craxton, M., Banker, A. T. *et al*. (1993) Preparation and fluorescent sequencing of M13 clones: microtiter methods. *Methods Enzymol.*, **218**, 173–87.

Ward, E. S. and Howe, C. J. (1989) Troubleshooting in chain-termination DNA sequencing. In *Nucleic acids sequencing, a practical approach* (ed. C. J. Howe and E. S. Ward), pp. 79–97. Oxford University Press, Oxford.

18

Assessment of gene transcription and expression control

NICHOLAS P. BATES and HELEN C. HURST

18.1 Introduction

Despite the fact that nearly all the cells in an individual organism have an identical genotype, huge differences are seen in their structure and function. This is achieved by the selective expression or repression of genes, so that those required to produce a particular phenotype are activated while the others are suppressed. Thus during embryogenesis, multipotent cells differentiate to form distinct tissue precursors as a result of changes in gene expression. A similar process continues to occur in renewable tissues within the mature organism where differentiated cells are derived from stem cell precursors through progressive changes in the pattern of expressed genes. These are long-term changes, but short-term responses also occur in response to external stimuli. This can be mimicked by the addition of serum to previously serum-deprived cells in tissue culture. This induces growth and division, and is accompanied by a rapid increase in expression of genes involved in the regulation of proliferation, such as c-*fos* (Fort *et al.*, 1987). Altered gene expression is associated with pathological conditions as well. For example, simple overexpression of a single gene, c-*erb*B-2, is sufficient to constitute one of the steps in the development of some breast adenocarcinomas (D'Souza *et al.*, 1993). As a result of the fundamental role gene expression plays, both in health and disease, the field is now being intensively researched.

The expression of a gene as a functional protein involves several distinct processes. It must first be transcribed by RNA polymerase to give an RNA primary transcript. This is then edited by the excision of intron sequences and exported from the nucleus, to act as a template for polypeptide production by ribosomes in the cytoplasm. After post-translational modification, and transport to its site of action, the mature protein can then influence the cellular phenotype. Alterations in overall gene expression may involve one or more of these processes, but the most important control point is the initiation of gene transcription, the subject of this chapter. Its importance is illustrated by the fact that as many as 10% of human genes may encode protein factors involved in the control of transcription.

Specialized areas of genomic DNA are involved in controlling the rate of gene transcription. These include the gene promoter region adjacent and 5′ to the transcription start site, and more distant enhancer sequences. The promoter contains a recognition site for RNA polymerase and its associated factors, which can bind and form the transcription initiation complex close to the start site. RNA polymerization is dependent upon the formation of the transcription inititation complex, but the individual proteins are ubiquitous, and are not thought to play a part in transcription regulation. This is achieved by specific factors, which bind to other sites within the promoter and enhancer regions, and interact with the initiation complex thus altering its stability and hence the frequency of transcription starts. The presence or absence of these specific factors determines whether genes controlled by them are transcribed or not (reviewed by Lobe, 1992). Binding of positively acting transcription factors will stimulate the transcription rate above the low basal level achieved by general transcription factors alone. Other specific factors may reduce transcription, but the overall effect of a given transcription factor depends upon the others that are present. One which weakly increases transcription when tested in isolation, for example, may act as an inhibitor when several factors are tested together, by competing with a more active transcription factor for an overlapping binding site. In fact, *in vivo* most genes are controlled by several different transcription factors, with interacting effects, and this must be borne in mind when analysing these processes *in vitro* model systems.

Overall gene expression at the protein level may be assessed by Western blotting, or by immunocytochemistry if information on the subcellular localization of the protein is required (see Chapter 11). At the RNA level, expression is assessed by Northern blotting. *In situ* hybridization may be used to determine the position of cells expressing a particular RNA within a tissue (see Chapter 9). These techniques are discussed in this book. In order to come to accurate conclusions about mechanisms of transcription control, the results from several different techniques need to be considered. It can be misleading to rely on the results of a single method without corroborative evidence obtained in an independent way. It is impossible to include all the methods that may be relevant within the scope of this chapter. Laboratory manuals dedicated to gene transcription are available, and the reader is referred to these for a more extensive treatment of the subject (Hames and Higgins, 1993; Latchman, 1993). The nuclear run-on technique, which is useful for assessing gene transcriptional activity under near physiological conditions, is well covered in the review by Marzluff and Huang (1985).

This chapter will concentrate on the assessment of gene promoter activity, and the identification of transcription factor binding sites. We will describe three fundamental techniques. Transient transfection of promoter/reporter gene constructs allows an assessment of the promoter or enhancer function of a cloned DNA fragment. The proteins involved in promoter or enhancer function may then be examined by identifying transcription factor binding sites using DNAase footprinting. Finally, the gel retardation assay allows the binding of a transcription factor to its DNA recognition site to be analysed in isolation and can lead to the identification of the particular factor. The evidence that a specific transcription factor binding site is required for promotor or enhancer activity may then be strengthened by mutation of specific binding sites and comparing the activity in functional transfection assays with the wild-type sequence. Together, these commonly performed techniques allow a great deal of information to be obtained about the normal transcriptional control of a gene.

18.2 Transient transfection

18.2.1 Introduction

The technique of transient transfection can be used to assess the promoter or enhancer activities of cloned fragments of a gene, which is helpful in determining the position of functionally important regulatory sites. Measurement of promoter activity involves cloning the test fragment upstream of a reporter gene, which is chosen to given a non-mammalian product that is readily assayed. This protocol uses the bacterial chloramphenicol acetyl transferase (CAT) gene but other reporters such as luciferase are also widely used and may in some cases be more efficient. The promoter/CAT construct (illustrated in Figure 18.1) is transiently transfected into appropriate cells in tissue culture, and the resultant reported gene expression will be proportional to the degree of promoter activity in the fragment being tested. The CAT gene without a promoter may be used as a negative control. Chloramphenicol acetyl transferase activity is detected by a simple biochemical assay. Inevitably, there are differences in the efficiency of transfection between plates of cells, and this is allowed for in this protocol by co-transfecting a β-galactosidase expression plasmid, pCH110 (Pharmacia). This bacterial enzyme activity is also easily assayed, and the results are used to correct the CAT assay results for variations in transfection efficiency.

It is important to select the cells that are transfected on the basis of their physiological relevance to the gene being studied rather than their ease of transfection. The pattern of transcription factor expression varies greatly between cell types, so results obtained in one cell line do not automatically apply to a different one. For example, the promoter activities of genes important in breast cancer should be assessed in breast cancer cell lines; it is tempting to use more easily transfected cell lines, but the results may not be very relevant. The protocol described here has been optimized for use with human breast cancer cell lines and uses the calcium phosphate precipitate technique to introduce foreign DNA into the cells (see Chapter 19). Other cultured cells vary in their growth rates, their ability to take up foreign DNA

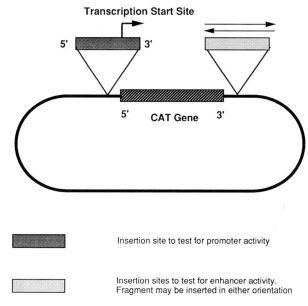

Figure 18.1. Diagramatic representation of reporter constructs for transfection assays. Test promoter fragments thought to contain transcription initiation sites (depicted by an arrow) can be cloned in front of the CAT coding region. Prospective enhancer sequences may be additionally cloned in either orientation further 5' or 3' to the promoter.

and their sensitivity to the toxic effects of transfection. Modification of the transfection protocol with these factors in mind may be needed for other cell types and alternative transfection protocols (DEAE dextran, lipofection, electroporation) may need to be considered (see Chapter 19). The quantities of reagents suggested are suitable for cells grown in 9 cm tissue culture dishes.

18.2.2 Protocol

Day 1
1. Split the cells to be transfected so that they are approximately 30% confluent. Two plates should be prepared for each set of experimental conditions. If similar results are obtained from each duplicate plate this confirms satisfactory transfection.

Day 2
1. Feed the cells with 9 ml of fresh medium. Return them to the incubator for 2 h before transfection.

2. Add the following to a sterile 5 ml tube for each pair of tissue culture plates, in the indicated order:

Sterile Millipore water   to give a final volume of 1 ml
Promoter CAT construct   10 μg
pCH110   6 μg
2 M CaCl$_2$ 124 μl
(This must be added last)

3. With an automatic pipettor, bubble the mixture above while adding 1 ml of 2x DNA precipitation buffer (5Prime-3Prime Inc; Catalogue Number 5302-878960), dropwise.
4. Allow to stand for 20 min. The solution may be slightly cloudy.
5. Add 1 ml of solution dropwise to each plate of cells.

Day 3
1. Microscopy should confirm the presence of innumerable calcium phosphate precipitates on the bottom of each plate.
2. Wash each plate twice in sterile phosphate buffered saline (PBS).
3. Add 10 ml of the usual growth medium.

Day 5
1. Remove medium from plates, and wash twice with cold PBS.
2. Add 200 μl lysis buffer (Nonidet P40, 0.65%; Tris/HCl, pH 8, 10 mM; EDTA, 1 mM; NaCl, 150 mM) to each plate, tip from side to side to ensure that all the surface is covered, and leave flat for 5 min.
3. Confirm cell lysis has occurred by microscopy.
4. Tip each plate for a few seconds to allow the lysate to collect before transferring it to a microcentrifuge tube.
5. Spin the lysate in a refrigerated microcentrifuge (14,000 r.p.m., 4 °C, 5 min)
6. Transfer the supernatant to a fresh tube, and discard the pelleted debris.
7. The lysate may now be frozen indefinitely at −20 °C, or used immediately for assay.

18.2.2.1 Assay for β-galactosidase activity
1. Mix: 400 μl lacZ buffer (Na$_2$HPO$_4$, 60 mM; NaH$_2$PO$_4$, 40 mM; KCl, 10 mM; MgCl$_2$, 1 mM; β-mercaptoethanol, 0.27% (v/v); store at 4 °C)
   100 μl cell lysate
   100 μl *o*-nitrophenyl β-D-galactopyranoside, 4 mg/ml (Sigma N1127. Make up at 4 mg/ml in 100 mM Na$_2$HPO$_4$/NaH$_2$PO$_4$ buffer, pH 7.0)
2. Incubate the samples, plus a lysis buffer only control, at 37 °C until the sample colour has changed to a pale lemon yellow.
3. Add 250 μl 1 M Na$_2$CO$_3$ to stop the reaction. The mixture will turn a deeper yellow.
4. Spin the samples in a microcentrifuge (14 000 r.p.m, 5 min).

5. Transfer the supernatant to a plastic spectrophotometer cuvette without disturbing the pellet.
6. Measure the absorbance at 420 nm (A_{420}), using the control incubation to zero the spectrophotometer.
7. The linear range of this assay is $0.2 < A_{420} < 0.8$. Sample falling outside this range should be repeated.

18.2.2.2 *Assay for chloramphenicol acetyl transferase activity*

1. Heat a 56 μl aliquot of each sample, and a lysis buffer control, at 68 °C for 5 min. This inactivates the CAT inhibitory proteins that are present in some mammalian cells.
2. Return to ice.
3. Make a mix containing the following per sample:

8 mM chloramphenicol	20 μl
0.5 mM acetyl CoA	18 μl
^{14}C-acetyl CoA (Amersham CFA 729)	2 μl
2 M Tris/HCl, pH 7.8	4 μl

4. Add 44 μl of the mix to each sample. A precipitate may form, but can be ignored.
5. Incubate at 37 °C for 2 h or longer. The length of time taken for the β-galactosidase assays to change colour gives a rough indication of the time for which these assays should be incubated.
6. Add 130 μl of cooled ethyl acetate to each sample.
7. Vortex the tubes, and then spin in a microcentrifuge (10 min, 14 000 r.p.m.).
8. Transfer 100 μl of the upper layer to a scintillation vial, containing 5 ml scintillant (such as Aquasol, NEN Research Products) taking care to avoid the lower layer.
9. Count each sample (^{14}C programme, 1 min).
10. Express the corrected result as: (sample activity − control activity)/β-galactosidase A_{420}.
11. The control activity should be less than 300 counts per min.
12. Samples achieving more than 15 000 counts per minute should be repeated after dilution.

Solutions
Chloramphenicol (Sigma C0378)

> Make a stock of 100 mM in ethanol
> Dilute to 8 mM in water for use
> Store at −20 °C

Acetyl CoA (Sigma A2897)
> Make a 0.5 mM solution in water
> Store in batches at −20 °C

18.2.2.3 Troubleshooting

The manipulations involved in this protocol are all fairly straightforward, and unlikely to cause problems. Having said that, transfections may fail for no apparent reason. The cells should be in good condition before attempting to transfect them, and not allowed to become severely overgrown. We have found that cell lines become less easy to transfect after multiple passages, and it is good practice to renew them from liquid nitrogen stocks at regular intervals. Some growth media, such as RPMI, are unsuitable for transfection by the calcium phosphate method. The pH of the precipitation buffer is critical to transfection efficiency, and small variations will lead to failure of the whole procedure. For this reason, we use a commercial product, which gives greater batch consistency than making up the reagent in the laboratory. This is obtainable from several manufacturers including Clontech, Invitrogen and Promega.

If a particular gene fragment is shown to have promoter activity, then it is usual to try to pinpoint the functional sequences more precisely using smaller fragments (deletion analysis). This may reveal the presence of both activating and silencing functions within the test promoter and should identify those fragments that will be informative to analyse by DNAase footprinting (see below). However, it should also be borne in mind the DNA is being investigated under conditions very different to those that exist *in vivo*. The overall length of promoter fragment being used is obviously important. Sequences 5′ to the fragment being used may play a role in promoter activity, and this will of course not be detected. In addition, no assessment is made of the influence that more distant enhancer sites may have on transcription. These are frequently found at sites other than within 5′ region of a transcription unit, for example in the first intron or entirely 3′ to the gene. Activity of these elements may be investigated by generating additional constructs around the CAT gene. This is illustrated in Figure 18.1 where additional fragments can be inserted in the construct to test for enhancer activity.

18.3 DNA footprinting

18.3.1 Introduction

Once a region of DNA has been shown to be functionally active as a promoter or enhancer, the position of transcription factor binding sites can be determined by DNA footprinting, which identifies the position at which proteins bind within a DNA sequence. The DNA fragment being tested, which acts as a probe, is radioactively labelled on one strand only. If it is then digested with an agent that randomly cleaves the probe, under conditions that allow an average of one cut per molecule, a series of DNA fragments of different sizes will be formed. This protocol uses deoxyribonuclease I (DNAase I), but other chemical agents can be used for more sophisticated analysis (Revzin, 1993). The digest is then run on a sequencing gel, which allows fragments differing by a single base pair in length to be distinguished. A ladder of bands will be seen (see Figure 18.2 lanes 2 and 11). Parallel incubations, where the probe is preincubated with a protein nuclear extract, will give a similar pattern of bands, except where proteins have bound to the DNA. Here, the DNA will be protected from digestion and the bands corresponding to that position will be missing from the gel, giving a gap, or footprint, in the ladder of fragments (see FP1 and FP2 indicated on Figure 2). The position of a footprint can be determined by running a sequencing reaction of the probe DNA on the same gel (lane 1, Figure 2) and directly comparing the two lanes. This gives both the position of the binding site within the promoter and its sequence.

18.3.2 Protocol

18.3.2.1 Preparation of a nuclear extract

A crude nuclear protein extract from the same cell line(s) used for functional transfection assays is ideal for these assays.

1. Remove media from cultured cells at about 80% confluency.
2. Wash each plate twice with cold PBS.
3. Add 1 ml of PBS to each plate, scrape off the cells and transfer them to a microcentrifuge tube.
4. Spin at full speed for one minute in a microcentrifuge, and discard the supernatant.
5. Add 300 to 500 μl of lysis buffer (Tris/HCl, pH 8.0, 20 mM; NaCl, 20 mM; Nonidet P40, 0.5%; just before use add DTT, 1 mM; PMSF, 0.5 mM), and leave on ice for 5 to 10 min.
6. Spin in the microcentrifuge for 1 min, and discard supernatant, leaving a nuclear pellet.
7. Resuspend the pellet in 50 to 500 μl of buffer C (HEPES, pH 8.0, 0.20 mM; glycerol, 25% (v/v); NaCl, 420 mM; $MgCl_2$, 1.5 mM; EDTA, 0.2 mM; DTT, 1 mM; PMSF, 0.5 mM), depending on its size.
8. Mix well and leave on ice for 30 min, or mix at 4 °C for 1 h for larger volumes.
9. Spin for 1 min to pellet nuclei. The supernatant is the crude nuclear extract, which may be used directly in the gel retardation assay.
10. Before using for DNAase I footprinting, the extract should be dialysed against buffer D (HEPES–KOH, pH 8.0, 20 mM; glycerol, 20% (v/v); KCl, 100 mM; $MgCl_2$, 1.5 mM; EDTA, 0.2 mM; lauryl diamine oxide, 0.1% (v/v); sodium molybdate, 1 mM; PMSF 0.5 mM; DTT, 0.5 mM).

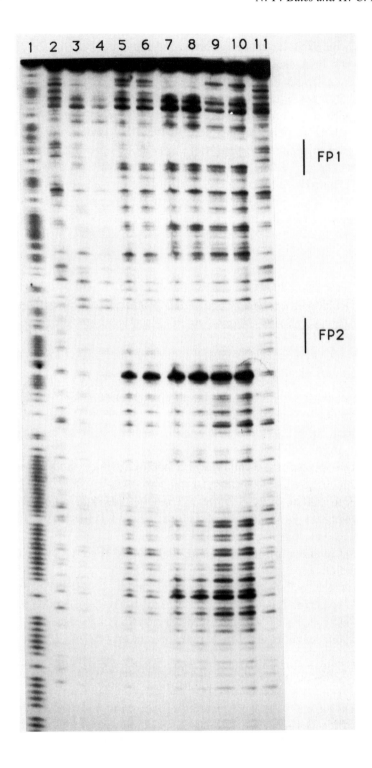

18.3.2.2 *Preparation of the DNA probe*

1. It is assumed that the promoter or enhancer sequences being tested have previously been subcloned into a suitable plasmid. Digest 100–150 μg of plasmid DNA with appropriate restriction enzymes to produce a promoter fragment of 200–500 base pairs that will label with $(\alpha^{32}P)$dATP only at one end. For example, an *Xba*I, *Eco*RI digested fragment will label only at the *Eco*RI site.
2. Run the digest on an agarose gel.
3. Cut out and electroelute the required restriction fragment and recover it either by precipitation or with a commercial DNA purification product.
4. Resuspend the pellet in 20–30 μl TE (10 mM Tris/1 mM EDTA, pH 8).
5. Use 5 μl to determine the concentration by spectrophotometry, then dilute to 0.5 pmole/μl.

18.3.2.3 *Labelling the probe*

We have found that using reverse transcriptase to end-label the probe produces higher activity than the Klenow fragment.

1. To label 1 μl (i.e. 500 fmoles) mix:
    4–5 μl $(\alpha\text{-}^{32}P)$dATP (10 μCi/μl)
    1 μl M-MLV reverse transcriptase (Life Technologies)
    2 μl 5× buffer
    1 μl 0.1 M DTT
    distilled water to 10 μl.
2. Incubate at 37 °C for 1 h.
3. Add 2 μl of 10 mM 'cold' dATP and incubate at 37 °C for 15 min.
4. Make up to 50 μl with distilled water.
5. Remove unincorporated $(\alpha\text{-}^{32}P)$dATP, for example by passing through a Sephadex G50 (Pharmacia) spun column.
6. Adjust final volume to exactly 50 μl, i.e. 10 fmoles/μl.
7. Count 1 μl in a scintillation counter (^{32}P, 1 min). Should have about 0.5 $\times 10^5$ c.p.m./μl.
8. Store at −20 °C.

Figure 18.2. DNAase 1 footprinting gel. A 350 bp promoter fragment has been end-labelled and digested with DNAase 1 either directly (lanes 2 and 11) or after incubation with crude nuclear extracts derived from a number of cell lines known to express the gene under study (lanes 3–10). Two footprints designated FP1 and FP2 can be clearly seen in lanes 5–10 but are not detected in lanes 2 and 3 where suboptimal DNAase 1 digestion has occurred. An A + G sequencing ladder of the promoter fragment has been run in lane 1.

18.3.2.4 Maxam and Gilbert sequencing reaction

1. Take 5 μl of the probe made above and add:
   5 μl distilled water
   3 μl 10% formic acid.
2. Incubate at 37 °C for 13 min.
3. Add:
   300 μl DNAase stop buffer (EDTA 10 mM; NaOAc, 0.3 M, pH 5.2).
   2 μl carrier tRNA (10 mg/ml)
   750 μl ethanol.
4. Precipitate on dry ice.
5. Spin and wash pellet in 70% ethanol.
6. Spin, and dry pellet.
7. Resuspend pellet in 100 μl of fresh 1/10 dilution of piperidine in distilled water.
8. Allow the piperidine to cleave the DNA at 95 °C for 30 min.
9. Add:
   60 μl 3 M sodium acetate.
   440 μl distilled water.
10. Mix well. This dilutes out the piperidine.
11. Add:
    2 μl carrier tRNA (10 mg/ml)
    650 μl iso-propanol.
12. Precipitate on dry ice.
13. Spin, and remove supernatant.
14. Wash pellet with 1 ml 70% ethanol.
15. Spin, and dry pellet.
16. Cerenkov count pellet (^{3}H for 1 min).
17. Resuspend pellet in formamide sequencing gel buffer at 1000 to 5000 counts/μl.

18.3.2.5 DNAase I incubations

1. Set up the following reactions on ice for each experimental condition, diluted in DNAase buffer (HEPES/KOH, pH 8.0, 20 mM; KCl, 50 mM; glycerol, 20% (v/v); DTT, 2 mM; MgCl$_2$, 2 mM; store at −20 °C).

Tube	Probe mix	Protein	DNAase buffer
1	5 μl	0 μg	To 45 μl total
2	5 μl	50 μg	To 45 μl total
3	5 μl	100 μg	To 45 μl total
4	5 μl	150 μg	To 45 μl total

These quantities are suggested as a starting point, but may need to be varied. Dilute the probe mix in DNAase buffer such that each tube receives 5–10 fmoles with a suggested starting point of 7 fmoles per tube.

2. Incubate at room temperature for 20 min to allow protein/DNA binding.
3. Meanwhile, make two dilutions of pancreatic deoxyribonuclease (Amersham, E2210Y) as follows.

    1 in 100: $2\mu l$ enzyme made up to 200 μl in DNAase buffer (this is 10 ng/ml and is used for experimental tubes containing protein).

    1 in 10000: 2 μl of the 1 in 100 dilution made up to 200 μl in DNAase buffer (this is 0.1 ng/ml and is used for control tubes containing DNA probe alone).

    These dilutions can be stored at $-20\,°C$ and used repeatedly.

4. *Perform steps 4 and 5 on one tube at a time*

    Once the 20 min binding incubation is complete, add 3–5 μl of the correct pancreatic DNAase dilution to the side of the tube, vortex and start timing.

5. After 30–60 s at room temperature, add 50 μl 2 × PK stop buffer (Tris/HCl, pH 7.5, 100 mM; SDS 2% (w/v); EDTA 20 mM; proteinase K 400 μg/ml; store at $-20\,°C$).

6. When step 5 has been completed for every tube, incubate them all at $37\,°C$ for 10 min.

5. Phenol/chloroform extract samples.
6. Ethanol precipitate (add 2 μl of 10 mg/ml carrier tRNA) and spin.
7. Wash in 70% ethanol, and air dry pellet.
8. Cerenkov count samples and resuspend in formamide DNA sequencing stop buffer at 1000 to 3000 counts/μl such that each sample has the same cpm/μl and at least 5 μl of each sample is available to load on the gel.
9. Load 5 μl of each sample on gel and an equal number of counts of the Maxam and Gilbert sequencing ladder on one side as a marker track.

18.3.2.6 Electrophoresis

1. Pour a sequencing gel (6% to 10% acrylamide/bisacrylamide (38:2) depending upon the size of the probe, 7 M urea, 1 × TBE)
2. Heat all samples at $95\,°C$ for 3 min.
3. Load 5 μl in each track (experimental tubes and A + G reactions).
4. Run at 33 mA until Bromophenol Blue is at bottom of gel, which takes approximately 90 min.
5. Dry the gel, and autoradiograph at $-80\,°C$. (The autoradiograph may need to be left for more than 24 h).

18.3.2.7 Troubleshooting

It is important that the probe is effectively labelled. It is very difficult to get useful information if the labelling reaction has not given the specified level of activity, and it is better to repeat this rather than continue with the experiment. The probe may be checked on a sequencing gel to ensure that it produces a single band. If the autoradiograph shows only very slowly migrating bands, corresponding to undigested probe (e.g. lanes 3 and 4, Figure 18.2), then the DNAase I digestion should be increased by either adding more enzyme or increasing the incubation time. Conversely, if the autoradiograph shows very little remaining full length probe and numerous small bands, indicating over-digestion, then the DNAase I digestion should be reduced in another reaction. If there are no obvious footprints on the autoradiograph, it is worth increasing the amount of protein in the binding reactions. There are often areas of increased DNAase I digestion on either side of a protein binding site, revealed as more intense bands adjacent to the footprint on the autoradiograph (e.g. below FP2 in lanes 5–10, Figure 18.2). This phenomenon may sometimes be seen under conditions that are not adequate to give a footprint. Unfortunately, DNAase I shows some sequence preference, and this may result in some gaps being seen in the ladder of fragments and consequently it can be difficult to decide if there are footprints in some sequences.

Like all *in vitro* assays, this technique does not always reflect the *in vivo* situation. In an intact cell, the DNA is highly structured, through the binding of structural proteins such as the histones not normally present in nuclear extracts. Because of this, proteins bound at distant sites along the DNA may normally be in contact and help each other to bind *in vivo*, but this will not be reproduced here.

18.4 The gel retardation assay

18.4.1 Introduction

Once a binding site has been identified using DNA footprinting, the gel retardation assay, or electromobility shift assay, provides a less cumbersome way of investigating its interactions with transcription factors of interest. In essence, a radioactively labelled DNA fragment containing the sequence of interest is incubated with the proteins being tested, and the mixture electro-phoresed through a non-denaturing acrylamide gel. DNA fragments that remain unbound migrate rapidly through the gel, while those that have bound to a protein move more slowly (see Figure 18.3). It is this retardation of migration that gives the assay its name. The position of the labelled DNA probe on the gel is revealed by autoradiography. The assay is best used with short fragments of DNA that represent a single transcription factor binding site. However, some proteins can only bind in cooperation with others. If these proteins are unavailable, or if they rely on a separate binding site that is not included in the probe, binding that does occur in the natural state will not be detected by the gel retardation assay.

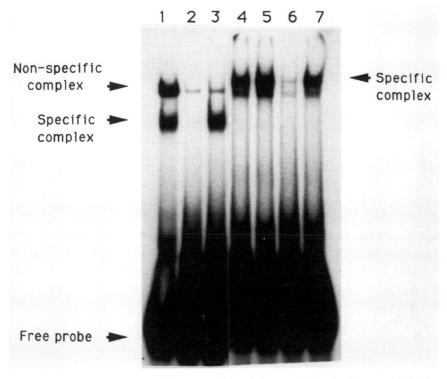

Figure 18.3. Electromobility shift assay. Two different 25 bp double-stranded oligonucleotide probes have been end-labelled and incubated with the same crude extract
and the resulting complexes separated in a non-denaturing gel (probe 1, lanes 1–3;
probe 2, lanes 4–7). In each case the excess of free probe runs with the dye front and
the protein–DNA complexes run more slowly. The two different probes bind quite
distinct proteins, which result in specific complexes of very different mobility (compare
lanes 1 and 4). These complexes are competed away by addition of an excess of the
same, unlabelled oligonucleotide to the incubation (compare lanes 1 and 2; lanes 4 and
6) but an excess of an unrelated oligonucleotide (as in lanes 3 and 5) cannot compete
away the specific complexes. Also an excess of an oligonucleotide with a single base
mutation in the centre (as used in lane 7 for probe 2) no longer competes as it does not
retain the core DNA recognition sequence for the nuclear protein that forms the
complex. This nuclear extract also contains a non-specific DNA-binding protein that
forms a complex of the same mobility on both probes as indicated and is equally
competed with both cognate and unrelated oligonucleotides (e.g. compare lanes 2
and 3).

The DNA binding specificity can be evaluated by competition assays when
an excess of unlabelled oligodeoxynucleotide is added. If the competitor
oligonucleotide also provides a protein binding site it should compete for the
binding protein and the retarded complex on the labelled probe will be
abolished. The binding requirements of a protein can be tested using a series of
mutated competitor oligonucleotides; mutations that abolish protein binding
will also abolish competition (illustrated in Figure 18.3). The nature of the

interacting protein may also be investigated by adding antibody to the incubation mixture. If the test antibody recognizes the bound protein, the DNA/protein complex may either be abolished, or its migration slowed further to give a supershifted band, depending on whether or not the antibody interferes with the DNA/protein interaction.

18.4.2 Protocol for the gel retardation assay

18.4.2.1 Introduction

The length of the double-stranded DNA probe has an important bearing on the results obtained. The best probe corresponds to the sequences shown to be protected in the footprinting assay, and is commonly 20 to 50 base pairs long. This may most conveniently be made by annealing two complementary oligonucleotides, as described in this protocol. Labelling with $(\alpha\text{-}^{32})$dATP is facilitated by ensuring that overhanging unpaired thymidine residues are present at the end of the probe, by appropriate design of the oligonucleotides or choice of restriction enzymes. This labelling protocol assumes that overhanging thymidine residues are present.

The protein to be tested for binding may be derived from crude nuclear extracts (see above) or from more purified preparations of these extracts or by *in vitro* synthesis from cloned cDNA templates of candidate cognate factors. Whichever source is used, the quantity of protein to add will need to be determined empirically.

18.4.2.2 Preparation of oligonucleotides probe

1. Dissolve each of two complementary oligonucleotides in 10 mM Tris/1 mM EDTA, pH 8.0 (TE) at a concentration of approximately 1 mg/ml.
2. Measure the absorbance at 260 nm (A_{260}) to determine the actual concentration.
3. Add equal amounts, approximately 10 µg, of each oligonucleotide to a tube containing 2.5 µl of 10× polynucleotide kinase buffer.
4. Make the volume up to 25 µl with distilled water.
5. Heat to 95 °C for 5 min, 65 °C for 10 min, 37 °C for 10 min and then allow to cool to room temperature.
7. Dilute the double-stranded oligonucleotide to a final concentration of 0.1 µg/µl.

18.4.2.3 Labelling the oligonucleotide probe

1. Set up the following reaction mix:

0.1 µg/µl double-stranded oligonucleotide	1 µl
DTT 0.1 M	1 µl

5 × M-MLV reverse transcriptase buffer	2 μl
(α-^{32}P)dATP (10 μCi/μl)	2 μl
M-MLV reverse transcriptase (200U)	1 μl
Water	3 μl

2. Incubate at 37 °C for 30 min.
3. Add 90 μl water to bring the total volume to 100 μl
4. Remove the unincorporated nucleotides (for example, by passing the product down a Sephadex G50 spun column).
5. This probe concentration may now be taken as 1 ng/μl.

18.4.2.4 Binding reaction

1. Set up tubes for the experiment on ice.
2. The final volume for each reaction is 25 μl.
3. All dilutions are made in gel retardation assay buffer (GRB) (HEPES, pH 7.9, 25 mM; EDTA, 1 mM; DTT, 5 mM; NaCl, 150 mM; glycerol, 10% (v/v); store at -20 °C).
4. Make the following additions to each tube in the indicated order:

GRB to give a final volume of:	25 μl
nuclear extract	1 to 5 μg
poly dA/poly dT (Pharmacia) (non-specific blocker)	1 μg
competitor DNA if required	10 to 100 fold excess over probe
Probe	0.1 ng

It is convenient to make a mix of the non-specific blocker and probe to add to all the tubes.
5. Incubate at room temperature for 20 min. If the factors being tested are very labile, incubation on ice may be preferable.

18.4.2.5 Electrophoresis

1. Load the binding reaction directly onto a 6–8% acrylamide gel (44:0.8) made with 0.5 × TBE.
2. Electrophorese at 200 V for 90 min in 0.5 × TBE running buffer.
3. Fix the gel for 10 min with 10% acetic acid.
4. Dry the gel on Whatman 3 MM paper under vacuum.
5. Autoradiograph. An overnight exposure is usually adequate.

18.4.3 Troubleshooting

If no specific complexes are seen, it may help to vary the amounts of protein and the non-specific blocker. Sometimes, a different non-specific blocker, such as poly dI/poly dC or sonicated salmon sperm DNA may give better results. Several non-specific DNA binding proteins are present in nuclear extracts, and these can bind to the ends or single-stranded extensions of oligonucleotide probes. This can give extra, non-specific complexes on the final gel as shown in Figure 18.3. These may be eradicated by including one to two micrograms of another double-stranded oligonucleotide, unrelated to the probe, to the blocking mix to provide an excess of DNA ends.

References

D'Souza, B., Berdichevsky, F., Kyprianou, N. and Taylor-Papadimitriou, J. (1993) Collagen-induced morphogenesis and expression of the α2-integrin subunit is inhibited in c-erbB-2-transfected human mammary epithelial cells. *Oncogene* **8**, 1797–806.

Fort, P., Rech, J., Vie, A. *et al*. (1987) Regulation of c-fos gene expression in hamster fibroblasts: initiation and elongation of transcription and mRNA degradation. *Nucl. Acids Res.* **15**, 5657–67.

Hames, B. D. and Higgins, S. J. (1993) *Gene transcription, A practical approach*. Oxford University Press, Oxford.

Latchman, D. S. (1993) *Transcription factors, A practical approach*. IRL Press, Oxford.

Lobe, C. G. (1992) Transcription factors and mammalian development. *Curr. Topics in Devel. Biol.* **27**, 351–83.

Marzluff, W. F. and Huang, R. C. C. (1985) Transcription of RNA in isolated nuclei. In *Transcription, a practical approach*, pp. 89–129. IRL Press, Oxford.

Revzin, A. (1993) *Footprinitng of nucleic acid–protein complexes*. Academic Press Inc., San Diego.

Gene manipulation: cellular level

19

Delivery of DNA into cells: production of competent cells and cell transfection

SIMON CASTLEDEN and RICHARD G. VILE

19.1 Introduction and background

DNA transfection has made it possible to study the function of cloned eukaryotic genes.

Recombinant plasmids are frequently employed in transfection experiments. With the availability of a wide spectrum of specialized and versatile eukaryotic cloning/expression vectors, investigators have been given powerful tools to look into mechanisms governing gene expression, and to try to identify genes participating in diverse processes, such as metabolism, the immune response, differentiation and development, the repair of DNA damage, and malignant transformation.

The success of any gene-transfer experiment depends largely on the quality of the donor DNA. Highly purified plasmid molecules ensure both consistent and optimal levels of gene expression in 'transient' assays, as well as reproducible frequencies of stable transfection. All stock solutions should be prepared using ultra pure water and the highest purity and grade of chemicals. Where possible, buffers and all glassware should be autoclaved to ensure inactivation of any contaminating deoxyribonucleases. DNA preparation, cloning and construction of recombinant DNA molecules are discussed in Chapter 14.

Irrespective of the method used to introduce DNA into cells, the efficiency of transient or stable transfection is determined largely by the cell type that is used. Different lines of cultured cells vary by several orders of magnitude in their ability to take up and express exogenously added DNA. Furthermore, a method that works well for one type of cultured cell may be useless for another.

The choice of promoter used is also very important, from the strong and promiscuous cytomegalovirus (CMV) or SV40 promoters to tissue-specific promoters.

19.1.1 Transient transfection

Only a small amount of the DNA actually integrates into the host genome, the rest being retained extrachromosomally prior to degradation. Expression from this unintegrated DNA can produce transient gene expression. At 48 to 72 h post-transfection, cells or the medium in which they are growing can be harvested to study the transient expression of the transferred genes. The actual time chosen depends very much on the doubling time of the cells and the specific characteristics of expression of the gene.

19.1.2 Stable transfection

The objective of stable, long-term transfection is to generate individual clones containing transferred DNA. It is therefore necessary to differentiate between those cells that have been transfected by the exogenous DNA and the bulk of non-transfected cells. This screening can be accomplished by drug selection when an appropriate drug resistance marker is included in the transfected DNA either as part of the plasmid, or as a separate plasmid as a co-transfection.

There are now several different drug selection markers that can be used. For example, the bacterial gene for aminoglycoside phosphotransferase confers resistance to the drug G418 (neomycin). Resistance to the drug hygromycin B is conferred by the hygromycin B phosphotransferase gene.

Typically, cells are maintained in non-selective medium for 24 to 72 h post-transfection, before being split into selection medium. The ideal is to let the cells divide only once, to allow integration and expression of the transfected gene. Thereafter, all selected clones will represent the progeny of a single transfection event. It is important to establish that the concentration of drug used is sufficient to eliminate parental, non-transfected cells thus leaving only the transfected ones alive. Initially, during the period of greatest cell killing, the selection medium should be changed two or three times in the first week, and then two to three times in the next couple of weeks, thereby removing dead cells and allowing individual clones to develop. These can then be individually picked and bulked up (Figure 19.1). Small squares of sterilized Whatman filter paper soaked in trypsin, placed on to the clone, and then lifted off and placed into fresh selective medium will transfer some of the cells, which can then continue to grow as unique clones.

Many methods for introducing DNA into cells have been developed. It should be noted that all the methods listed in this chapter have worked with standard cell lines, but may need adjustment for optimization in a particular cell line.

19.1.3 Bacterial transformation

Transfection of DNA into competent bacteria provides a method of establishing permanent 'hard copies' of a plasmid as well as amplifying the copy. Recombinant DNA molecules can be introduced into *Escherichia coli* cells, and

subsequent clones can be screened for the recombinant plasmid. DNA and

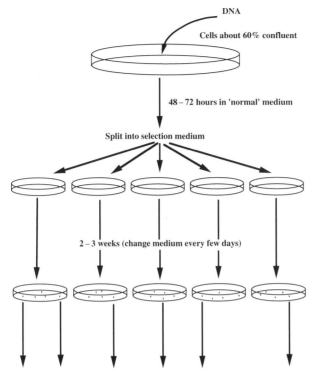

Figure 19.1. Generation of individual cell clones.

treated bacteria are mixed together allowing some of the cells to take up DNA molecules (see Chapter 14). The bacteria are plated out onto agar containing the selective drug, on which only bacteria containing the DNA will form colonies (Figure 19.2). The plasmid has a drug resistance gene such as the ampicillin resistance gene, driven by a bacterial promoter within its backbone. Encouraging bacteria to take up DNA is very inefficient, and thus only very few bacteria can be transformed. Those that are transformable must first be made competent using physical and/or chemical manipulation (Protocol 1).

19.2 Transfection of eukaryotic cells

Introducing naked DNA into eukaryotic cells can be achieved using a number of methods, both physical and viral. The most frequently used methods include co-precipitation with calcium phosphate, or with other divalent cations, the use of polycations or lipids to complex with DNA or encapsidation with liposomes. In addition, exposure of cells to rapid high-voltage pulses (electroporation), introduction of DNA into cells by microinjection, 'shooting' DNA into cells on microprojectiles, and various types of viral vectors can all be used as transfer techniques. The actual size of the genetic cassette is determined by the nature

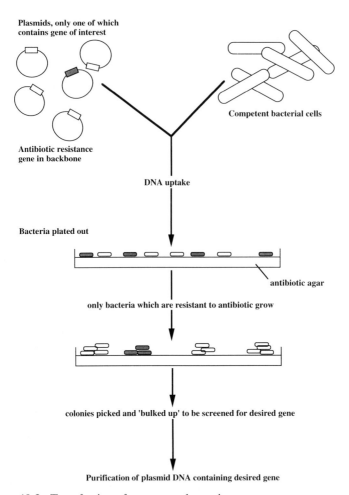

Figure 19.2. Transfection of competent bacteria.

of the transfer technique. Naked plasmids can be any size, however, the bigger they are, the less the likelihood of an efficient transfer. Retroviral and adenoviral vectors are currently limited to up to about 6 kb of insert. The adenoviral genome is around 36 kb, and therefore has potential for much greater insert sizes with manipulation. The efficiency of physical transfection methods *in vitro* are typically now 1–10% in suitable recipient cells. Viral methods, however, can increase this efficiency under ideal circumstances to 100% of target cells.

19.2.1 *Calcium phosphate transfection*

When DNA, calcium chloride and a phosphate buffer are mixed at a neutral pH, a calcium phosphate–DNA complex precipitates out of solution. This complex is then adsorbed to cell membranes and some is taken up by the cells by endocytosis. Phagosomes containing the precipitate are transported to various organelles within the cell, including the nucleus. This procedure was first described by Graham and van der Eb (1973), although subsequently many modifications to the standard methods have been made. Treatment of cells post-transfection with various chemicals, such as a glycerol shock, have been shown to result in a 10 to 100 fold increase in efficiency, particularly in transient expression. However, often only 1–10% of cells are stably transfected (Protocol 2).

19.2.2 DEAE-Dextran transfection

The technique of DEAE-Dextran transfection was first described by McCutchan and Pagano (1968). It is thought that the polymer may bind to DNA thereby inhibiting the action of nucleases and promotes binding to cells. The endocytosis of the DNA is thus promoted. DEAE-Dextran transfection differs in three main ways to calcium phosphate co-precipitation. (a) Less DNA is required for DEAE-Dextran protocols. In fact, levels of DNA that are approaching those used for calcium phosphate transfection, can be inhibitory. (b) There are a limited number of cell lines that can be transfected efficiently with DEAE-Dextran. (c) Generally only transient transfections with cloned genes are possible, but stable transfection of cells is less common. The two main variations adopted to increase the efficiency of DEAE-Dextran transfections are the time for which the cells are exposed to the transfection mixture, and the concentration of the DEAE-Dextran in the mixture (Protocol 3).

The efficiency of transfection with calcium phosphate or DEAE-Dextran for certain cell types can be improved after exposure to a number of chemicals. Dimethyl sulfoxide (DMSO) and glycerol are two of the most efficient, but others include chloroquine and sodium butyrate. All are toxic to cells, and the concentration and exposure times have to be optimized for each cell type. Glycerol and DMSO are used post-transfection. The cells are exposed to 10–20% glycerol or DMSO solutions for 30 s up to a few minutes. The mechanism, though not really understood, is likely to modify the cell membrane structure and thus improve DNA uptake. Chloroquine is added with the transfection mixture, and appears to enhance DNA binding and inhibit intracellular lysosomes. Sodium butyrate, again added post-transfection, may activate chromatin structures of newly transfected genes. Overall, one of the most important factors to improve efficiency of transfection is to ensure that the cells are growing well and are healthy, and that the DNA has been prepared to the highest standards.

19.2.3 Liposomal transfection

Liposomes are lipid vesicles with artificial membranes that form an increasingly valuable part of the repertoire of DNA delivery methods. They have been commonly used as carriers of pharmaceutical agents. Liposome preparation, targeting and delivery are beyond the scope of this chapter; however, most of these procedures involve encapsulation of DNA within the liposomes, followed by fusion of the liposomes with the cell membranes. There are many sorts of liposomes, but one particular genre has become popular as transfection mediators, a reagent that consists of polycationic liposomes composed of a positively charged lipid called DOTMA, (N[1-(2,3-dioleoyloxy)propyl]-N,N,N-triethylammonium), and was designed by the application of the technology of molecular shape and hydrophobicity to form stable cationic bilayers (Felgner *et al.*, 1987) (Figure 19.3). This has been shown to be far more convenient, efficient and efficacious than with other physical methods. The mechanism of action is not completely understood, but the evidence suggests the following model. DOTMA forms liposomes in an aqueous environment either with or without phospholipids. The active cationic liposome itself consists of sonicated vesicles containing equal parts of DOTMA and dioleoylphosphatidylethanol-

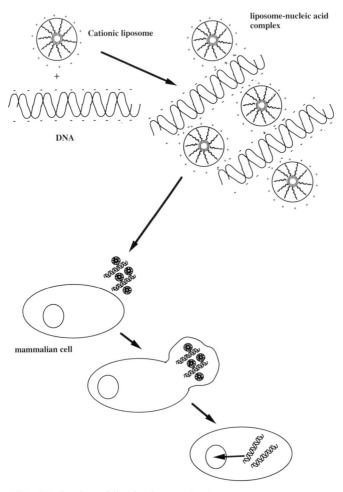

Figure 19.3. Mechanism of lipofectin transfection.

amine in water. Spontaneously formed DNA/DOTMA complexes occur following mixing with aqueous DNA solutions. The polynucleotide is almost entirely trapped within the complex. The efficiency of the reagent is critically controlled by the ratio of DOTMA to DNA. This is to ensure that the final complexes have a net positive charge to allow them to interact with the cell membranes (Protocol 4).

19.2.4 Electroporation

Electroporation is, as the name suggests, the production of pores in the cell membrane by use of an electric current. In this simple and rapid procedure high-voltage pulses across the cells in specially designed cuvettes cause nanometre-sized pores. DNA enters the cells either by means of these holes, or as a result of the redistribution of the membrane components as the pores reseal (Knight, 1981). The efficiency of the technique relies on a number of factors. The time for which the electric pulse is passed, and the strength of the electric field are particulry important, as is the temperature at which the experiment takes place. Transient or stable transfectants can be produced using electroporation. Insertion by electroporation can result in gene copy numbers ranging from one to twenty. Large lengths of genomic DNA, up to 65 kb in length, can be transfected by this method (Protocol 5).

19.2.5 Microinjection

Direct microinjection of DNA into nuclei is both technically complicated and for small scale use only. It has the advantage of avoiding exposure of the DNA to the contents of the cell prior to integration into the nucleus. Its application in biochemical analysis is very limited.

19.2.6 Biolistics

Biolistics (biological ballistics) is a method of DNA delivery in which small particles coated in DNA are shot into cells. The biolistic process was originally developed by Sanford *et al*. (1987) for plant cell transfection, but is now used for all sorts of cells.

19.2.7 Protoplast fusion

Protoplast fusion is a delivery technique in which protoplasts formed from bacteria carry the plasmid DNA into mammalian cells. Chloramphenicol is used to amplify the plasmid DNA by growing the bacteria in it. The bacteria are then treated with lysozyme which removes the cell wall to form protoplasts. These are spun down onto a monolayer of mammalian cells and fusion is encouraged by the addition of polyethylene glycol (PEG). Bacterial and plasmid DNAs are both passed into the mammalian cells. The cells are then washed and incubated for 1 to 2 days in normal medium containing a selection drug which will eliminate any surviving bacteria.

19.2.8 Retroviral vectors

Retroviral vectors are capable of infecting virtually every dividing cell in a target population. Retroviruses are diploid positive-stranded RNA viruses that replicate through an integrated DNA intermediate. The incoming RNA of the virus is reverse-transcribed into DNA by a virally encoded reverse transcriptase that is carried as a protein in each viral particle. The viral DNA is integrated pseudo-randomly into the host genome. The normal cellular transcription machinery is then utilized to make RNA copies of the genome and a spliced version that encodes the major *gag, pol* and *env* proteins, which are all part of the viral particle. The virus then assembles around the two RNA genomes and buds off from the cell carrying a piece of the outer lipid membrane and embedded envelope proteins.

There are many ways in which retrovirus vector infection might potentially damage host cells, including insertion of the vector into an essential gene of the host cell, the activation of a silent proto-oncogene by introduction of viral promoter or enhancer sequences, or the rescue of infectious virus from the defective vector by recombination with cellular sequences. Retroviruses have now been extensively modified so that they can integrate into host cell genomes and express foreign genes efficiently and stably without harming host cells, and have been developed as transducing viral vectors.

In the genome structure of the DNA provirus, the *cis* acting promoter and packaging sequence are located at the 5' and 3' ends in and around the long terminal repeats (LTRs). The structural genes (*gag*, *pol* and *env*) occupy the central 8 kb region, thereby making it relatively easy to cut them out and replace them with the gene of interest (Vile, 1991). A modification to these vectors has been to include part of the *gag* sequence after all, to include the full encapsidation signal, which gives a 5 to 10 fold increase in viral titre.

The vector is introduced into a packaging cell line that has been engineered to produce *gag*, *pol* and *env* protein in trans, but no RNA molecules that encode complete retroviral genomes nor RNA molecules that may be encapsidated. The introduction of a vector into a packaging line gives rise to a producer cell line that makes only replication-defective vector particles capable of introducing their genes into target cells without the capacity for replication. The most useful and well used model vectors for the efficient introduction of foreign genes into target mammalian cells have been derived from murine and avian retroviruses (Protocol 6).

Mechanisms of infection, replication, integration, and gene expression from these replication-defective vectors have been well reviewed (Varmus, 1988) (Figure 19.4). Retroviral vectors are capable of infecting a broad class of cell types. Cell replication and DNA synthesis are required for provirus integration

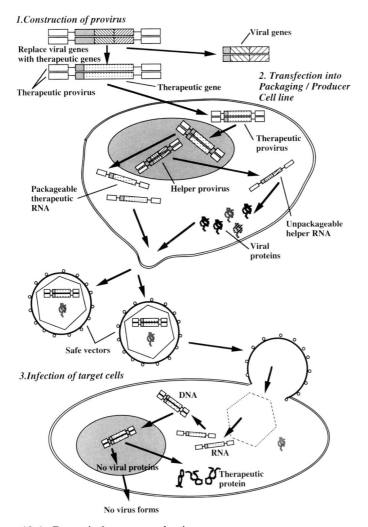

1.Construction of provirus

Replace viral genes
with therapeutic genes

Viral genes

Therapeutic provirus

Therapeutic gene

2. Transfection into
Packaging / Producer
Cell line

Therapeutic
provirus

Packageable
therapeutic
RNA

Helper provirus

Unpackageable
helper RNA

Viral
proteins

Safe vectors

3.Infection of target cells

DNA

RNA

No viral proteins

Therapeutic
protein

No virus forms

Figure 19.4. Retroviral vector production.

and this effectively restricts efficient use of retroviral vectors to replicating cells.

An advantage of retroviral vectors over other gene transfer tools is the presumed structural and functional stability of the integrated form of the retroviral vector or provirus. Although many retroviral sequences are stable, proviruses, like other transgenes can show high frequencies of instabilities. Vector design, the nature of the target cell, the presence or absence of selection pressure, and the nature of the expressed genes can contribute to vector instability, although the mechanism is not understood.

Retroviral vectors generally integrate into random sites in the cell genome, although some specific preferred sites have been noticed. Random integration occasionally leads to insertional mutagenesis through the interruption of vital cellular genes or through the insertion of retroviral regulatory sequences that modulate the expression of flanking cellular genes. Problems of promoter interference and the likelihood of insertional mutagenic events have been reduced by designing retroviral vectors that are devoid of their own promoter and enhancer sequences and are therefore transcriptionally disabled. Such disabled vectors express vector-encoded genes exclusively from promoters introduced together with the gene, but these can be associated with reduced viral titres.

The development of efficient helper cell lines and elucidation of the retroviral packaging signal required for encapsidation of virus RNA into particles has aided the preparation of high-titre retroviral vectors.

19.2.9 *Adenoviral vectors*

Adenoviral vectors are another viral vector that can be used to transfect cells. The reasons for the development of adenoviral vectors included the promise of higher titres than retroviral vectors, the expected benign clinical profile, and their likely ability to result in high levels of transgene expression.

Adenoviruses are non-enveloped icosohedral double-stranded DNA viruses with the ability to infect many cell types. Efficient transduction has been observed in both replicating and non-replicating cells. Viral replication is in the nucleus of the cell without integration into the host DNA, although this can happen on occasion. DNA replication is mediated by both viral and cellular proteins. Replication is initiated by a viral protein covalently linked to the 3′ end of both strands and eliminates the need for a nucleic acid primer. The genome is about 36 kb in size (Horowitz, 1991).

After binding to a target cell through the capsid 'fibre' protein, the virus is taken up into endosomes where the low pH leads to its dissociation and release of DNA from the endosomes. The DNA then makes its way to the nucleus where the early proteins are transcribed, leading to DNA replication and transcription of the late genes that give rise to the capsid proteins. The viruses are assembled in the cytoplasm and the host cells are lysed to release the virus.

The usual method of making a vector is to substitute the gene of interest with its own promoter sequence in place of the early E1A and E1B genes, although substitutions and deletions in the E3 and late regions have also been used. In the E1, E3 doubly deleted vectors, up to 7.5 kb can be inserted. Substitution of the E1 genes is particularly attractive as the E1 proteins have oncogenic and lytic properties. Substitution can be performed by recombination or by molecular manipulation (Figure 19.5).

It is possible to make adenoviral vectors by co-transfecting adenoviral constructs carrying appropriate adenovirus DNA sequences and a plasmid containing the foreign gene sequence into special producer cells which contain the E1 genes (called 293 cells) (Graham and Prevec, 1991). This leads to a small percentage (usually 1% to 10%) of recombinant adenoviral molecules

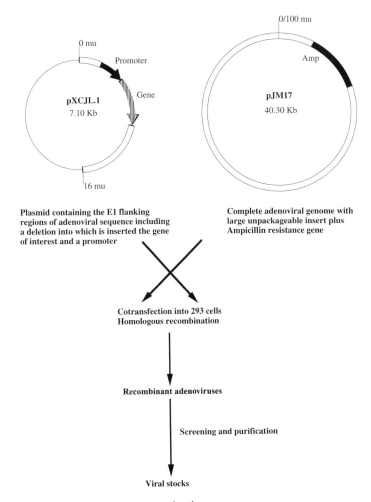

Figure 19.5. Adenoviral vector production.

with the foreign gene inserted in the vector by homologous recombination. The desired recombinants are then produced by the 293 cells and can be screened for the presence of the desired gene. The supernatant from these cells, which are themselves killed by the virus product, can then be used to infect recipient cells (Protocol 7).

19.2.10 Assessing efficiency

'Reporter genes' can be used to monitor transfection efficiency. These genes, which should not be endogenous to the host cell, are part of the plasmid DNA and the protein that is the product of the reporter gene should be easily assayed. The reporter gene can be part of the plasmid under test, or can be co-transfected with the study DNA. Alternatively the reporter plasmids can be separately transfected on to similar cell lines under the same conditions. Efficiency can usually be assayed after only 2 or 3 days.

Three of the reporter genes that are in common use include (a) the growth hormone gene, the secreted protein release being monitored using a radio-immunoassay; (b) chloramphenicol acetyltransferase (CAT); and (c) β-galactoside. Cells expressing the enzyme can be visualised *in situ* by staining with X-Gal substrate, which results in blue cells.

19.3 Protocols

Protocol 1: Transformation of bacteria

Reagents, buffers and solutions

L. Broth (Luria Bertani Broth)

To 950 ml of deionized water add:

Bacto-tryptone	10 g
Bacto-yeast extract	5 g
NaCl	10 g

Shake until all the solids have dissolved.
Adjust pH to 7.0 with 5 M NaOH.
Make up to 1 litre with water.
Autoclave to sterilize.

Transforming Buffer I (TFBI)

100 mM	KCl
50 mM	$MnCl_2$
35 mM	CH_3COONa
10 mM	$CaCl_2$

15% (v/v) glycerol
Adjust pH to 5.8 (acetic acid or NaOH)
Autoclave to sterilize

Transforming Buffer II (TFBII)

75 mM	$CaCl_2$
10 mM	KCl
10 mM	MOPS (3-[*N*-Morpholino]propane-sulfonic acid)

15% (v/v) glycerol
Adjust pH to 6.8 (acetic acid or NaOH)
Store and use at 4 °C

Tris–EDTA (TE)

10 mM	Tris pH 8.0
1 mM	EDTA pH 8.0

Cells

Inoculate a small number of *Escherichia coli* bacteria (e.g. HB101) into 2 or 3 ml of L. Broth, and shake overnight at 37 °C to grow them up to exponential phase. Reinoculate some of this exponentially growing culture into 100 ml of L. Broth in a conical flask and shake for 1 to 2 h at 37 °C until the culture has an absorbance of about 0.6–0.7. (When the flask is held up to the light and tipped slightly, the culture should be opaque enough to cloud the vision through the base of the flask).

DNA

Usually transforming bacteria is used to amplify a ligation product or a small amount of plasmid. For a plasmid with a concentration of around 1 $\mu g/\mu l$ in TE or sterile water, about 2 μl are used. For a ligation, about one-half of the ligation mixture is used.

Protocol

1. Split the 100 ml of growing bacteria into two conical tubes and spin at 1000 g for 6 min to form a pellet.
2. Place the tubes on ice.
3. Resuspend one pellet in 10 ml of TFBI, and then resuspend the second pellet in the same suspension. Leave on ice for about 40 min.
4. Spin down the TFBI suspension at 1000 g for 6 min to form a pellet.
5. Keeping the tube and pellet cool at all times, resuspend the pellet in 2 ml of cold TFBII. Leave to incubate on ice for 1 to 2 h. The longer bacteria are left to become competent, up to a point, the more efficient the transformation will be.
6. Measure 100 μl of the suspension into a precooled 1.5 ml microfuge tube on ice, for each transformation to be done.
7. Add the DNA to the competent bacteria and mix gently. Leave on ice for 30 min. Preheat a heat block to 42 °C.
8. Place the tubes into the heat block for 1 min only. This is to heat shock the bacteria, increasing the chance of DNA being taken up.
9. Place the tubes back onto ice for 2 or 3 min.
10. Add 400 μl of L. Broth to each tube and shake at 37 °C for about 1 h.
11. Plate the growing bacteria onto agar plates containing the appropriate selection drug.
12. After the plates have dried, incubate inverted at 37 °C overnight.
13. Pick any colonies into L. Broth containing the selection drug and shake overnight at 37 °C, and prepare DNA from these cultures to identify which contain the correct plasmid.

Protocol 2: Calcium phosphate transfection

Reagents, buffers and solutions

2 M CaCl$_2$ solution
2× HEPES-buffered saline (2 × HBS)

50 mM	HEPES, pH 7.1 [HEPES: 4-(2-hydroxyethyl)-1-piperazine ethanesulfonic acid]
280 mM	NaCl
1.5 mM	Na$_2$HPO$_4$

Sterile deionized water
Reagents stored at −20 °C.
Reagents available as a kit, Profection calcium phosphate transfection system supplied by Promega
Tris-EDTA (TE)

10 mM	Tris, pH 8.0
1 mM	EDTA, pH 8.0

Cells

Put up cells such that they are about 60–70% confluent when transfected. For this protocol, figures are quoted for transfecting two T25 flasks or a 100 mm plate.

DNA

The DNA should be at a concentration of about 1 mg/ml, dissolved in TE or water.

Protocol

1. The three components should be thawed and warmed to room temperature.
2. Into pre-labelled microfuge tubes put the following mixture.

10–30 µl	DNA (or 28 µl plus 2 µl of co-transfection plasmid)
62 µl	CaCl$_2$

   Sterile water make up to 500 µl
3. 500 µl HEPES buffered saline is then added dropwise to the solution.
4. Incubate at room temperature for 15–30 min.
5. Add 500 µl of mixture to the tissue culture medium of each flask of cells to be transfected.

6. Incubate for 24 h at 37 °C.
7. Wash gently three times with serum-free medium.
8. Incubate in normal medium for 24–48 h.
9. Harvest cells or supernatant to assay for gene activity in transient transfection; or for stable transfection, split recipient cells into selection medium and leave for up to 2–3 weeks, changing the medium two or three times during this period whilst colonies develop.

Troubleshooting

The pH of the buffer is critical and several samples in the range of pH 6.8–7.2 should be tested. (This means that the infused CO_2 may be required to be decreased to 3–5% in the 24 h prior to and following transfection.) In addition, the amount of DNA is also important, usually 5–50 μg of DNA per 100 mm plate.

If there are dead cells they should be removed prior to transfection and after at least 4 h in fresh medium transfection can be commenced.

Protocol 3: DEAE-Dextran

Reagents, buffers and solutions

DEAE-Dextran at 10 mg/ml in PBS
1× Phosphate buffered saline (PBS)

137 mM	NaCl
2.7 mM	KCl
4.3 mM	Na_2HPO_4
1.47 mM	KH_2PO_4

The final pH should be 7.3
Reagents available as a kit, DEAE-Dextran transfection system from Promega.
Tris–EDTA (TE)

10 mM	Tris pH 8.0
1 mM	EDTA pH 8.0

Cells

Put up cells such that they are about 30–50% confluent when transfected. For this protocol, figures are quoted for transfecting a T25 flask or a 60 mm plate.

DNA

The DNA should be at a concentration of about 1 mg/ml, dissolved in TE or water.

Protocol

1. Prewarm 10 ml of 1 × PBS to 37 °C. This is the wash solution. Also warm the DEAE-Dextran solution to 37 °C.
2. Prepare the transfection mixtures as follows:
   (a) Filter sterilize 0.4 ml of 1 × PBS per T25 flask or 60 mm plate.
   (b) Dilute the DNA in 1 × PBS and add DEAE-Dextran. Dilute the DNA to a final volume of 326 μl and add 17 μl of 10 mg/ml DEAE-Dextran, per T25 or 60 mm plate.
3. Remove the medium from the cells and wash the cells twice with the warmed PBS.
4. Add the DNA/DEAE-Dextran mixture: 343 μl per T25 or 60 mm plate. Dispense evenly over the cells. The concentration of DEAE–Dextran exposed to the cells is 0.5 mg/ml.
5. Incubate for 30 min at 37 °C.
6. Add 3.5 ml of normal growth medium to the flask and incubate at 37 °C in a CO_2 incubator.
7. Leave cells for 48–72 h.
8. Harvest cells or supernatant to assay for gene activity in transient transfection; or for stable transfection, split recipient cells into selection medium and leave for up to 2–3 weeks, changing the medium two or three times during this period whilst colonies develop.

Troubleshooting

Maximum transfection efficiency occurs when cells are 30–50% confluent. Cells should be split at least 12 h prior to transfection. The time taken for transfection is also important. Some delicate cells may be damaged by times in excess of 4 h whereas other cell types may be relatively impermeable to foreign DNA and require in excess of 24 h.

Protocol 4: Lipofection

Reagents, buffers and solutions

Lipofectin is a DOTMA:DOPE liposome formulation, made by Gibco BRL. It comes as a 1 mg/ml preparation.

Serum-free tissue culture medium.
Tris–EDTA (TE)

| 10 mM | Tris, pH 8.0 |
| 1 mM | EDTA, pH 8.0 |

Cells

Cells should be about 60% confluent. For this protocol figures are quoted for transfection of a single T25 flask or 60 mm tissue culture plate.

DNA

The DNA should ideally be at a concentration of about 1 mg/ml, dissolved in TE or water.

Protocol

1. Prepare the following solutions in microfuge tubes:
   Solution A: For each well of cells to be transfected, dilute 1–20 μg of DNA into a *final* volume of 100 μl serum-free medium.
   Solution B: For each transfection, dilute 2–50 μl of Lipofectin reagent into a *final* volume of 100 μl of serum-free medium.
2. Combine the two solutions, mix gently, and incubate at room temperature for 10–15 min. The solution may appear cloudy, however this will not impede the transfection.
3. While the Lipofectin reagent–DNA complexes are forming, wash the cells twice with 2 ml of serum-free medium to ensure removal of any serum that may react with the Lipofectin.
4. For each transfection, add 0.8 ml of serum-free medium to each microfuge tube containing the Lipofectin reagent–DNA complexes. Mix gently, overlay the complex onto cells, and swirl the plates to ensure uniform distribution.
5. Incubate the cells for 5–24 h at 37 °C in a CO_2 incubator.
6. Replace the DNA containing medium with 2 ml of normal growth medium containing serum and incubate cells at 37 °C in a CO_2 incubator for another 48–72 h.
7. Harvest cells or supernatant to assay for gene activity in transient transfection; or for stable transfection, split recipient cells into selection medium and leave for up to 2–3 weeks, changing the medium two or three times during this period whilst colonies develop.

Protocol 5: Electroporation

Reagents, buffers and solutions

1× Phosphate buffered saline (PBS)

137 mM	NaCl
2.7 mM	KCl
4.3 mM	Na_2HPO_4
1.47 mM	KH_2PO_4

The final pH should be 7.3

HEPES-buffered saline (HeBS)

20 mM	HEPES
137 mM	NaCl
5.4 mM	KCl
1.3 mM	Na_2HPO_4
6 mM	glucose

The final pH should be 7.4

Phosphate-buffered sucrose

272 mM	sucrose
7 mM	sodium phosphate, pH 7.4
1 mM	$MgCl_2$

Tissue culture medium with serum

Tris-EDTA (TE)

10 mM	Tris, pH 8.0
1 mM	EDTA, pH 8.0

The protocols outlined below are designed for use with the Bio-Rad Gene Pulser, but will also be applicable to other similar appliances.

PBS, HeBS and phosphate buffered saline are all electroporation buffers. Some cells are killed easily and the high voltages required to efficiently electroporate in PBS or HeBS are fatal. Phosphate-buffered sucrose can be used at lower voltages.

Cells

Electroporation can be carried out on both suspension and adherent cells. The methods are very similar for both, once adherent cells are trypsinized and in a suspension. Cells should be growing in mid to late log phase. For stable transfections, 5×10^6 cells are required, with transients requiring up to 4×10^7 cells, depending on the strength of the promoter.

DNA

For stable transfections, DNA should be linearized, but for transient expression plasmid DNA is sufficient. It is important that the linearized DNA is purified from the digestion mix, before use.

Notes

The capacitance and voltage for each pulse must be optimized for each cell type used. For high salt, low resistance buffers such as PBS and HeBS the capacitor setting should be 25 μF in the first instance, increasing or decreasing the voltage to optimize the transfection. This is usually when about 40–70% of the cells remain viable. Transient expression of a gene can be used to set the conditions for stable transfection, because the parameters are very similar. The voltage used can be reduced by using lower salt solutions such as phosphate-buffered sucrose.

It has also been shown that using low voltage, high capacitance settings with longer pulses was more efficient for sensitive cells. The voltage should be started at 250 V, 960 μF and changed up to 350 V or down to 100 V stepwise to optimize efficiency.

For this protocol everything is kept as close to 0 °C as possible. It has been shown that recovery and viability can be improved by working at room temperature. It is well worth experimenting with temperature as a further criterion.

Protocol

1. Adherent cells should be trypsinized to remove them from the plate surface. The trypsin should be inactivated with serum-containing medium, and the cells should be spun down (1000 × g for 5 min, 4 °C). Suspension cells should also be harvested by centrifuging (1000 × g for 5 min, 4 °C).

2. The cells should be washed by resuspension in ice cold electroporation buffer, and recentrifuging. This is particularly important for the adherent cells, to ensure removal of trypsin. Repeat if necessary.

3. Finally, resuspend the cells in electroporation buffer to a density of 1×10^7 cells/ml for stable transfections. This concentration may need to be increased up to 8×10^7 cells/ml.

4. Measure 0.5 ml of cell suspension into electroporation cuvettes and put on ice.

5. Add DNA to the cell suspension in the cuvette. For transient expression, 10–40 μg of plasmid DNA is required and for stable transfections 1–10 μg of linearized DNA is sufficient. Carrying out a co-transfection is not recommended.

6. The DNA and the cell suspension should be mixed by gently flicking the cuvette. The cuvette should then be placed on ice for 5 min.

7. Place the cuvette into the holder in the electroporation apparatus at room temperature and shock.

8. Return the cuvette to ice for 10 min.

9. Dilute cells in 20 times their volume of normal medium and wash out the cuvette to remove all transfected cells.

10. Incubate cells at 37 °C for 48 h.
11. Harvest cells or supernatant to assay for gene activity in transient transfection; or for stable transfection, split recipient cells into selection medium and leave for up to 2–3 weeks, changing the medium two or three times during this period whilst colonies develop.

Protocol 6: Retroviral vectors

Reagents, buffers and solutions

Polybrene 1000× solution (4 mg/ml)
Tris–EDTA (TE)

10 mM	Tris, pH 8.0
1 mM	EDTA, pH 8.0

Cells

Change the medium on the producer line 24 h before the infection is to be done. At the same time seed the cells to be infected, so that they will be 50–60% confluent 24 h later.

Protocol

1. Take the supernatant off the producer line and filter through a 0.45 μm filter to remove any cells and debris.
2. Add viral supernatant (at dilutions of 1, 1:10, and 1:100) along with polybrene (at 4 μg/ml) to the medium on the cells which are to be infected. Leave for between 4 and 16 h.
3. Change the medium for new normal growth medium and incubate for 24–48 h.
4. Harvest cells or supernatant to assay for gene activity in transient transfection; or for stable transfection, split recipient cells into selection medium and leave for up to 2–3 weeks, changing the medium to remove dead cells whilst colonies develop.

Protocol 7: Adenoviral vectors

Reagents, buffers and solutions

1× phosphate-buffered saline (PBS)

137 mM	NaCl
2.7 mM	KCl
4.3 mM	Na_2HPO_4
1.47 mM	KH_2PO_4

The final pH should be 7.3

Cells

Recipient cells should be about 80% confluent. Medium containing adenovirus particles can be taken straight off 293 producer cells, or frozen at $-70\,°C$ for use later.

Protocol

1. The medium is removed from the recipient cells, and the supernatant from the adenoviral producer cells added to the cells in as little serum-free medium or PBS as possible.
2. The virus is allowed to adsorb for 30–60 min at 37 °C.
3. The cells are re-fed with normal tissue culture medium containing foetal-calf serum and incubated at 37 °C overnight. The medium is then changed and the cells left for another 24–48 h, when the cells or supernatant are assayed for gene activity.

References

Felgner, P. L., Gadek, T. R., Holm, M. *et al.*, (1987) *Proc. Natl Acad. Sci.* **84**, 7413–17.
Graham, F. L. and Prevec L. (1991) *Gene transfer and expression protocols*, Humana Press Inc., New Jersey, 109–128.
Graham, F. L. and van der Eb, A. J. (1973) *Virology* **52**, 456–67.
Horwitz, M. S. (1991) *Fundamental virology*. Raven Press, New York, 1679–721.
Knight, E. D. (1981) *Techn. Cell Physiol.* **113**, 1–20.
McCutchan, J. H. and Pagano, J. S. (1968) *J. Natl Cancer Inst.* **41**, 351–6.
Sanford, J. C. (1987) *J. Particle Sci. Technol.* **5**, 27–37.
Varmus, H. E. (1988) *Science* **240**, 1427–35.
Vile, R. G. (1991) *Practical molecular virology*. Humana Press Inc, New Jersey. 1–16.

20

Gene manipulation: antisense RNA technology

P. L. IVERSEN and ELIEL BAYEVER

20.1 Introduction and background

An antisense approach to gene manipulation is being viewed with increasing interest by the scientific community. In addition to research laboratory use, antisense technology has recently been adapted for clinical use, with a number of human trials now ongoing or being planned (Cohen and Hogan, 1994).

Simply, this approach leads to the interruption, either temporarily or permanently, of gene expression. It is generally thought of as a way to prevent gene expression resulting in an unwanted cellular activity but could, conversely, interrupt or suppress expression of negative regulation gene products activating a desired gene expression activity. In any event, the underlying tenet is to interfere with the mRNA translation process, and a number of technologies are evolving that satisfy these conditions.

20.2 Phosphorothioate-backbone oligonucleotides

Currently, the most popular means of this gene manipulation strategy uses synthetic phosphorothioate-backbone oligodeoxyribonucleotides, or oligo-nucleotides (ODNs). ODNs may be synthesized complementary to DNA (antigene) or RNA (antisense). Antisense is an approach employed to attenuate selected protein synthesis. When the ODN binds to mRNA, complementary base-pair hydrogen bonding occurs, preventing translation. Furthermore, mRNA–ODN heterodimer degradation is manifested, at least to some extent due to the cells' RNAase H activity, as this enzyme degrades the RNA portion of an RNA:DNA heterodimer.

20.2.1 Nuclease resistance

Unmodified DNA ODNs are susceptible to $5'$ and $3'$ exonucleases and a variety of endonucleases that eliminate sequence-specific interactions between the ODN and cellular nucleic acids. These degradative enzymes are ubiquitous (Emlen and Mannik, 1978) and unmodified ODNs have a half-life of less than an hour in fetal calf serum (Wickstrom, 1986). Phosphorothioate ODNs, however, are nuclease-resistant (Eckstein, 1975; 1979; 1984), which originally led to their use as potential antiviral agents based upon their ability to induce interferons (DeClercq *et al.*, 1970).

The ability to synthesize ODNs with different 'backbones' has resulted in the development of agents with specific properties. The phosphorothioate 'backbones', where one of the non-bridging oxygen atoms in each of the internucleotide phosphate linkages is replaced by sulphur atom, have the following potentially advantageous properties: They are (a) relatively stable against endo- and exo-nucleases, producing a molecule with longer biological activity than the naturally occurring, congener phosphodiester; (b) readily soluble in aqueous solutions; (c) taken up by a wide variety of cell types; (d) becoming relatively simple and inexpensive to synthesize; and (e) show negligible depression of melting temperature (T_m) compared to normal DNA and hybridize well to target mRNA. Nuclease resistance is particularly advantageous because much longer intracellular stability prevails than with the less resistant backbone. The sulphur atom substitution has a larger ionic radius than oxygen, which may confer other novel biological properties. Additionally, the sulphur appears to be more charged relative to the oxygen, which could alter the interactions of the phosphorothioate with the active site of enzymes utilizing a DNA substrate.

This backbone modification does not appear to interfere significantly with the ability of these nucleic acids to form hydrogen bonds essential for base pairing. Therefore, these ODN analogues can bind to native and complementary mRNA sequences. One can, therefore, assume that (1) the ODN can hybridize to mRNA *in vivo* and (2) the action of the ODN is selective for the target. In summary, phosphorothioate ODNs successfully satisfy the first critical issue in therapeutic development, that of nuclease resistance.

20.2.2 Cell uptake, distribution and exocytosis of ODNs

Mammalian cells internalize ODNs and their derivatives by an endocytotic mechanism (Vlassov and Yakubov, 1991; Crooke, 1993). ODN derivatives clearly can affect specific nucleic acid functions *in vitro* and inside cells (Cohen, 1989; Hélène and Toulmé, 1990; Degols, Leonetti and Lebleu, 1991). Cells incubated with radiolabelled ODNs result in radioactive material binding to the cells as well as its accumulation in the cell nuclei (Vlassov *et al.*, 1986a; Zamecnik *et al.*, 1986). Relatively rapid uptake occurs, as cells exposed to a 20 μM concentration of a 20-mer produced an intracellular 1.5 μM concentration within 15 min (Goodchild *et al.*, 1988). Autoradiography of cells reveals both cytoplasm and nucleus labelling (Iversen, 1993). Further, ODNs with an alkylating group, an aromatic nitrogen mustard at the 5' terminal phosphate, provide direct evidence of the intracellular localization (Karpova *et al.*, 1980; Vlassov *et al.*, 1986*b*).

Detailed studies on the cell uptake mechanism have been performed with fluorescently labelled ODNs. ODNs are taken up by cells in a saturable manner compatible with endocytosis, and maximal binding is achieved within 2 h of incubation. The process is slowed by decreased temperature or inhibitors of endocytosis such as deoxyglucose, cytochalasin B and sodium azide. Further, 20% of the material was taken into nuclei and 50% was associated with mitochondria, lysosomes and other vesicular structures (Loke *et al.*, 1989; Neckers, 1989; Iversen, 1993). Once in the cytoplasm, but not in vesicles, the ODN is rapidly transported to the nucleus (Chin *et al.*, 1990; Leonetti *et al.*, 1991). Different cell lines bind different amounts of ODNs, and the efficiency of ODN internalization depends on the cell growth conditions. In addition to greater internalization in rapidly growing cells, ODN association with dead cells may be 50 times greater than that with living cells (Zhao *et al.*, 1993). Increasing the cell monolayer density from 8×10^4 to 5×10^5 cells/cm^2 resulted in a threefold decrease in maximal binding of the oligomer per cell (Ceruzzi and Draper, 1989). Reports such as that of Milligan *et al.* (1993) conclude that ODN analogues do not efficiently permeate into the cell. Hence, the reported literature contains conflicts, and a broader perspective is necessary regarding ODN cellular uptake.

Since ODN binding is saturable, a cell surface receptor capable of mediating cellular uptake is suggested. The binding is trypsin-sensitive, indicating the transport involves a protein (Emlen *et al.*, 1988). An ODN with an *N*-hydroxy-succinimidyl (NHS) ester attached to the 'Denny-Jaffe' cross-linking reagent and an ODN with an aminolinker group at the 5' end are incubated in cells in the dark. Ultraviolet light exposure then activates the cross-linking reagent revealing an 80-kDa protein, the putative receptor (Geselowitz and Neckers, 1992; Neckers, 1993). DNA binding is most likely mediated by this receptor in platelets, leukocytes and lymphocytes (Diamantstein and Blitstein-Willinger, 1978; Ohlbaum, Csuzi and Antoni, 1979; Dorsch, 1981; Bennett, Gabor and Merritt, 1985). This putative receptor is linked to cellular functions and pathophysiology (Bennett, Peller and Merritt, 1986; Bennett, Kotzin and Merritt, 1987). Finally, chloroquine, monensin and phenylarsine do not diminish internalization, conflicting with a traditional receptor uptake mechanism (Wu-pong, Weiss and Hunt, 1992).

ODN modification can alter cellular uptake and distribution kinetics. Uptake has been enhanced through the conjugation of the 3'-end of the ODN with poly-L-lysine (Stevenson and Iversen, 1989; Leonetti, Degols and Lebleu, 1990*a*) or by antibody-targeted liposomes (Leonetti *et al.*, 1990*b*). Further ODN backbone structural modifications may alter ODN uptake selectivity. Finally, the non-ionic methylphosphonates do not bind to the same cellular receptor as the ionic ODNs (Miller and Ts'o, 1987).

In summary, ionic ODN cellular uptake is most likely mediated by a cell surface protein capable of recognizing single-stranded DNA (ssDNA). Uptake is saturable, temperature- and energy-dependent, and endocytosis-inhibiting drugs prevent internalization. The ODNs are primarily distributed within subcellular vesicles and the nucleus. Uptake is different between different cell types, is dynamic throughout the cell cycle, and is different for different ODN sequences. Finally, much of the ODN that becomes associated with a cell is not available for potential therapeutic interaction with cellular nucleic acids and proteins.

20.2.3 Sequence specificity and non-sequence specific effects

Non-sequence-specific ODN effects in which homopolymeric ribonucleotides interfered with the replicase $Q\beta$ were the first to be appreciated (Haruna and Spiegelman, 1966). Non-specific effects were further observed in the inhibition of DNA polymerase (Erickson and Grosch, 1974; Stebbing, Grantham and Carey, 1976; Gao *et al.*, 1988) and reverse transcriptase (Majumdar *et al.*, 1989). Finally, an RNAaseH mechanism can also result in non-sequence specificity as described for phosphorothioate ODNs in the xenopus oocyte (Woolf *et al.*, 1992). This lack of specificity is the result of RNAaseH cleavage of RNA:DNA heteroduplexes containing imperfect complementary binding. Hence, a segment only 10 bases in length in an ODN 14 bases long might act as five different 10-mer. While these data suggest that an investigator should exercise caution in the interpretation of data, the studies were conducted at 25 °C, which would favour the stability of shorter-length hybrids.

20.2.4 Selection of the appropriate gene product

Because a therapeutic action is desired, the appropriate gene target must be intimately involved in disease maintenance or pathogenesis. For example, the *bcr-abl* mRNA is a rational target for antisense inhibition, since this chromosomal rearrangement (translocation of exons encoding the oncogene *bcr* with exons encoding *abl*) is directly involved in CML pathogenesis. Murine cells are transformed by the *bcr-abl* gene *in vitro* (Daley, Van Etten and Baltimore, 1990). Conversely, CML cell proliferation is inhibited by *bcr-abl* antisense ODNs *in vitro* (Skorski *et al.*, 1991; Szczylik *et al.*, 1991). Further, the same b2a2 targeted antisense strategy resulted in elimination of leukemic cells *in vitro* (deFabritis *et al.*, 1993). Finally, antisense phosphorothioate ODNs to *bcr-abl* inhibit Ph$^+$ leukemia growth in mice *in vivo* (Skorski *et al.*, 1994). Antisense ODNs that target this translocation should be specific for the malignant clone; hence, no sequence-specific toxicity would be expected. In summary, antisense ODNs which target the *bcr-abl* junction in CML Ph$^+$ cells effectively inhibit leukemic cell growth and eliminate leukemic cells, providing the rationale for utility in the treatment of human disease.

20.2.5 *Availability of 'drug-quality' ODNs*

A recently-completed phase I clinical trial in collaboration with Lynx Thera-
peutics, Inc., in Hayward, CA, utilized phosphorothioate ODNs complementary to the p53 mRNA to investigate their potential therapeutic use as
intravenously administered agents in acute myeloid leukemia (Bayever *et al.*,
1993). Provided for this trial were phosphorothioate ODNs of 'drug quality',
the concept of which is discussed in the Manufacture section, p. 379.

20.2.6 *Evaluation of* **in vivo** *pharmacokinetics and toxicology*

A variety of ODNs have been evaluated in animals, but generalized conclusions are still difficult to make. However, the following statements are justified:

- Unmodified, phosphodiester ODNs have very short half-lives *in vivo*, in
  the range of a few minutes. Single-stranded phosphodiester ODNs
  (ssODN) have a shorter half-life than double-stranded ODNs (dsODN).
  Finally, a critical chain length of approximately 15 bases can be observed,
  and shorter chains have shorter half-lives (Chused, Steinberg and Talal,
  1972; Dorsch *et al.*, 1975; Emlen and Mannick, 1978; 1982; 1984;
  Zamecnik *et al.*, 1986; Goodchild, Zamecnik and Kim, 1991; Griffin
  et al., 1993).
- ssODN is localized preferentially in the liver and kidneys, while dsODN
  tends to stay in circulation to be broken down by nucleases. The blood–
  brain and blood–testicle barriers effectively reduce the accumulation of
  phosphorothioate ssODN in the brain and testes (Iversen, 1993). A
  dsODN can be degraded to structures that become ssODN, which then
  behave as native ssODN (Emlen and Mannick, 1978; 1982; 1984).
- Phosphodiester ODNs conjugated with a 5′ end cholesteryl moiety increases the half-life approximately tenfold (van Berkel *et al.*, 1991).
  Conversely, a 3′ phosphopropyl amine phosphodiester conjugation does
  not significantly increase ODN stability (Zendegui *et al.*, 1992).
- Methylphosphonate-modified backbone ODNs are resistant to nucleases
  in blood serum and are accumulated in the liver and kidneys (Miller and
  Ts'o, 1987; Sarin *et al.*, 1988; Colvin, 1990).
- ODNs bind non-specifically to a significant amount of serum protein.
  This binding may be a determinant of plasma half-life, as phosphorothio-
  ates are highly protein-bound and have relatively long half-lives while
  methylphosphonates are not protein-bound and have relatively short
  half-lives (Colvin, 1990; McCormack *et al.*, 1990). The ODN serum
  protein affinity may be modified by specific ODN base composition
  (Griffin *et al.*, 1993) or through ODN conjugation with a ligand-like
  cholesterol to enhance interactions with low density lipoproteins (van
  Berkel *et al.*, 1991).
- The elimination half-life of phosphorothioate ODNs is from a few hours

to days, a significantly longer time period than that of unmodified ODNs (Agrawal, Temsamani and Tang, 1991; Goodarzi, Watabe and Watabe, 1992; Karamyshev *et al.*, 1993; Iversen *et al.*, 1994).

– Deaths have been reported of mice injected with 160 mg/kg doses of phosphodiester ODNs (Agrawal, 1991) and following rapid intravenous bolus injections into monkeys (Cornish *et al.*, 1993).

20.2.7 Demonstration of **in vitro** and **in vivo** efficacy of phosphorothioate ODNs

In order to establish the potential utility, feasibility and determine their relatively non-toxic characteristics, antisense phosphorothioate ODNs have been utilized to modulate selected gene expression *in vivo*. Ducks infected with duck hepatitis B virus treated with antisense ODNs directed at the 5' region of the pre-S gene resulted in complete inhibition of viral replication and gene expression (Offensperger *et al.*, 1993). Mice with HTLV-I induced tumors were given intraperitoneal injections of ODNs antisense to HTLV-I-*tax*, which inhibited the *tax* expression by 90% in established tumors (Kitajima *et al.*, 1992). Smooth muscle cell proliferation and accumulation was inhibited with antisense *c-myb* (Simmons *et al.*, 1992). Human leukemia–SCID mouse chimeras with K562 cells treated with phosphorothioate ODNs antisense to *c-myb* increased survival by more than 3.5-fold over untreated or unrelated sequence controls with significantly less disease infiltration into the central nervous system and the ovary (Ratajczak *et al.*, 1992). Finally, human leukemia–SCID mouse chimeras with BV173 cells treated with phosphorothioate ODNs antisense to the *bcr-abl* junction (1 mg/day for nine consecutive days) reduced clonogenic leukemia cells, decreased *bcr-abl* mRNA transcripts and more than doubled survival compared with mice treated with sense ODNs or 6-base mismatched antisense ODNs (Skorski *et al.*, 1994).

20.3 Other backbone oligonucleotides

Three sites within a mononucleotide may be modified: the base, the sugar, or the phosphate. Chemical modifications of the base and/or sugar may impair Watson–Crick base pairing, resulting in less efficient hybridization and thus less antisense potency. 'Backbone' chemical modifications are the most common synthesis alterations because these frequently produce nuclease-resistant ODNs with little change to the hybridization characteristics. The backbone modifications can be classified as ionic, non-ionic and complex. The most popular ionic backbone is the phosphorothioate modification, where replacement of one of the non-bridging oxygens on the phosphate with a sulphur results in a phosphorothioate ODN. Replacement of both non-bridging oxygens with sulphur results in a phosphorodithioate. The most common non-ionic modification is to the methylphosphonate backbone in which one of the non-bridging oxygens is replaced with a methyl group. Complex backbone modifications can be extensive and include such changes as the replacement of the phosphodiester group with a peptide bond.

20.3.1 Ribozymes

A more recent, but promising, development for antisense gene manipulation has been the use of ribozymes. Ribozymes are polymeric RNA molecules that catalyse the hydrolysis of single-stranded RNA (Uhlenbeck, 1993). This method of modulating gene expression will require proper target selection and suitable ribozyme delivery. These are currently being delivered either encapsulated in liposomes or integrated with retroviral vectors. The latter have the advantage of potentially producing constant high levels of ribozymes, but cell targeting and transfer efficiency remain a problem (see below).

20.3.2 Retroviruses

Retroviruses are a popular method of transfer of exogenous genes to target cells (see Chapter 19). Compared with other methods, retroviral transfer results in successful transfer of exogenous sequences to a higher percentage of target cells than other gene transfer techniques. Single copies of proviral DNA are inserted at random with respect to the host genome so that exogenous transcriptional units carried within the vector remain intact. A higher cell viability follows retroviral infection compared with many other methods, and murine retroviruses, the most commonly used retroviral vectors, are capable of infecting a wide range of host species and cell types under certain circumstances. Once the target cells are infected, a DNA copy of viral RNA containing the gene of interest, in this case producing an antisense RNA molecule, is inserted into the genome as part of the retroviral life cycle. The viruses, which are able to infect cells of a wide host range, are altered to replace sequences essential for viral growth (GAG, POL, and ENV) with the genes to be transferred. The defective viral RNA is packaged in viral-producing cell lines that contain the deleted sequences but not the packaging sequence, resulting in helper-free stocks of virus particles capable of a single infection. These retroviral vectors have been used as vehicles for transfer of both antisense RNA and ribozymes. Other viruses such as adenoviruses and adeno-associated viruses are also being developed for gene transfer.

20.3.3 Triplex

ODNs can bind to specific sequences of double-stranded DNA, forming triple helices. This phenomenon may be useful in inhibition of gene transcription with the advantage that a diploid cell contains only two copies of selected targets. The potential number of triplex sites is limited to purine (pyrimidine)-rich sequences (Hélène, 1993).

20.3.4 Aptomer double-stranded

Two complementary ODNs that will form a dsDNA segment may mimic the genomic DNA and bind to critical trans-acting proteins, resulting in modulation of gene expression by interfering with endogenous control of transcription, an approach employed successfully *in vitro* (Bielinska *et al.*, 1990; Wu *et al.*, 1990) as well as *in vivo* (Tracewell, Desjardins and Iversen, 1995).

20.4 Manufacture

20.4.1 Chemical structure and nomenclature

ODN structure is assembled of nucleic acids comprising adenine, cytosine, guanine and thymine (Figure 20.1(a)). The employed nomenclature – 5′d{ApCpGpT}3′ – represents a 4-mer oligonucleotide (Figure 20.1(b)) four bases long composed of deoxyadenosine at the 5′ end followed by deoxycytosine, deoxyguanine and deoxythymidine.

20.4.2 Issues related to synthesis and purification

Reverse phase HPLC is the most frequently recommended purification method for a 5′-dimethoxytrityl (DMT) terminal oligodeoxynucleotide, a method providing for the separation of the relatively hydrophobic 5′-DMT from other products generated during synthesis, including shorter ODNs that arise from synthesis errors. Once the HPLC-purified material is recovered, the 5′-DMT group is removed and separated from the ODN. A valuable reference for the novice is the *Applied Biosystems User Bulletin*, Issue Number 13 – revised 1 April, 1987 (Applied Biosystems, Foster City, CA).

20.4.2.1 Characterize the presence of n − 1, n − 2, etc., and n + 1 materials in the synthetic product

ODNs are generally between 15 and 30 nucleotides in length to ensure efficiency, potency, and specificity based upon stochastic and thermodynamic considerations. To evaluate the possible multiple effects of an array of different length ODNs would be difficult. The best method for characterization of an ODN is polyacrylamide gel electrophoresis (20% acrylamide, 7 M urea)

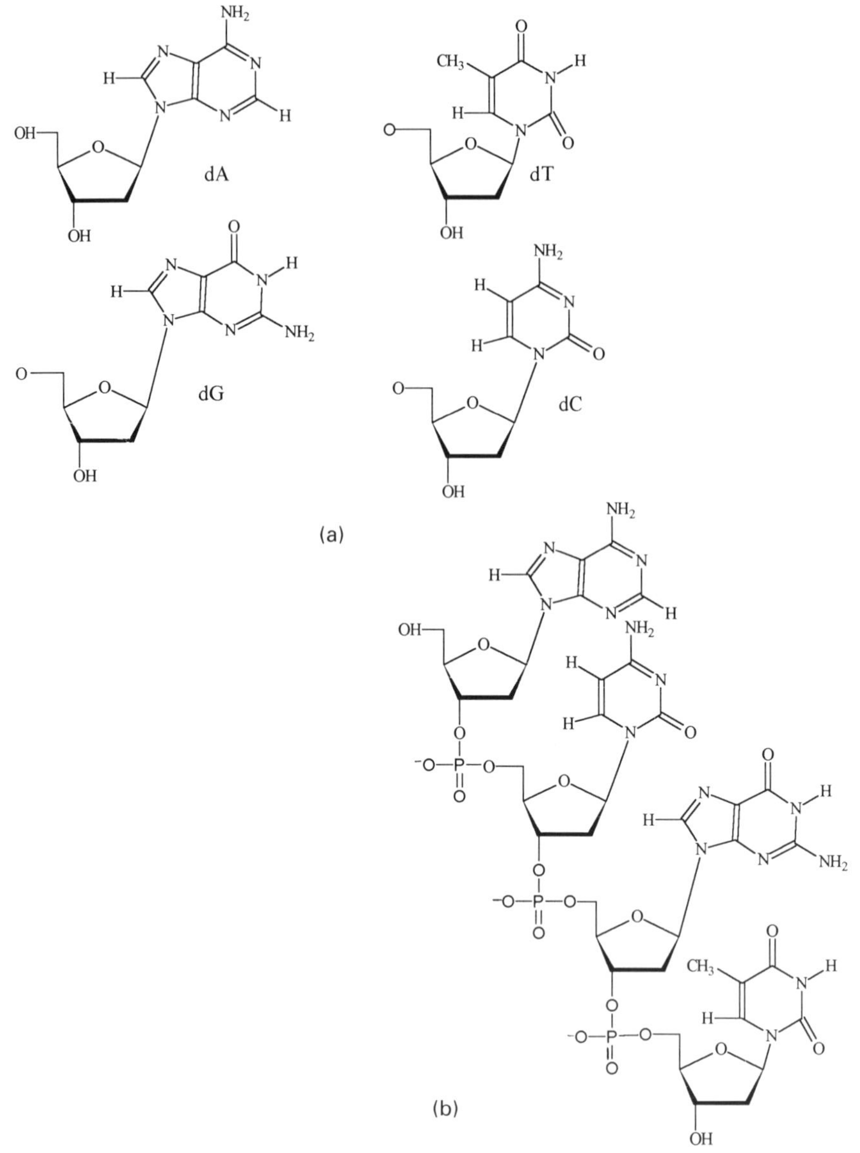

Figure 20.1. A 4-mer oligonucleotide (b) and its component nucleotides (a).

(Efcavitch, 1990) followed by visualization by staining in 0.001% (w/v) Stains-All (Aldrich) (Dahlberg, Dingham and Peacock, 1969). To load a sufficient amount of material so that minor products can also be visualized is best. Interestingly, the phosphorothioate ODNs stain purple in the Stains-All and phosphodiester ODNs appear blue. If sufficient material is loaded onto the gel, a quantitative evaluation of the length purity can be determined by densitometry of the stained gel.

20.4.2.2 *The ODN should not contain modified nucleotide bases*

Nucleic acid analogues represent known antimetabolites, hence sequence-independent toxicity. In addition, the base modifications may reduce interference with hydrogen bonding, resulting in a reduction in potency and/or efficacy. Finally, modified bases may cause the ODN to become antigenic. The simplest method to determine the presence of modified bases would be the analysis of the ultraviolet spectrum of the sample in the range of 200 to 400 nm. The molar extinction coefficient can be calculated as the sum of the molar absorption for each residue of adenine (1.54×10^4), cytosine (7.30×10^3), guanine (1.17×10^4) and thymine $(8.80 \times 10^3 \text{ M}^{-1}\text{cm}^{-1})$.

20.4.2.3 *Evaluate the ODN for homogeneity of the phosphorothioate backbone*

To ensure not too many non-phosphorothioate 'defects' exist in the backbone is important. Because the sulphur provides the basis for tissue and fluid stability in the case of a phosphorothioate ODN, non-sulphur linkages may reduce the biological residence time and subsequent biological activity. ^{31}P-NMR is a frequently employed method that provides information concerning the homogeneity of substituents bonded to the phosphate backbone. The use of NMR is not a particularly quantitative method, and strong anion exchange methods have been employed for analysis of phosphorothioate backbone ODNs (Bergot and Egan, 1992).

20.4.2.4 *Ensure the synthetic material is sterile and endotoxin free, or apyrogenic*

Sterility tests can be performed as described in *The Extra Pharmacopoeia* (Martindale, The Pharmaceutical Press). Pyrogens are evaluated by analysis in the rabbit pyrogen test and the limulus lysate tests, also described in the *Pharmacopoeia*.

20.4.2.5 *Ensure organic reagents employed in synthesis are removed*

The removal of organic reagents employed in synthesis may best be accomplished by proton NMR; however, discussion of the procedural details and results interpretation is beyond the scope of this chapter.

20.5 Troubleshooting

20.5.1 *Controls*

What is a good control? This aspect of antisense research seems to be replete with pundits proclaiming to know the absolute best control (see Chapter 22).

Perhaps a more practical approach should include the notion that multiple controls are necessary. The investigator should provide controls for the *same ODN chemistry* (e.g. backbone, sugar or nucleotide base modifications) and the selected sequence of nucleic acids. First, the length of the ODN should be optimum, that is, length versus activity should be established for a selected target site. The limits to this matrix include understanding that nucleic acid hybridization will not occur for ODNs with fewer than approximately 12 bases at 37 °C, and ODNs longer than 25 are generally no more active than 18-mers. The control should be the *same length* as the active ODN.

ODN entry into cells appears to be sensitive to the sequence of nucleic acids, so a control with completely different sequence or different base composition may not be equivalent in cellular bioavailability. Hence, a control with the exact *same base composition* and the opposite sequence is afforded by a reverse-order control. What was at the 5′ end becomes the 3′ end; for example, 5′-ATCG-3′ becomes 5′-GCTA-3′. The sense-strand control is not a particularly good control, as the base composition is not always equivalent to the antisense compound. The reverse-order control does not control for the relationship of a base sequence to a given end of the ODN, that is, preserve the *same base sequence motifs*. Since this relationship may be important, a scrambled control is recommended, for example, 5′-ACTGGG-3′ could be rearranged to 5′-ACGTGG-3′.

Additional sequences antisense to the same target mRNA at different nucleotide positions with different nucleotide sequences provide valuable information as positive controls. The investigator should not expect to observe equivalent potency or efficacy from all antisense sequences that target the same transcript, but a consistent qualitative response avoids result interpretation bias by target selection.

The selected ODN should be evaluated by *computer search* for potential inadvertent sequence identity with non-targeted genes. This is not a particularly powerful approach, as only an insignificant per cent of the genome is known. This may control for multiple actions of the ODN. One way to solve the problem of insufficient information about non-targeted genes is to *find a cell or cell line that does not express the target gene* or only contains mutant versions of that gene. This control is particularly useful, as the anticipated response to the ODN is no change.

20.5.2 Cell culture conditions (oxidants, glutamine)

In vitro cell culture conditions vary with the cell type to be assayed. Constituents added to the medium may augment or suppress ODN activity. Glutamine, for example, is shown to abrogate cell death in acute myeloid leukemia cells exposed to an ODN complementary to p53, OL(1)p53, which is dependent on DNA damage resulting from oxygen radical formation for toxicity. These same cells failed to respond to this ODN in low oxygen culture conditions unless oxidants were also present. As noted below, concerning *in vitro* vs. *in vivo* systems, support cells or other cell types grown in conjunction with the cells of interest may enhance ODN activity or may inhibit the growth of the target cells so that ODN response cannot be clearly determined.

20.5.3 Assessment of response

Choosing more than one assay system to monitor response to the ODN is important. For example, a single or limited timepoint viability assay may be misleading in cases where an ODN does not cause cell death as predicted but may lead to differentiation or halt proliferation of the cells. To ascertain cellular response, a time course that includes viability, proliferation and differentiation assays in parallel is useful. If an ODN blocks cell proliferation, it may trigger apoptosis, which can be determined through DNA ladders, flow cytometry for the apoptotic cell peak (A_0), and electron microscopy. A biological response may occur later than subcellular events, so a separate time course must be determined to best analyse changes in protein, mRNA and other metabolic events. The experimental design is greatly enhanced if an internal control gene is monitored in addition to the targeted gene product. Caution is required in the selection of the internal control protein, as the protein half-life should be close to that of the targeted protein and proteins regulated by the cell cycle are expected to modulate with the passage through the cell cycle.

Fluorescent or radioactive labelling can be used to determine uptake and internal localization of the ODN (Chin *et al.*, 1990; Iversen, Crouse and Perry, 1992*a*) as well as cell lineage and cell cycle specific ODN binding (Krieg *et al.*, 1991; Iversen, Mata and Ebadi, 1992*b*; see also Introduction). Similarly, changes in target protein function may be monitored by monoclonal antibody, immunohistochemistry or Western blot.

Response *in vivo* is consistent with that which might be determined for any pharmacological agent. For example, an investigator should monitor bone marrow, blood and urine, and conduct animal autopsy with histological examination. In one case involving monkeys, continuous monitoring of cardiovascular parameters was employed (Cornish *et al.*, 1993).

20.5.4 **In vitro** *versus* **in vivo** *assays*

Both *in vitro* and *in vivo* assays must be regarded as model systems with unique advantages and disadvantages. In the cause of *in vitro* culture, these may be performed on long-established cell lines or on primary cells and tissues obtained directly from the patient population in question.

Upon initiation of a primary cell culture, numerous cell types may be present. Some may not survive in tissue culture for extensive periods of time. This could lead to the possibility of studying a subpopulation of the target cells that may not be indicative of the response of the entire complement of cells seen in the patient, so cell type should be confirmed. Unlike cell lines, primary cell cultures tend to have a rather limited lifespan, thus impairing assays that require long-term monitoring to reveal responses to ODNs.

In vivo model systems permit the testing of ODNs in a multi-cellular, microenvironmentally interactive unit where subtle changes, such as inter-actions between growth or inhibitory factors, may alter the outcome. Though perhaps cost limiting, the best alternative for some studies may be non-human primates, in which the target sequence retains a high degree of homology to human.

20.5.5 Biological response not due to antisense effect

The mere fact that a response is noted following exposure to an antisense ODN is not evidence of an antisense mechanism. Care should be taken when interpreting antisense studies that do not contain a series of controls to demonstrate specific gene inhibition (Wagner, 1994). If the biological response to the ODN is indirect, such as change in cell growth rate or cell viability, then the response may be due to a near infinite number of possible ODN-based interactions. Well-controlled assays (described above) are essential in establish-ing the antisense mechanism of action. In some cases, a variety of biological effects can be produced, but inclusion of controls may rule out the antisense mechanism. In other cases, some effects may be due to differences in cell lines, culture conditions or ODN purity.

The genesis of non-antisense effects may be due to ODN binding to small molecules (Ellington and Szostak, 1990) or proteins (Bock *et al.*, 1992; Gao *et al.*, 1992; Yakubov *et al.*, 1993), activation of transcription factors (Perez *et al.*, 1994), or hybridization to mismatch sequences (Woolf, Melton and Jennings, 1992). Additional, poorly understood, non-antisense mechanisms have been reported for ODN intervention in the processes of viral penetration, uncoating and replication (Yakubov *et al.*, 1993; Azad *et al.*, 1993). Finally, the degradation and metabolism of nucleosides and nucleoside analogues may interfere with cellular differentiation and proliferation (Kamano *et al.*, 1992; Rathbone *et al.*, 1992).

Antisense attenuation of a protein does not ensure that biological response is due to an antisense mechanism of action. Since the ODN can bind small molecules and proteins, participate in mismatch hybridization, and activate transcription factors as described above, it should not be surprising that ODNs produce multiple effects. Just because targeted protein synthesis is diminished in response to an antisense ODN does not exclude the possibility that the biological response is due to lower cellular abundance of the targeted protein.

Exclusion of this possibility may require the use of multiple ODN chemistries to isolate potential non-sequence specific effects. For example, phosphorothioate ODNs may activate SP1 in a non-sequence-dependent manner, so if SP1 responses produce a portion of the biological response, then use of a phosphodiester ODN may be employed to provide an antisense effect without producing overlapping non-sequence-dependent effects.

References

Agrawal, S. (1991) Antisense oligodeoxynucleotides: a possible approach for chemotherapy of AIDS. In *Prospects for Antisense Nucleic Acid Therapy of Cancer and AIDS* (ed. E. Wickstrom), pp. 143–58. Wiley-Liss, New York.

Agrawal, S., Temsamani, J. and Tang, J. Y. (1991) Pharmacokinetics, biodistribution and stability of oligodeoxynucleotide phosphorothioates in mice. *Proc. Natl. Acad. Sci. USA* **88**, 7595–9.

Azad, R. F., Driver, V. B., Tanaka, K. *et al.* (1993) Antiviral activity of a phosphorothioate oligonucleotide complementary to RNA of the human cytomegalovirus major immediate-early region. *Antimicrob. Agents Chemother.* **37**, 1945–54.

Bayever, E., Iversen, P. L., Bishop, M. R. *et al.* (1993) Systemic administration of a phosphorothioate oligonucleotide with a sequence complementary to p53 for acute myelogenous leukemia and myelodysplastic syndrome: initial results of a phase I trial. *Antisense Res. Devel.* **3**, 383–90.

Bennett, R. M., Gabor, G. T. and Merritt, M. J. (1985) DNA binding to human leukocytes: evidence for a receptor-mediated association, internalization and degradation of DNA. *J. Clin. Invest.* **76**, 2182–90.

Bennett, R. M., Kotzin, B. L. and Merritt, M. J. (1987) DNA receptor dysfunction in systemic lupus erythematosus and kindred disorders. *J. Exp. Med.* **166**, 850.

Bennett, R. M., Peller, J. S. and Merritt, M. J. (1986) Defective DNA-receptor function in systemic lupus erythematosus and related diseases: evidence for an autoantibody influencing cell physiology. *Lancet* **25**, 186.

Bergot, B. J. and Egan, W. T. (1992) Separation of synthetic phosphorothioate oligodeoxynucleotides from their oxygenated (phosphodiester) defect species by strong-anion-exchange high-performance liquid chromatography. *J. Chromatogr.* **599**, 35–42.

Bielinska, A., Shivdasani, R., Zhang, L. and Nabel, G. (1990) Regulation of gene expression with double-stranded phosphorothioate oligonucleotides. *Science* **250**, 997–1000.

Bock, L. C., Griffin, L. C., Latham, J. A. *et al.* (1992) Selection of single-stranded DNA molecules that bind and inhibit human thrombin. *Nature* **355**, 564–6.

Ceruzzi, M. and Draper, K. (1989) The intracellular and extracellular fate of oligodeoxyribonucleotides in tissue culture systems. *Nucleosides and Nucleotides* **10**, 155–66.

Chin, D. J., Green, G. A., Zon, G. *et al.* (1990) Rapid nuclear accumulation of injected oligodeoxyribonucleotides. *New Biol.* **2**, 1091.

Chused, T. M., Steinberg, A. D. and Talal, N. (1972) The clearance and localization of nucleic acids by New Zealand and normal mice. *Clin. Exp. Immunol.* **12**, 465.

Cohen, J. S. (ed) (1989) *Oligodeoxynucleotides antisense inhibitors of gene expression.* CRC Press, Boca Raton, Florida.

Cohen, J. S. and Hogan, M. E. (1994) The new genetic medicines. *Scientific American* **271**, 77–82.

Colvin, O. M. (1990) Disposition and metabolism of oligodeoxynucleoside methylphosphonate following a single iv injection in mice: Short Communication. *Drug Metab. Disp.* **18**, 815–18.

Copple, B. L., Gmeiner, W. M. and Iversen, P. L. (1995) Reaction between metabolically activated acetaminophen and phosphorothioate oligonucleotides. *Toxicol. Appl. Pharmacol.* **133**, 53–63.

Cornish, K. G., Iversen, P. L., Smith, L. J. *et al.* (1993) Cardiovascular effects of a phosphorothioate oligonucleotide with sequence antisense to p53 in the conscious rhesus monkey. *Pharmacol. Comm.* **3**, 239–47.

Crooke, R. M. (1993) Cellular uptake, distribution and metabolism of phosphorothioate, phosphodiester, and methylphosphonate oligonucleotides. In *Antisense research and applications* (ed. S. T. Crooke and B. Lebleu), pp. 427–49. CRC Press, Boca Raton, Florida.

Dahlberg, A. E., Dingman, C. W. and Peacock, A. C. (1969) Electrophoretic characterization of bacterial polyribosomes in agarose-acrylamide composite gels. *J. Mol. Biol.* **41**, 139–69.

Daley, G. Q., Van Etten, R. A. and Baltimore, D. (1990) Induction of chronic myelogenous leukemia in mice by the P210bcr/abl gene of the Philadelphia chromosome. *Science* **247**(4944), 824–30.

DeClercq, E., Eckstein, F., Sternbach, H. and Merigan, T. C. (1970) The antiviral activity of thio-phosphate substituted polynucleotides *in vitro* and *in vivo*. *Virology* **42**, 421–8.

deFabritis, P., Amadori, S., Calabretta, B. and Mandelli, F. (1993) Elimination of clonogenic Philadelphia-positive cells using *BCR-ABL* antisense oligodeoxynucleotides. *Bone Marrow Transpl.* **12**, 261–5.

Degols, G., Leonetti, J.-P. and Lebleu, B. (1991) Antisense oligonucleotides as pharmacological modulators of gene expression. In *Targeted drug delivery. Handbook of experimental pharmacology* (ed. R. L. Juliano), Springer-Verlag, Berlin.

Diamantstein, T. and Blitstein-Willinger, E. (1978) Specific binding of poly(I).poly(C) to the membrane of murine B lymphocyte subsets. *Eur. J. Immunol.* **8**, 896–9.

Dorsch, C. A. (1981) Binding of single-stranded DNA to human platelets. *Thrombosis Res.* **24**, 119.

Dorsch, C. A., Chia, D., Levy, L. and Barnett, E. V. (1975) Persistence of DNA in the circulation of immunized rabbits. *Rheumatology* **212**, 161.

Eckstein, F. (1975) Investigation of enzyme mechanisms with nucleoside phosphorothioates. *Angew. Chem. Int. Ed.* **14**, 160–6.

Eckstein, F. (1979) Phosphorothioate analogs of nucleotides. *Accts. Chem. Res.* **12**, 204–10.

Eckstein, F. (1984) Investigation of enzyme mechanisms with nucleoside phosphorothioates. *Ann. Rev. Biochem.* **54**, 367–402.

Efcavitch, J. W. (1990) The electrophoresis of synthetic oligonucleotides. In *Gel electrophoresis of nucleic acids – a practical approach*, 2nd edn (ed. D. Rickwood and B. D. Hames), pp. 125–49. I.R.L. Press, Oxford.

Ellington, A. D. and Szostak, J. D. (1990) *In vitro* selection of RNA molecules that bind specific ligands. *Nature* **346**(6287), 818–22.

Emlen, W. and Mannik, M. (1978) Kinetics and mechanisms for removal of circulating single-stranded DNA in mice. *J. Exp. Med.* **147**, 684.

Emlen, W. and Mannik, M. (1982) Effect of preformed immune complexes on the clearance and tissue localization of single stranded DNA in mice. *Clin. Exp.*

Immunol. **40**, 264.

Emlen, W. and Mannik, M. (1984) Effect of DNA size and strandedness on the *in vivo* clearance and organ localization of DNA. *Clin. Exp. Immunol.* **56**, 185–192.

Emlen, W., Rifai, A., Magilvey, D. and Mannik, M. (1988) Hepatic binding of DNA is mediated by a receptor on nonparenchymal cells. *Am. J. Pathol.* **133**, 54–60.

Erickson, R. J. and Grosch, J. C. (1974) The inhibition of avian myeloblastosis virus deoxyribonucleic acid polymerase by synthetic polynucleotides. *Biochemistry* **13**, 1987–93.

Gao, W.-Y., Han, F.-S., Storm, C. *et al.* (1992) Phosphorothioate oligonucleotides are inhibitors of human DNA polymerases and RNase H: Implications for antisense technology. *Molec. Pharmacol.* **41**, 223–9.

Gao, W., Stein, C. A., Cohen, J. S. *et al.* (1988) Effects of phosphorothioate oligodeoxyribonucleotides on herpes simplex virus type-2 induced DNA-polymerase. *J. Biol. Chem.* **264**, 11521–4.

Geselowitz, D. A. and Neckers, L. M. (1992) Analysis of oligonucleotide binding, internalization and intracellular trafficking utilizing a novel radiolabeled cross-linker. *Antisense Res. Devel.* **2**, 17.

Goodarzi, G., Watabe, M. and Watabe, K. (1992) Organ distribution and stability of phosphorothioated oligodeoxyribonucleotides in mice: Short Communication. *Biopharm. Drug Dispos.* **13**, 221–7.

Goodchild, J., Letsinger, R. L., Sarin, P. S. *et al.* (1988) Inhibition of replication and expression of HIV-1 in tissue culture by oligodeoxynucleotide hybridization competition. In *Human retroviruses, cancer and AIDS: Approaches to prevention and therapy* (ed. D. Bolognesi), pp. 423–38. Alan R. Liss, New York.

Goodchild, J., Zamenick, P. C. and Kim, B. (1991) The clearance and degradation of oligodeoxynucleotides following intravenous injection into rabbits. *Antisense Res. Devel.* **1**, 153–60.

Griffin, L. C., Tidmarsh, G. F., Bock, L. C. *et al.* (1993) *In vivo* anticoagulant properties of a novel nucleotide-based thrombin inhibitor and demonstration of regional anticoagulation in extracorporeal circuits. *Blood* **81**, 3271–6.

Haruna, B. I. and Spiegelman, S. (1966) Selective interference with viral RNA formation *in vitro* by specific inhibition with synthetic polyribonucleotides. *Proc. Natl. Acad. Sci. USA* **56**, 133.

Hélène, C. (1993). Control of gene expression by triple helix-forming oligonucleotides: the antigene strategy. In *Antisense Research and Applications* (ed. S. T. Crooke and B. Lebleu), pp. 375–85. CRC Press, Boca Raton, Florida.

Hélène, C. and Toulmé, J.-J. (1990) Specific regulation of gene expression by antisense, sense and antigene nucleic acids. *Biochem. Biophys. Acta.* **1049**, 99–125.

Iversen, P. (1993) *In vivo* studies with phosphorothioate oligonucleotides: rationale for systemic therapy. In *Antisense research and applications* (ed. S. T. Crooke and B. Lebleu), pp. 461–70. CRC Press, Boca Raton, Florida.

Iversen, P., Crouse, D. and Perry, G. (1992*a*) Binding of antisense phosphorothioate oligonucleotides to murine lymphocytes is lineage specific and inducible. *Antisense Res. Devel.* **2**, 223–233.

Iversen, P. L., Mata, J. E., Tracewell, W. and Zon, G. (1994) The pharmacokinetics of an antisense phosphorothioate oligodeoxynucleotide against *rev* (art/trs) from human immunodeficiency virus (HIV) Type 1 in the adult male rat following single injections and continuous infusion. *Antisense Res. Devel* **4**, 43–52.

Iversen, P., Mata, J. and Ebadi, M. (1992*b*) Synthetic antisense oligonucleotide probes the essentiality of metallothionein gene. *Biological Signals* **1**, 293–9.

Kamano, H., Tanaka, T., Yamaji, Y. *et al.* (1992) *E. coli* gpt gene expression effects on K562 human leukemia cell proliferation and erythroid differentiation altered by mycophenolic acid. *Biochem. Int.* **26**, 537–43.

Karamyshev, V. N., Vlasov, V. V., Zon, D. *et al.* (1993) Distribution and stability of oligonucleotide derivatives in mouse tissues. Translated from *Biokhimiya*, **58**, 590–8. Plenum Publishing Corporation.

Karpova, G. G., Knore, D. G., Ryte, A. S. and Stepanovich, L. E. (1980) Selective alkylation of poly(A) tracts of RNA inside the cell with the derivative of ethylester of oligothymidylate bearing 2-chloroethylamino group. *FEBS Lett.* **122**, 21–4.

Kitajima, I., Shinohara, T., Minor, T. *et al.* (1992) Human T-cell leukemia virus type I tax transformation is associated with increased uptake of oligodeoxynucleotides *in vitro* and *in vivo*. *J. Biol. Chem.* **267**, 25881–8.

Krieg, A. M., Gmelig-Meyling, F., Gourley, M. F. *et al.* (1991) Uptake of oligodeoxy-ribonucleotides by lymphoid cells is heterogenous and inducible. *Antisense Res. Devel.* **1**, 161–71.

Leonetti, J.-P., Degols, G. and Lebleu, B. (1990*a*) Biological activity of oligonucleo-tide–poly-(L-lysine) conjugates: mechanism of cell uptake. *Bioconj. Chem.* **1**, 149.

Leonetti, J.-P., Machy, P., Degols, G. *et al.* (1990*b*) Antibody-targeted liposomes containing oligodeoxyribonucleotides complementary to viral RNA selectively inhibit viral replication. *Proc. Natl. Acad. Sci. USA* **87**, 2448.

Leonetti, J.-P., Mechti, N., Degols, G. *et al.* (1991) Intracellular distribution of microinjected antisense oligonucleotides. *Proc. Natl. Acad. Sci. USA* **88**, 2702.

Loke, S. L., Stein, C. A., Zhang, X. H. *et al.* (1989) Characterization of oligonucleo-tide transport into living cells. *Proc. Natl. Acad. Sci. USA* **86**, 3474–8.

Majumdar, C., Stein, C. A., Cohen, J. S. *et al.* (1989) Stepwise mechanism of HIV reverse transcriptase: primer function of phosphorothioate oligonucleotides. *Biochemistry* **28**, 1340.

McCormack, J. J., Bigelow, J. C., Chrin, L. R. and Mathews, L. A. (1990). High-performance liquid chromatographic analysis of phosphorothioate analogues of oligonucleotides in biological fluids. *J. Chromatog.* **533**, 133–140.

Miller, P. and Ts'o, P. O. P. (1987) A new approach to chemotherapy based on molecular biology and nucleic acid chemistry: Matagan (masking tape for gene expression). *Anticancer Drug Design* **2**, 117–123.

Milligan, J. F., Matteucci, M. D. and Martin, J. C. (1993) Current concepts in antisense drug design. *J. Med. Chem.* **36**, 1923–37.

Neckers, L. M. (1989) Antisense oligonucleotides as a tool for studying cell regulation: mechanism of uptake and application to the study of oncogene function. In *Oligodeoxynucleotides antisense inhibitors of gene expression* (ed. J. S. Cohen), pp. 211–32. CRC Press, Boca Raton, Florida.

Neckers, L. M. (1993) Cellular internalization of oligodeoxynucleotides. In *Antisense research and applications* (ed. S. T. Crooke and B. Lebleu), pp. 451–60. CRC Press, Boca Raton, Florida.

Offensperger, W. B., Offensperger, S., Walter, E. *et al.* (1993) *In vivo* inhibition of duck hepatitis B virus replication and gene expression by phosphorothioate modified antisense oligodeoxynucleotides. *EMBO J.* **12**, 1257–62.

Ohlbaum, A., Csuzi, S. and Antoni, F. (1979) Binding of exogenous DNA by human lymphocytes and by their isolated plasma membranes. *Acta Biochim. Biophys. Acad. Sci. Hung.* **14**, 165–76.

Perez, J. R., Li, Y., Stein, C. A. *et al.* (1994) Sequence-independent induction of Sp1 transcription factor activity by phosphorothioate oligodeoxynucleotides. *Proc. Natl. Acad. Sci. USA* **91**, 5957–61.

Ratajczak, M. S., Kant, J. A., Luger, S. M. *et al.* (1992) *In vivo* treatment of human leukemia in a scid mouse model with *c-myb* antisense oligodeoxynucleotides. *Proc. Natl. Acad. Sci. USA* **89**, 11823–7.

Rathbone, M. P., Middlemiss, P. J., Gyshers, J. W. *et al*. (1992) Purine nucleosides and nucleotides stimulate proliferation of a wide range of cell types. *In Vitro Cell Dev. Biol.* **28A**, 529–36.

Sarin, P. S., Agrawal, S., Civeira, M. P. *et al*. (1988) Inhibition of acquired immunodeficiency syndrome virus by oligodeoxynucleoside methylphosphonates. *Proc. Natl. Acad. Sci. USA.* **85**, 7448–51.

Simmons, M., Edelman, E. R., Dekyser, J.-L. *et al*. (1992) Antisense *c-myb* oligonucleotides inhibit intimal arterial smooth muscle accumulation *in vivo*. *Nature* **359**, 67–70.

Skorski, T., Nieborowska-Skorska, M., Nicolaides, N. C. *et al*. (1994) Suppression of Philadelphia leukemia cell growth in mice by *bcr-abl* antisense oligodeoxynucleotides. *Proc. Natl. Acad. Sci. USA* **91**, 4504–8.

Skorski, T., Szczylik, C., Malaguarnera, L. and Calabretta, B. (1991) Gene-targeted specific inhibition of chronic myeloid leukemia cell growth by *bcr-abl* antisense oligodeoxyribonucleotides. *Folia Histochem. Cytobiol.* **29**, 85.

Stebbing, N., Grantham, C. A. and Carey, N. H. (1976) Anti-viral activity of single-stranded homopolynucleotides against encephalomyocarditis virus and semliki forest virus in adult mice without interferon induction. *J. Gen. Virol.* **30**, 21.

Stevenson, M. and Iversen, P. L. (1989) Inhibition of human immunodeficiency virus type I-mediated cytopathic effects by poly(L-lysine)-conjugated synthetic antisense oligodeoxyribonucleotides. *J. Gen. Virol.* **70**, 2673–82.

Szczylik, C., Skorski, T., Nicolaides, N. C. *et al*. (1991) Selective inhibition of leukemia cell proliferation by *bcr-abl* antisense oligonucleotides. *Science* **253**, 562.

Tracewell, J., Desjardins, J. and Iversen, P. L. (1995) *In vivo* modulation of the rat cytochrome P450 1A1 by double-stranded phosphorothioate oligonucleotides. *Toxicol. Appl. Pharmacol.* **135**, 179–184.

Uhlenbeck, O. C. (1993) Using ribozymes to cleave RNA's. In *Antisense research and applications* (ed. S. T. Crooke and B. Lebleu), pp. 83–96. CRC Press, Boca Raton, Florida.

van Berkel, T. J. C., Doan, T. L., Smidt, C. P. and Falco, S. (1991) Association of antisense oligonucleotides with lipoproteins prolongs the plasma half-life and modifies the tissue distribution. *Nucl. Acids Res.* **19**, 4695–700.

Vlassov, V. V., Godovikov, A. A., Kobertz, N. D. *et al*. (1986a) Nucleotide and oligonucleotide derivatives as enzyme and nucleic acid targeted irreversible inhibitors. Biochemical aspects. In *Advances in enzyme regulation* **24** (ed. G. Weber), pp. 301–20. Pergamon, London.

Vlassov, V. V., Gorokhova, O. E., Ivanova, E. M. *et al*. (1986b) Interaction of alkylating oligonucleotide derivatives with mouse fibroblasts. *Biochem. Kletka* **2**, 323–7.

Vlassov, V. and Yakubov, L. A. (1991) Oligonucleotides in cells and in organisms: pharmacological considerations. In *Prospects for antisense nucleic acid therapy of cancer and AIDS*. (ed. E. Wickstrom), pp. 243–66. Wiley-Liss, New York.

Wagner, R. W. (1994) Gene inhibition using antisense oligodeoxynucleotides. *Nature* **372**, 333–5.

Wickstrom, E. (1986) Oligonucleotide stability in subcellular extracts and culture media. *J. Biochem. Biophys. Meth.* **13**, 97.

Woolf, T. M., Melton, D. A. and Jennings, C. G. (1992) Specificity of antisense oligonucleotides *in vivo*. *Proc. Natl. Acad. Sci. USA* **89**, 7305–9.

Wu, H., Holcenberg, J., Tomich, J. *et al*. (1990) Inhibition of *in vitro* transcription by specific double-stranded oligodeoxyribonucleotides. *Gene* **89**, 203–9.

Wu-pong, S., Weiss, T. L. and Hunt, C. A. (1992) Antisense c-myc oligodeoxyribonucleotide uptake. *Pharm. Res.* **9**, 1010–17.

Yakubov, L., Khaled, Z., Zhang, L. M. *et al*. (1993) Oligodeoxynucleotides interact with recombinant CD4 at multiple sites. *J. Biol. Chem.* **268**, 18818–23.

Zamecnik, P. C., Goodchild, J., Taguchi, Y. and Sarin, P. S. (1986) Inhibition of replication and expression of human T-cell lymphotropic virus type III in cultured cells by exogenous synthetic oligonucleotides complementary to viral RNA. *Proc. Natl. Acad. Sci. USA* **83**, 4143–6.

Zendegui, J. G., Vasquez, K. M., Tinsley, J. H. *et al*. (1992) *In vivo* stability and kinetics of absorption and disposition of 3′ phosphopropyl amine oligonucleotides. *Nucl. Acids Res.* **20**, 307–14.

Zhao, Q., Matson, S., Herrera, C. J. *et al*. (1993) Comparison of cellular binding and uptake of antisense phosphodiester, phosphorothioate, and mixed phosphorothioate and methylphosphonate oligonucleotides. *Antisense Res. Devel.* **3**, 53–66.

Troubleshooting

21

Microbial contamination

KAREN HENDERSON

Occasionally, microorganisms will appear in cell cultures or in nucleic acid samples and solutions. Sources of infection include tissue culture reagents, air currents and the operator. The most commonly occuring contaminants include the following.

21.1 Fungus

Morphology

Yeasts normally exist in the unicellular form and are globose, ellipsoidal or occasionally cylindrical particles of 5–15 μm. Some remain attached to each other following division, forming chains. The filamentous fungi consist of a mass of branched hyphae, which are usually septate.

Detection

Macroscopically a heavy infection of yeast may result in the medium becoming slightly cloudy and acidic. Filamentous fungi produce fluffy, fuzzy particles in, or floating on, the medium. Both unicellular and filamentous fungi can be clearly seen using the phase contrast microscope. Growth in fungal medium enables further analysis.

Prevention

Good sterile technique, particularly in relation to airflow, and use of anti-mycotics (e.g. amphotericin B) in media and water baths.

21.2 Bacteria

Morphology

Rods and cocci – the majority having a diameter in the range 0.2–1.5 μm.

Detection

Macroscopically, culture medium becomes cloudy and acidic over the course of 12 h or so. Under the phase contrast microscope, the spaces between cell colonies have a granular appearance and at ×400 clumps and single rods and cocci become apparent. Not all bacteria are motile but motility, when it does occur, helps to distinguish bacterial cells from cell debris. Though it is seldom necessary, the presence of bacteria can be confirmed by using conventional stains and light microscopy. Growth in bacteriological media provides a means of further analysis.

Prevention

Good sterile technique and broad spectrum antibiotics such as gentamycin, penicillin and/or streptomycin.

21.3 Mycoplasma

Morphology

Size 100 nm–1 μm, capable of passing through sterilizing filters.

Detection

Undetectable by phase contrast and light microscopy. These microorganisms live within the cultured cells and occasionally their presence results in a deterioration in the health of the culture, e.g. reduced proliferation rates. Means of detection include growth in mycoplasma broth and use of the DNA stain, bisbenzimide (a protocol is given below).

Prevention

Good sterile technique and anti-mycoplasma agents such as gentamycin and tylocine.

21.4 Viruses

Morphology

20–300 nm.

Detection

Various means including electron microscopy and co-cultivating the cell line with test cell lines suseptible to a range of viruses.

Prevention

Good sterile technique.

21.5 Tests and technique

The type of infection should be noted, the culture discarded and all reagents checked for contamination. It is sometimes possible to salvage infected cultures using antibiotics but this should only be attempted when the cells are irreplaceable. If a particular type of contaminant appears repeatedly it may be necessary to identify the organism further in order to enable subsequent eradication. Contamination tests should be carried out at regular intervals.

The use of antibiotics in tissue culture is not a substitute for aseptic technique and in fact many antibiotics have been found to be cytotoxic. Periodic discontinuation of antibiotic use should be carried out occasionally to prevent antibiotic resistance and to test for cryptic contaminations that would otherwise be masked. Bacterial and fungal contamination is, in general, easy to detect and cultures should be screened on a daily basis. It should be noted, however, that the diptheroids (*Corynebacterium*) are very slow growing and can easily be missed, particularly when culture medium is changed on a very frequent basis. Mycoplasma detection is more difficult and though cells may appear to be healthy, mycoplasma can and do affect cell properties and so routine screening for mycoplasma, approximately once a month and on new lines entering the tissue culture laboratory, is essential and a procedure for this is outlined below.

Test for mycoplasma

Bisbenzimide is a carcinogen and is heat and light sensitive

1. Grow cells on coverslips to 50% confluency in six-well plates
2. Remove half of the culture medium and add an equal volume of fixative (1 part glacial acetic acid to 3 parts methanol), 2 min.
3. Decant, add 5 ml fresh fixative and allow to stand for a further 5 min. Repeat.
4. Aspirate and air dry.
5. Stain in 5 ml Hoechst 33258 solution (0.25–0.5 $\mu g/ml$ bisbenzimide in distilled water), 15–30 min.
6. Wash 3× distilled water.
7. Air dry.
8. Mount coverslip, cell side down, in mounting solution (1 part glycerol: 1 part pH 5.5 buffer – 22.2 ml citric acid (0.1 M) and 27.8 ml disodium phosphate (0.2 M))
9. Examine using fluorescence microscope-emission at 440 nm, excitation 330/380 nm.

10. Mycoplasma are visible at a magnification of ×500, as small uniformly shaped fluorescent bodies throughout the cytoplasm. Cell nuclei also fluoresce.

Alternatively, cultures can be tested indirectly by incubating the spent medium with an indicator cell line such as 3T6, which is then used in the Hoechst test. This strategy is particularly useful for suspension cultures.

22

Safety aspects of genetic manipulation work

STEVE LEGON

22.1 Background

Historically it has been the case that powerful new technologies have been introduced without careful consideration of the problems that they might cause for society. This was not the case with genetic manipulation. When the possibility of creating and propagating recombinant DNAs became a reality in the early 1970s the press and public were ready to believe that all manner of horrors were about to be created. At a more cerebral level, many scientists were also concerned about the consequences of creating genetically modified organisms that could not have been produced by 'natural' means.

Counterparts of the viral oncogenes were just being characterized in the human genome and little was known about mammalian gene structure. Perhaps there was a cancer-causing gene that, when incorporated into the genome of *E. coli*, would give rise to a 'cancer bug'. In retrospect this appears naïve but it would be equally naïve to pretend that our present understanding of the human genome is such that we can rule out the possibility of accidentally creating recombinants that would be harmful either to mankind or to the environment. A number of situations can be foreseen in which a significant risk can be identified and, in most advanced countries, regulatory authorities have set up mechanisms to ensure that genetic manipulation proposals are examined and approved before work can commence. Because regulations differ widely in different countries, this section concentrates on the general issues and principals involved in risk assessment and containment. For specific advice, the reader should contact his or her Safety Office for the procedures applicable to the institution and country. The reader should be aware that there is still considerable public sensitivity on this issue making strict compliance with the relevant guidelines essential for the continued preservation of the confidence of both legislators and public. The following section identifies some of the problem areas. The intention is to demonstrate that, despite an impressive safety record, there are situations in which apparently benign experimentation can result in the creation of recombinants with harmful potential. The concluding section then gives brief guidance on the principals of physical and biological containment that are applied to limit the risks to the workers themselves and to the public at large.

22.2 Potential hazards

After more than two decades of work there has still been no documented case of any health problem that has been caused by a genetically modified organism. Anyone considering working in this area should be aware that their health may suffer but this is far more likely to be as a result of their exposure to the mundane hazards of laboratory life (toxic chemicals, broken glassware, defective electrophoresis equipment, etc) rather than the genetically modified organisms which they might create. Most workers in the field are aware of this and there has been a tendency to believe that the work is automatically 'safe' but the following examples illustrate the potential of genetic modification to create microorganisms with the potential to be harmful. Careful assessment and adequate containment of experiments is still necessary.

22.2.1 Modification of host range

Genetic manipulation techniques give us the power to produce chimaeric organisms that can be invaluable as research tools. The ability to 'swap' genes between, for example, closely related viruses, allows one to investigate the role of each gene in determining the phenotype. The aims of such an experiment may be mundane but the risks associated with such an approach are considerable. Even where the genes concerned are well characterized there is always the possibility that the recombinant will have unforeseen properties, the worst of which might be that the host range of the virus might be extended. Viruses evolve naturally to cross species barriers (HIV, for instance) but recombinant DNA technology has the power to accelerate the process dramatically.

22.2.2 Cytotoxic antibody constructs

Genetically modified organisms have the potential to produce a variety of proteins that have the potential to kill. Constructs of this nature are being produced in many centres for the best of reasons. Typically a bacterium is programmed to produce a recombinant protein in which the gene for a toxin is joined to the variable region of an antibody. The target antigen may be expressed on a cancer cell giving the recombinant protein potential as an anti-tumour agent. The problems arise in identifying antibodies that are specific to the tumour. If normal subjects express the marker at low levels then one has to know which cells these are. If all cells express the antigen at low levels then small amounts of the recombinant protein produced in the gut would be unlikely to cause problems. If a minor cell population expresses the antigen at high levels then this cell population might be at risk from a small amount of the recombinant protein. One has the example of HIV to point to the dangers of targeting what is quantitatively an insignificant cell population.

22.2.3 Drug resistance

The ability of pathogens to evolve drug resistance is a major problem limiting the effectiveness of antibiotic treatments. This is a natural process brought about by the selective pressure imposed on the pathogens by the treatment. No responsible researcher would propose to introduce an antibiotic resistance gene into a pathogen that was normally treated with that antibiotic. However, one has to be particularly careful that a seemingly innocuous experiment might carry with it the risk of generating an antibiotic-resistant strain of the pathogen. Consider the following experiment, in which a gene is cloned from a pathogen into a 'safe' *E coli* vector, is manipulated to inactivate it and is then reintroduced back into the pathogen to 'knock out' the endogenous copy of the gene. On the face of it this is a very safe form of manipulation that results in the loss of a gene function – an event that will happen many times in the course of any natural infection. However, typical bacterial plasmids contain genes encoding antibiotic resistance and any procedure that calls for the excision of a cloned sequence from such a plasmid will inevitably result in a low level of contamination by DNA fragments including the antibiotic-resistance gene. Consequently, there will be a low but finite chance of transferring the antibiotic resistance to the pathogen. The researcher may well be a clinician who, after washing his/her hands and exchanging one white coat for another, then goes to the hospital ward to examine a patient being treated for this disease with the antibiotic in question.

22.2.4 Oncogenes

The risks associated with the handling of oncogenes are difficult to assess. Even if a construct were engineered to express an activated oncogene from a cDNA sequence it is still difficult to see how this would result in any hazard if this protein were to be expressed by bacteria in the gut. If there is a risk from handling potentially oncogenic sequences then it would be associated with the DNA rather than the protein product. The integration and expression of an activated oncogene sequence takes a normal cell an important step towards malignant transformation. Although many further steps are required it would clearly be undesirable if significant numbers of cells in the body were to be exposed to this risk. Naked DNA can be taken up and expressed by cells when injected into the body and it would therefore be prudent to avoid accidental exposure to DNA preparations of expressible oncogene sequences. Some mammalian viral vectors are specifically designed to insert gene sequences into cells and experiments using such vectors with oncogene sequences require careful containment. Less obvious is the possible risk associated with handling oncogene sequences in *E coli*. For any risk to be realized one has first to suppose that the gut is colonized by bacteria carrying the oncogene. If that were to happen, what is the frequency with which DNA fragments might be taken up by gut cells? One would suppose that the frequency is very low but there is no experimental data available to support this assumption.

22.3 Containment measures

Containment measures can be divided into biological containment and physical containment. Biological containment seeks to limit the genetically modified sequences to a disabled host/vector system with no capacity to transfer the modification to more viable organisms whereas physical containment prevents the organism from having access to the world outside.

It is necessary to understand the type of risk that is to be contained before appropriate measures can be devised. One has to distinguish between genetically modified organisms that might have a selective advantage over their natural counterparts (e.g. extension of host range) and those that might harm those who are in contact with them (e.g. cytotoxic antibody) but that could not survive in the long term outside the laboratory. The former need secure containment whereas, in the latter case, one has only to limit exposure to acceptable levels. The distinction is important as absolute containment of a microorganism is difficult and expensive to achieve whereas reducing unnecessary exposure is simply a matter of adopting sensible work practices – the specific measures being related to the degree of risk that exposure is thought to entail.

22.3.1 Biological containment

The most important form of containment is biological containment. If the genetic modifications are conducted using disabled microorganisms and cloning vectors that cannot be transferred to wild type organisms than it is very difficult to imagine circumstances in which the potential hazards can be realized. In the simplest case of propagating sequence in an *E. coli* plasmid, the bacterium will almost certainly be a derivative of the K12 strain that has been maintained in culture for many decades. Strain K12 is a multiple auxotroph and has acquired other mutations that make it unable to compete with wild type strains in the gut. The plasmids that are available are almost always derived from a plasmid called pAT153, which lacks the sequences necessary for it to be mobilized during bacterial conjugation. Although these biological containment features are no longer widely promoted, it is none the less true that the host strains and vectors available today have been designed with a view to maintaining biological containment such that even where there is a potential selective advantage for the modified organisms, there is very little risk of the modification being transferred to wild type organisms. Consequently, measures of physical containment can be relaxed unless specific hazards can be foreseen or if, for any reason, it is necessary to work with microorganisms that can survive in the environment.

22.3.2 Physical containment

Physical containment measures that prevent the escape of *any* microorganisms are impractical for routine work. However, in the absence of adequate biological containment, they are essential if modifications are proposed that might provide a selective advantage for the recombinants over naturally occurring forms, since the escape of even a single organism could lead to its displacing naturally occurring forms. The containment measures seen in typical genetic manipulation laboratories are designed to limit the escape of microorganisms and not to prevent escape altogether. These measures reduce exposure of the operator and other workers to acceptable levels and are appropriate where there may be an immediate hazard but there is no possibility of the recombinant becoming established in the environment. For most types of experiment there is no perceived risk and 'good microbiological practice' is accepted as providing adequate protection. Perhaps the most important measures of physical containment are the avoidance of aerosols, safe disposal of waste, measures to deal with spills and, above all, adequate supervision to control the occasional individual who decides to ignore the rules. The reader is advised to consult with the appropriate authorities at local and national level for the specific containment measures that are required in their situation.

23

Miscellaneous troubleshooting

JANUSZ A. Z. JANKOWSKI

23.1 Tissue storage and preparation

23.1.1 Tissue fixation

When processing fresh tissue, particularly during homogenization, several pathogens are present that are highly infective, including HIV, hepatitis B and tuberculosis and all samples should be treated as high-risk specimens. Some tissues, however, are still potentially infective even after fixation, such as samples from patients with Kreuzfeld–Jacob disease.

Mark the proximal and distal edges of the tissue so that subsequently it can be confidently orientated. Roll the mucosal surface on to the outside tissue specimen and place into cardice (-70–$80\,°C$). Long-term storage should be in either a $-80\,°C$ freezer or in a liquid nitrogen tank ($-170\,°C$).

23.1.2 Tissue processing samples for histopathology morphological assessment

When processing large specimens such as tumours divide them into equal portions and section each of the blocks by routine H&E (haematoxylin and eosin) staining to ensure that all specimens have similar morphology.

If the tissue is to be paraffin embedded use neutral buffered formalin, slow acting formal saline or rapidly acting formal mercury, which causes less tissue shrinkage. Carnoy's is a rapid fixative and preserves carbohydrates well for staining such as Periodic acid–Schiff. Cytological samples are best stored in Bouin's fluid but this penetrates poorly and prolonged fixation can cause problems.

23.1.3 Tissue processing for molecular biological studies

Tissue can be stored indefinitely in liquid nitrogen and when required for use $0.5\,cm^3$ ($\sim1\,g$) of tissue is put in $2\,ml$ of liquid nitrogen and ground to a fine powder. This step should be done quickly because when the tissue thaws it is liable to degradation. Alternatively, grind pieces in Dounce homogenizers in the presence of the appropriate protein, RNA or DNA tissue buffer in an alcohol bath in ice at $-10\,°C$. Avoid creating bubbles as this may oxidize specimens, therefore use a circular grinding action rather than plunging with the homogenizer. Pour the frozen grains of tissue into an autoclaved 15 ml conical tube and store at $-70\,°C$ until required and wash the homogenizer in between each tissue extraction with alcohol or DEPC-treated water.

23.1.4 Storage of molecules following extraction

Protein may be stored at $-20\,°C$ for 1 month but longer storage at $-70\,°C$ is suggested. In addition, samples should be separated into 200 μl aliquots containing 1 μg/μl of protein so as to prevent freeze–thaw degradation.

RNA should always be stored at $-70\,°C$ or at $-180\,°C$ when very small samples are available. An addition of 0.5% (w/v) SDS or 1 mM EDTA will also further retard the actions of contaminating RNases

Genomic DNA should be stored at $4\,°C$ or at $-20\,°C$ if it is lyophilized. Storage of DNA solutions at less than $0\,°C$ will result in DNA shearing with ice crystals.

23.2 Failed experiments – general guidance

23.2.1 Failed RNA extraction

Problem	Cause
Increased DNA	Inadequate tissue homogenization
	Inadequate DNA shearing by aspiration with needle
Increased protein	Insufficient centrifugation to separate phases
	Inefficient phenol/chloroform extraction
Decreased RNA	Time taken to process sample and 'protect RNA'
	Tissue (pancreas has excess RNAascs)
	Temperature of storage
	Technique unpractised

23.2.2 Failed PCR or PCR errors

Unwanted reactions can occur with the following:

1. Too much initial DNA used, which joins together resulting in either no bands or a smear.
2. Primer–dimers electrophorese at 40–60 bp, which is usually due to use of too much primer.
3. Mispriming leads to unexpected bands (i.e. background bands) and is usually due to too low annealing temperatures.

4. Misincorporation, especially if dGATC bases are in an unbalanced concentration.
5. Carry-over DNA such as circular plasmid DNA may be amplified more extensively than linear DNA, resulting in bands of inappropriate size.
6. Contamination may occur from amplicons (amplified DNA strands) from previous reactions. This can be avoided by using the tissue culture hood until the problem resolves.

For the sake of validation:

1. Ensure PCR is in the exponential phase with pilot reactions (PCR dynamics).
2. Ensure cDNA has been made by testing with GAPDH primers.

23.2.3 Sequencing errors

1. Hairpin loops occur, especially in G-C rich regions and bands on the sequencing gel above the loop are abnormally spaced out. This can be prevented by three strategies:
   (a) Use c7 dGTP (deaza analogue of dGTP), which forms weaker G-C bonds.
   (b) Use dITP, which allows more rapid chain extension.
   (c) Use a combination of Sequenase/Taq DNA polymerase, which allows reactions to occur at a higher temperature, thereby smoothing out secondary structures that cause sequencing artefacts.
2. Watson–Crick compressions of C-C bases or Hoogstein compressions of G-G bases:
   (a) Use deaza analogues c7 dGTP or dITP (less strong).
   (b) Use increased annealing temperature $> 65\,°C$ (usually $40\,°C$).
   (c) Use Taq.
   (d) Use 20–40% (v/v) formamide in acrylamide gel, which breaks down hydrogen bonds.
3. Deletions or recombinations can occur, therefore use cells that prevent this, i.e. JM 109. Rec+ has a K-restriction system that stops homologous repeats and SURE1 cells prevent spontaneous recombinations.
4. Inappropriate terminations of the sequencing reaction are due to poor hybridization of primer. This can usually be limited by cleaning up the DNA template or by increasing the primer concentration twofold.

23.2.4 Poor growth of cultured cells

1. Culture the reagents separately to check for growth of bacterial cells.
2. Examine infection microscopically for cocci and rods, which are usually contaminants from the skin.

3. If there is no growth, it is possible that mycoplasma, fungi or yeast are present.
4. Filter all products or irradiate serum, gels or plasticware.
5. Ensure the CO_2 concentration in the incubator is appropriate, usually 5–10%.

23.3 Selection of adequate control experiments for gene analysis

Immunocyto-chemistry	positive control to ensure correct topographical immunoreactivity
	negative control to ensure no non-specific localization
Western blot	positive control to prove no degradation of protein
	negative control to prove no non-specific binding
Northern/ Southern	positive control to assess correct hybridization
	negative control to exclude background hybridization
PCR controls	positive control to ensure correct amplification of transcripts
	negative control to ensure no spurious amplification or contamination

23.4 Quality control

In general, temperature changes may vary widely in some laboratories without air conditioning, by as much as 15 °C. This will severely interfere with gel setting, PCR experiments and gene sequencing techniques.

Type 1 water from laboratory distillers and class 'A' certified glassware purity and sterility is required. Water and glassware should be checked each month by culture and HPLC testing.

Pipette calibration should be assessed by the balance/gravimetric method. Briefly weigh out a number of samples of distilled water at room temperature (18–22 °C) using a calibrated electronic microgram balance – 1 mg should equal 1 μl throughout the volume range of the pipette. Using this procedure it should be possible to decide whether the pipette is accurate (i.e. within 10–20% of ideal volume) or precise (each measurement within 5–10% of each other).

Spectrophotometers should also be calibrated with a DNA and protein standard of known concentrations to test the tungsten halogen lamps for the UV and visible spectras respectively. In addition a standard curve should be generated to establish the linear range of the spectrophotometer. pH meters should also be calibrated using predetermined acidic and alkaline solutions.

Gene manipulation and applications to molecular medicine

24

Molecular genetics and developmental disorders

PAUL BUSS

24.1 Introduction

As the molecular revolution continues, clinicians have to assimilate large volumes of information and decide on its clinical application. The strides made so far in our understanding of processes underlying inherited conditions have largely resulted from linkage studies. Utilizing additional laboratory techniques, our understanding of the molecular basis of normal developmental processes has improved and new avenues for research have evolved. This is particularly the case with regards to congenital malformation.

This chapter aims to outline some recent areas of advance in specific inherited diseases resulting in changes to the counselling and management of affected individuals and their families.

24.2 Huntington's disease (HD) (autosomal dominant)

24.2.1 Clinical aspects

Clinicians who encounter Huntingtons Disease recognize it as an awesome clinical entity. It represents the archetypal neurodegenerative disease. Most often clinically silent during childhood years and adolescence, the disease manifestations of minor mood-swings and tic-like movements in the third decade herald the development of severe symptoms. When fully developed, the patient, usually in early middle age, may be handicapped by rigidity, choreoform movements, severe psychological disease and dementia (Harper, 1991).

The pathological changes of cell death in the basal ganglia (the caudate nucleus and putamen primarily) lead to a typical presentation of relentless progression, sometimes rapid, but occasionally for up to 20 years, prior to death. In most juvenile presenters the gene is inherited paternally (Karien et al., 1993). The disease prevalence in European populations is estimated at between 1 in 10 and 14 000 (Harper, 1993; Karien et al., 1993). Attempts to provide treatment for the condition are largely unsuccessful at altering disease course, although phenothiazines and psychotropic drugs have been widely used. Families with an affected member have awaited identification of the disease gene, in the hope that it will give more accurate predictive testing, additionally providing insight into the biochemical disorder leading to neuronal loss and hence a rational treatment.

24.2.2 Genetics

Initially localized to 4p16.3, long-term collaborative research, using disequilibrium and haplotyping, identified and localized an expanded trinucleotide CAG repeat in the gene *IT15* (Huntington's Disease Collaboration Group, 1993). The gene itself is composed of 67 exons and encodes two mRNA transcripts of differings sizes and expressed in differing ratios in a number of tissues. Foetal brain expresses the longer transcript (13.5 kb) whereas the reverse is the case in adults. The role of the protein provisionally named 'Huntingtin' in the role of neuronal integrity, however, remains unclear.

24.2.3 Clinical implications

Identification of the disease gene allows more accurate counselling to those affected and their families. This includes more information with regards to the possible age of onset in presymptomatic cases (Craufurd and Dodge, 1993; Harper, 1993), although data correlating triplet repeat length with age of onset must be regarded cautiously (Kremen *et al.*, 1993). Informing someone that they may have a severe incurable disease that will possibly affect them in several years' time is not easy for the clinician, should not be undertaken lightly, and may not be wanted (Decruynaere, Evere-Kiebooms and Van den Berghe, 1993). In such circumstances potentially affected cases may opt for uncertainty, in the hope that in due course scientifically based therapy will be forthcoming. A very difficult counselling situation may arise when an intermediate repeat length is identified (Benjamin *et al.*, 1994). Such issues have arisen with cases considering prenatal testing. A recent survey indicated low uptake rates among parents – with most declining prenatal testing in the hope that a cure will be found in time for offspring (Adam, Wiggins and Whyte, 1993).

24.3 Cystic fibrosis (CF) (autosomal recessive)

24.3.1 Clinical aspects

In the Northern hemisphere this disease represents a cause of major long-term morbidity and mortality in children and young adults. The old term 'mucoviscidosis' adequately identifies the clinical problems encountered. One in 2000 children in the United Kingdom suffers from cystic fibrosis and the carrier frequency is estimated at approximately 1 in 25 (Williamson, 1993).

Ten percent of children will present in the newborn period with obstruction of the gastrointestinal tract by viscid meconium (meconium ileus). Most, however, present insidiously with either pancreatic insufficiency and poor weight gain (in spite of good calorie intake) or recurrent chest infections during early childhood. The identification of an affected child and the communication of this news to parents is a daunting clinical task. Cystic fibrosis is established as a 'dreaded' childhood disease, and although current therapies have been able usefully to extend life well beyond the second decade for many, the routine of hospitalization and regular physiotherapy, with complications including diabetes and cirrhosis, means considerable handicap.

Diagnosis is classically via measurement of increased sweat sodium (the sweat test) in affected cases. Following the advent of gene localization and then identification, hope that this would mean confident molecular diagnosis has been complicated by the identification of several hundred mutations.

24.3.2 Genetics

The gene, identified as recently as 1989 (Kerem *et al.*, 1989; Rommens *et al.*, 1989) codes for a protein named cystic fibrosis transmembrane conductance regulator (CFTR), an integral cell membrane protein. Mutations of the CFTR gene leads to abnormalities of cAMP-mediated active chloride transport across luminal surfaces. This results in intracellular water influx and dehydration of cellular surfaces, which become viscous.

The gene itself is 250 kb long encoding a 6.5 kb mRNA sequence corresponding to a 1480 amino acid protein. The most common site for mutation of the gene is in exon 10 – mutation of a phenylalanine triplet at position 508. This mutation probably acts by reducing the bioavailability of CFTR, retaining it within the endoplasmic reticulum rather than altering function at the cell membrane. Approximately two-thirds of cases homozygotically possess this mutation, although wide racial differences exist. In the UK over 90% of cystic fibrosis cases will possess at least one delta 508 mutation (Green *et al.*, 1993). Over 250 mutations within the gene have been described (Miedzybrodzka *et al.*, 1994). In general, genotype is not an accurate predictor of disease presentation (Al-Jader *et al.*, 1992).

24.3.3 Clinical implications

The identification of the CFTR gene has opened up major avenues for further research. It has also enabled accurate diagnosis to be established where clinical impressions suggest CF but where repeated sweat tests are unclear. Also it is useful in the newborn period or if other medical conditions, where sweat testing may be inaccurate, are encountered. Prenatal testing for the gene mutation is now performed but with psychological cost for parents (Mennie *et al.*, 1993). Uncertainty also arises because of the large number of mutations (in Wales we currently routinely test for six mutations covering 85% of affected cases). The issue of carrier testing (using mouthwash samples) in both the general population and in affected individuals' siblings, has been debated (Williamson, 1993; Raeburn and Marteau, 1994; Balfour-Lynn, Madge and Dinwiddie, 1985), leading to concern with regards to the ethical issues surrounding such programmes, particularly if undertaken commercially without safeguards of clearly defined counselling and supportive programmes (Clinical Genetic Society (UK), 1994). Most families with affected children, when questioned, directly recognize some benefit from earlier knowledge of the diagnosis although there is not a considerable body of evidence that earlier diagnosis affects clinical outcome. A GP-based survey indicated that carrier testing would be taken up if offered (Watson *et al.*, 1991) but concerns have been voiced that this may lead to increased prenatal testing and increased requests for termination of pregnancy (Borgo *et al.*, 1992).

Nevertheless, families of affected children with cystic fibrosis awaited the molecular discovery with great anticipation, the isolation of the sequence being a prerequisite for the development of gene therapy. *In vitro* studies initially demonstrated the requirement for just a single copy of retrovirally transvected CFTR-cDNA to restore the active chloride mechanism (Rich *et al.*, 1990).

Theoretically, retrovirus vectors carry a possibility of incorporation into the host genome, with repercussions on oncongenic potential. Thus the most favoured vector for human trials of gene therapy are adenoviruses. Rat experiments have revealed successful CFTR transvection with transient expression via this method (Coutelle *et al.*, 1993). Recently, human expression of CFTR in four affected individuals administered gene therapy, with few systemic or pulmonary effects, revealed promising results (Crystal *et al.*, 1994). Caplen *et al.* (1995) demonstrated transfer of CFTR to nasal epithelium *in vivo* in six subjects with cystic fibrosis using cationic liposomes.

24.4 Fragile-X syndrome (X-linked)

24.4.1 Clinical aspects

Fragile-X syndrome represents the most common inherited cause of learning disability. The disease affects approximately 1 in 1250 males but females may also be affected (estimated at around 1 in 2000) (Webb, 1991). The classical clinical features include characteristic facial dysmorphisms (broad forehead and prominent ears) – more distinctive in older cases. Autistic features have been described in up to 3% of cases.

Unfortunately, as the clinical signs develop with maturity the opportunity for successful educational intervention declines. Establishing the diagnosis requires laboratory confirmation. The dramatic advances in our knowledge of the molecular basis of fragile-X syndrome has nevertheless aided clinicians, patients and their families considerably (Slaney *et al.*, 1995).

24.4.2 Genetics

The gene *FMR-1* is located at the Xq 27.3 locus. The molecular defect is one of hyperexpansion of CGG trinucleotide repeats (Verkerk *et al.*, 1991) producing abnormal promoter DNA hypermethylation with disturbed protein synthesis. In normals, between 5 and 45 repeats are present and occasionally (in approximately 1% of cases) a premutation exists, which represents a fragile-X carrier state. This occurs when amplification of the triplet repeat exceeds 150 triplets in size.

Affected individuals may have an amplification of as much as 5000 base pairs. The amplified repeat is unstable in that in differing tissues and within cells of the same tissue different numbers of copies of the repeat may occur (Sutherland and Richards, 1993).

When transmitted by males, the repeat length remains virtually unchanged. Transmission by females causes gradual increase in mutation length. This occurrence of transmission lengthening and the association of varied lengths of mutation with clinical expression have been grouped under the umbrella term 'dynamic mutations' (Sutherland and Richards, 1993). In addition, the female state is of interest in that they may be symptomatic if carrying a sufficiently long amplified repeat – but the process of X-inactivation may well reduce its clinical significance (Turk, 1995).

24.4.3 Clinical implications

In practice, the diagnosis of fragile-X is established in the laboratory. Frequently clinicians are faced with large numbers of children with non-specific developmental delay. It is these children and their families that will benefit most from the earlier accurate diagnosis and counselling that molecular testing offers. The concept of presymptomatic or carrier testing (of what is as yet an incurable disorder) remains contentious and prenatal testing also has considerable ethical problems – particularly as we now understand that possession of the amplified repeat does not necessarily mean clinical expression as significant developmental delay.

Since the fragile-X gene was identified, three other 'fragile' regions have been identified and labelled Fra X-D, -E and -F. Clinicians need to be aware of this, particularly if a family is possibly mildly affected or manifests few dysmorphic features. Conveying this to the laboratory may enable identification of one of these uncommon variants.

24.5 Insulin-dependent diabetes mellitus (IDDM) (polygenic)

IDDM affects approximately 1 in 300 European children. It was first recognized to be a specific result of insulin deficiency 70 years ago. It represents an autoimmune disease of pancreatic islet cells resulting in their destruction and either no or suboptimal insulin responses to normal stresses to pancreatic function, with the development of symptoms. These result from a reduced intracellular bioavailability of glucose with a raised plasma glucose. Increased oxidation of fats leads to utilization of ketones as an alternative fuel. Commonly IDDM presents in childhood with lassitude, weight loss, polyuria and thirst – in some cases it may present as ketoacidosis and is life-threatening. The diagnosis is relatively easy to establish but the consequences are lifelong for the child and the family. Insulin replacement therapy is necessary and dietary restriction plays a vital part in management. Long-term complications (renal and eye disease) occur frequently and there is an increased incidence of congenital malformation in infants of diabetic mothers – the genetic component of which is yet to be established.

24.5.1 Genetics

Unlike the genes for Huntington's chorea, fragile-X or cystic fibrosis, identified using a combination of classical and modern techniques, diabetes is a complex genetic disease. Animal models and family studies for many years have demonstrated a genetic component to IDDM with many studies revealing linkage to major histocompatability (MHC) loci DR3 or DR4 (6p21.3) with 90% of cases possessing at least one copy (Goodfellow and Schmitt, 1994). A locus closely linked to the MHC (*IDDM1*) has already been implicated by Davies *et al.* (1994). Additional susceptibility loci have been identified in regions close to the insulin locus (Lucassen *et al.*, 1993; Davies *et al.*, 1994; Hashimoto *et al.*, 1994) and the IGF2 locus on chomosome 11 (11p15) (Goodfellow and Schmitt, 1994). Bennett *et al.* (1995) identified the *IDDM2* susceptibility gene as a variable number tandem repeat (VNTR) and allellic variation within the VNTR may be important in the development of IDDM. Another susceptibility locus, *IDDM3* (15q26) has been recently identified and *IDDM4* (11q13) close to FGF3 (Field, Tobias and Magnus, 1994). Other genes possibly implicated in producing increased susceptibility to disease development include genes localized to the immunoglobulin heavy chain region (14q32) and T cell receptor genes located on chomosome 7 (7q35) (Goodfellow and Schmidt, 1994). Identification of these loci has been achieved via the systematic searching of chomosomes for the IDDM gene – using linkage to highly polymorphic microsatellites in families with more than one affected member.

24.5.2 Clinical implications

At present these are not clear. Newly diagnosed diabetics and their families can be reassured that rapid improvements in our understanding of IDDM will lead to better information with regards to aetiology. Although it appears the HLA linked loci have the strongest linkage to susceptibility there is now strong evidence for a polygenic multiple-loci effect. How environmental factors and infective agents interact with this is uncertain. Undoubtedly the association between the susceptibility genes and clinical outcome remains a major challenge to present methods of analysis. With evidence that careful control of diabetes can reduce complications (Diabetes Control and Complications Research Group, 1993) it may in future be possible to detect which patients are most at risk of complications.

24.6 Congenital malformation (mostly sporadic)

24.6.1 Clinical aspects

The clinical burden on health care professionals from congenital malformation is likely to rise in the forthcoming years. Presently congenital malformation is the commonest cause of childhood death between 1 and 5 years of age and the second commonest cause of death in the first year of life in the UK. Estimates of the birth incidence of congenital malformations are as high as 5% (Baird *et al.*, 1988).

Parents expect perfect children and the thought of possible foetal abnormalities is often sidelined in the preparation for a normal pregnancy and birth. The neonatologist is frequently faced with the highly emotive situation of dealing with questions from parents of a child with a birth defect that may sometimes be life threatening and require operative intervention. The questions that all parents ask are why? and how? For some defects classical embryology will suffice as an explanation but for other defects this is not so. Some malformations can be explained by chromosomal, environmental, viral, chemical or multiple factors but in most instances they arise sporadically. Recent years have witnessed major advances in identification and characterization of genes specifically involved in the process of vertebrate embryonic organization. Wilkie, Amberger and McKusick (1994) have identified 139 disease loci and 65 genes implicated in the genesis of malformation. Exciting avenues for research are therefore developing. It is virtually certain some malformations will arise from the malexpression of genes involved in the processes of embryonic patterning, regionalization and organogenesis.

24.6.2 Genetics

Early developmental experiments, on drosophila, identified the presence of genes encoding positional information in developing embryos. Since, workers have elucidated the presence of such homeobox (*HOX*) genes in higher animals. In humans, specific *HOX*-containing sequences are found in four clusters: *HOXA* (chromosome 7); *HOXB* (chromosome 17); *HOXC* (chromosome 12) and *HOXD* (chromosome 2) (Glover and Hames, 1990).

Over forty HOX motif-containing genes have been identified – some lie outside the clusters. Genes of the *PAX* (paired homeobox) family and the genes *MSX1* and *MSX2* appear to be more likely candidates for human disease. Current work is still concentrating on normal domains of expression of *HOX*-containing genes in the developing embryo using RNA *in situ* hybridization. Extensive work on *HOX* has identified ordered specific *HOX* expression in limbs (Dolle *et al.*, 1989; Davidson *et al.*, 1992), face and skull (Balling *et al.*, 1991; Chisaka and Capechi, 1991) and central nervous system (Morris-Kay *et al.*, 1991). In the cases of limbs and facial skeleton specific alteration of *HOX* expression either via retinoic acid implantation or targeted mutagenesis can lead to predictable *HOX* transcript maldistribution and malformation.

The techniques of targeted/insertional mutagenesis have led to mouse models with consistent malformations, revealing several candidates for such conditions. Recently *MSX1* (HOX 7) has been implicated in palatal and tooth anomalies (Satokata and Maas, 1994). Developmentally this gene is expressed at the interface of ectoderm and mesoderm (implicated in the development of both structures). Genes of the *PAX* type have already been implicated in human disease.

Mutations of *PAX3* have been identified as the cause of Waardenburg's syndrome (Tassabehji *et al.*, 1992), an autosomal dominant condition combining pigmentary anomalies with deafness and occasional neural tube defects. *PAX6* has been implicated in Peters' anomaly of the eye and diseases affecting anterior eye structures (Glaser *et al.*, 1994; Hanson *et al.*, 1994). Additionally, mutations of *PAX3* have been identified in the striated muscle tumour of childhood, alveolar rhabdomyosarcoma – underpinning the role of genes of oncogenic potential in normal development (Galili *et al.*, 1993). Many such experimental malformations are predictable from our knowledge of *HOX* expression in developing tissues. *HOX* genes simply represent a fraction of those involved in the developmental process. Recently mutations of the fibroblast growth factor receptor genes 1 and 2 have been implicated in Aperts' syndrome (Wilkie *et al.*, 1995), Pfieffers' (Muenke *et al.*, 1994) and Crouzons' syndromes (Rutland *et al.*, 1995) – abnormalities affecting craniofacial skeleton and limbs. In addition, other 'developmental' loci must also be involved in regionalization.

There is evidence for a complex interaction between *HOXD* and *HOXA* loci in the developing limb bud where interaction between *MSX1* and *MSX2* is responsible for maintaining a pool of undifferentiated cells at the tip of the growing bud. Several regions of the long arm of chromosome 7 (where currently no *HOX* genes have been localized) have been implicated in human digital pattern anomalies (Buss, 1994; Heutink *et al.*, 1994; Tsukurov *et al.*, 1994).

24.6.3 Clinical implications

These are as yet unclear. As more specific data on roles of these genes in normal development accrues, counselling can only become more precise.

Occasionally malformations (i.e. cardiovascular) are inherited and in these families predictive testing may become possible. The identification of a mal-expressed gene is simply the beginning with regards to the development of interventional programmes. It is heartening for clinicians dealing with the consequences of congenital malformation to see that in many areas basic scientists and clinicians are exchanging expertise in order to identify potential candidate loci (Winter and Tickle, 1993).

References

Adam, S., Wiggins, S. and Whyte, P. (1993) Five year study of prenatal testing for Huntingtons disease, attitudes and psychological assessment. *J. Med. Genet.* **30**, 549–56.

Al-Jader, L., Merdedith, L., Ryley, H. C. *et al.* (1992) Severity of chest disease in cystic fibrosis patients in relation to their genotype. *J. Med. Genet.* **29**, 883–7.

Baird, P. A., Anderson, T. W., Newcombe, H. B. *et al.* (1988) Genetic disorders in children and young adults: a population study. *Am. J. Hum. Genet.* **42**, 677–93.

Balfour-Lynn, I., Madge, S. and Dinwiddie, R. (1985) Testing carrier status in siblings with cystic fibrosis. *Arch. Dis. Child.* **72**, 167–8.

Balling, R., Mutter, G., Gruss, P. and Kessel, M. (1991) Craniofacial abnormalities induced by ectopic expression of the gene Hox 1.1 in transgenic mice. *Genes Devel.* **5**, 1167–77.

Benjamin, C. M., Adam, S., Wiggins, S. *et al.* (1994) Proceed with care: direct predictive testing for Huntingtons disease *Am. J. Hum. Genet.* **55**, 606–17.

Bennett, S. T., Lucassen, A. M., Gough, S. C. L. *et al.* (1995) Susceptibility to human type 1 IDDM at IDDM2 is determined by tandem repeat variation at the insulin gene microsatellite locus. *Nature Genet.* **9**, 284–91.

Borgo, G., Fabiano, T., Perobelli, T. *et al.* (1992) Effect of introducing prenatal diagnosis on the reproductive behaviour of families at risk for cystic fibrosis. *Prenat. Diag.* **12**, 821–30.

Buss, P. W. (1994) Cleft hand/foot: clinical and developmental aspects. *J. Med. Genet.* **31**, 726–30.

Caplen, M. J., Alton, E. W. F., Middleton, P. G. *et al.* (1995) Liposome mediated CFTR transfer to the nasal epithelium of patients with cystic fibrosis. *Nature Med.* **1**, 39–44.

Chisaka, O. and Capechi, M. (1991) Regionally restricted defects from targeted disruption of mouse Hox 1.5 homeobox gene. *Nature* **350**, 433–9.

Clinical Genetic Society (UK) (1994) The genetic testing of children. Report of the working party. *J. Med. Genet.* **31**, 785–97.

Coutelle, C., Caplan, N., Hart, S. *et al.* (1993) Gene therapy for cystic fibrosis. *Arch. Dis. Child.* **68**, 437–43.

Craufurd, D. and Dodge, A. (1993) Mutation size and age of onset in Huntingtons disease. *J. Med. Genet.* **30**, 1001–11.

Crystal, R. G., McEvaney, N. G., Rosenfeld, M. A. *et al.* (1994) Administration of an adenovirus containing the human CFTR cDNA to the respiratory tract of individuals with cystic fibrosis. *Nature Genet.* **8**, 42–7.

Davidson, D. R., Crawley, A., Hill, R. E. and Tickle, C. (1992) Position dependent expression of two related homeobox genes in developing vertebrate limbs. *Nature* **352**, 429–31.

Davies, J. L., Karaguch, Y., Bennett, S. T. *et al*. (1994) A genome wide search for human type 1 diabetes susceptibility genes. *Nature* **371**, 130–6.

Decruyneare, M., Evere-Kiebooms, G. and Van den Berghe, H. (1993) Perception of predictive testing for Huntingtons disease in young women preferring uncertainty to certainty. *J. Med. Genet.* **30**, 557–61.

Diabetes Control and Complications Trial Research Group (1993) The effect of long term intensive treatment on the development and progression of long term complications in insulin dependent diabetes mellitus. *N. Eng. J. Med.* **329**, 977–85.

Dolle, P. and Ispisua-Belmonte, J. C. (1989) Coordinate expression of the murine Hox 5 complex homeobox containing genes during limb pattern formation. *Nature* **342**, 767–72.

Field, L. L., Tobias, R. and Magnus, T. (1994) A locus on Chromosome 15q26 (IDDM3) produces susceptibility to IDDM. *Nature Genet.* **8**, 189–93.

Galili, N., Davis, R. J., Fredericks, W. J. *et al*. (1993) Fusion of a fork head domain gene to PAX3 in the solid tumour alveolar rhabdomyosarcoma. *Nature Genet.* **5**, 230–5.

Glaser, T., Jepeal, L., Edwards, J. *et al*. (1994) PAX 6 gene dosage effect in a family with congenital cataracts, aniridia, anophthalmia and central nervous system defects. *Nature Genet.* **7**, 463–7.

Glover, D. M. and Hames, B. D. (1990) *Genes and embryos*. IRL/Oxford University Press, Oxford.

Goodfellow, P. N. and Schmitt, K. (1994) From the simple to the complex. *Nature* **371**, 104–5.

Green, M. R., Weaver, L. T., Heeley, A. F. *et al*. (1993) Cystic fibrosis detected by neonatal screening, incidence, genotype and early natural history. *Arch. Dis. Child.* **68**, 464–8.

Hanson, I. M., Fletcher, J. M., Jordan, T. *et al*. (1994) Mutations at the PAX6 locus are found in the heterogenous anterior segment malformations including Peters anomaly. *Nature Genet.* **6**, 168–72.

Harper, P. S. (1991) *Huntingdons disease* W. B. Saunders, London.

Harper, P. S. (1993) A specific mutation for Huntingtons disease. *J. Med. Genet.* **30**, 975–7.

Hashimoto, L., Habita, L., Beressi, J. P. *et al*. (1994) Genetic mapping of a susceptibility locus for insulin dependent diabetes mellitus on chromosome 11q. *Nature* **371**, 161–4.

Heutink, P., Zguricas, J., Oosterhout, L. V. *et al*. (1994) The gene for triphalangeal thumb maps to the subtelomeric region of chromosome 7q. *Nature Genet.* **6**, 287–92.

Huntingtons disease collaborative research group. (1993) A novel gene containing a trinucleotide repeat that is expanded and unstable near the Huntingtons disease chromosome. *Cell*, **72**, 971–83.

Karien, E. D. R., Pia, A. M. K. G., Mette, I. S. *et al*. (1993) Dynamic mutation in Dutch Huntingtons disease patients. Increased paternal repeat initially extending within the normal size range. *J. Med. Genet.* **30**, 996–1002.

Kerem, B. S., Rommens, J. M., Buchanan, J. A. *et al*. (1989) Identification of the CF gene; genetic analysis. *Science* **245**, 1070–80.

Kremen, B., Squitieri, F., Telenius, H. *et al*. (1993) Molecular analysis of late onset Huntingtons disease. *J. Med Genet.* **30**, 991–5.

Lucassen, A. M., Julier, C., Beressi, J. P. *et al*. (1993) Susceptibility to IDDM maps to a 4.1 kb segment of DNA spanning the insulin gene and associated variable number tandem repeat. *Nature Genet.* **4**, 305–10.

Mennie, M. E., Compton, M., Gilfillan, A. *et al*. (1993) Prenatal screening for cystic fibrosis. Psychological effects on carriers and their partners. *J. Med. Genet*. **30**, 543–8.

Miedzybrodzka, Z. H., Yin Z., Kelly, K. F. *et al*. (1994) Evaluation of laboratory methods for cystic fibrosis carrier screening: reliability, sensitivity, specificity and costs. *J. Med. Genet*. **31**, 545–50.

Morris-Kay, G., Murphy, P., Hill, R. E. and Davidson, D. R. (1991) Effects of retinoic acid on Hox 2.9 and krox 20 expression in the hindbrain of mouse *EMBO J*. **10**, 2981–99.

Muenke, M., Schell, U., Hehr, A. *et al*. (1994) A common mutation in the FGFR1 gene in Pfieffer syndrome. *Nature Genet*. **8**, 269–73.

Raeburn, J. A. and Marteau, T. (1994) Screening for carriers of cystic fibrosis. *Br. Med. J*. **309**, 1428–30.

Rich, D. P., Anderson, M. P., Gregory, R. J. *et al*. (1990) Expression of cystic fibrosis transmembrane conductance regulator corrects defects in chloride channel regulation in cystic fibrosis airway epithelial cells. *Nature* **347**, 358–63.

Rommens, J. M., Ianuzzi, M. C., Kerem, B. S. *et al*. (1989) Identification of the cystic fibrosis genes; chromosome walking and jumping. *Science* **245**, 1059–65.

Rutland, P., Pullyn, L. J., Reardon, W. *et al*. (1995) Identical mutations in the FGFR2 gene cause both Pfieffer and Crouzon syndrome phenotypes. *Nature Genet*. **9**, 173–8.

Satokata, I. and Maas, R. (1994) MSX1 deficient mice exhibit cleft palate and abnormalities of craniofacial and tooth development. *Nature Genet*. **6**, 348–50.

Slaney, S. F., Wilkie, A. O. M., Hirst, M. C. *et al*. (1995) DNA testing for Fragile X syndrome in school for learning difficulties. *Arch. Dis. Child*. **72**, 33–7.

Sutherland, G. R. and Richards, R. I. (1993) Dynamic mutations on the move. *J. Med. Genet*. **30**, 978–81.

Tassabehji, M., Read, A. P., Newton, V. E. *et al*. (1992) Waardenburgs syndrome. Patients have a mutation in the human homologue of the PAX 3 paired homeobox gene. *Nature* **355**, 635–6.

Tsukurov, O., Boehmer, A., Flynn, J. *et al*. (1994) A complex bilateral disease locus maps to chromosome 7q36. *Nature Genet*. **6**, 282–6.

Turk, J. (1995) Fragile X syndrome. *Arch. Dis. Child*. **72**, 3–5.

Verkerk, A. J., Pieretti, M., Sutcliffe, J. S. *et al*. (1991) Identification of a gene (FMR1) containing a CGG repeat coinciding with a breakpoint cluster region exhibiting length variation in fragile X syndrome. *Cell* **65**, 905–14.

Watson, E. K., Mayell, E., Chapple, J. *et al*. (1991) Screening for carriers of cystic fibrosis through the primary health care services. *Br. Med. J*. **303**, 304–7.

Webb, T. (1991) The epidemiology of fragile X syndrome. In Davies KE: In *The fragile X syndrome* (ed. K. E. Davies), pp. 40–54. Oxford University Press, Oxford.

Wilkie, A. O. M., Amberger, J. S. and McKusick, V. A. (1994) A gene map of congenital malformations. *J. Med. Genet*. **31**, 507–17.

Wilkie, A. O. M., Slaney, S. F., Oldridge, M. *et al*. (1995) Apert syndrome results from localised mutations of FGFR2 and is allellic with Crouzons syndrome. *Nature Genet*. **9**, 165–72.

Williamson, R. (1993) Universal community carrier screening for cystic fibrosis. *Nature Genet*. **3**, 195–9.

Winter, R. and Tickle, C. (1993) Syndactylies and polydactylies – embryological overview and suggested classification. *Eur. J Hum. Genet*. **1**, 96–104.

25

Oncology of solid tumours: future directions for clinical research

JANUSZ A. Z. JANKOWSKI

25.1 Introduction to molecular management strategies of cancer

Over the next half decade the impact of molecular biology on clinical medicine will increase dramatically. The changes in basic science technology such as improved automation, cheaper machines and more reliable computer software will enable the rapid application of techniques such as multiplex PCR. This PCR-based technique can identify several subtle genetic errors, as occur in most epithelial cancers, increase the accuracy of prognostic evaluation and improve the efficacy of therapy. The examples given below are of gastro-intestinal tumours but none the less they are applicable to many solid epithelial tumours.

25.2 Diagnostic assessment

25.2.1 Surveillance of colonic cancer

Many centres advocate surveillance programmes for patients with a previous history of polyps and ulcerative colitis or a positive family history of colon cancer. At present diagnosis is unsatisfactory because current methods are inaccurate; assessments of faecal occult blood (FOBs) lack specificity and have poor sensitivity. Colonoscopy is highly sensitive and specific for established colonic cancers but when screening premalignant lesions the augmentation of histological analysis of pinch biopsies is mandatory. Histological analysis has much variability in the classification of colonic dysplasia. At present there are no alternatives as other biomarkers are inadequate, including DNA aneu-ploidy, glycoproteins such as increased CEA (carcinoembryonic antigen) serum levels (> 10 ng/ml), APC (Cawkwell et al., 1993) and c-Ki-ras mutations (Meltzer, Mane and Wood, 1990) as they offer little if any advantage over conventional assessment. The identification of Ki-ras mutations in colorectal cytology was, however, able to show that molecular techniques have a place in aiding such surveillance techniques by identifying minute transformed cell populations in faecal samples (Sidransky et al., 1992).

25.2.2 Surveillance of oesophageal cancer

Recently our group has been able to adapt the technique of RT-PCR to assess the growth factor receptor c-erbB3 in Barrett's mucosa of the oesophagus. Barrett's oesophagus represents one of the most challenging enigmas in clinical oncology. Much has been learnt about the pathophysiology, conventional surveillance and therapeutic regimens in recent years but despite this patient survival has not improved substantially (Jankowski and McGarrity, 1993). The reasons for this are partly due to the heterogeneous nature of Barrett's metaplasia and the absence of one outstanding curative therapy (Jankowski *et al.*, 1994). Contemporary surveillance techniques such as the use of histological biopsies can at best discriminate between high-grade dysplasia and invasive carcinoma only if multiple biopsies are taken from each patient, as has been suggested by a recent publication (Levine *et al.*, 1994).

The use of molecular markers to diagnose oesophageal malignancy has increased over the last 5 years. In particular, the detection of multiple p53 missence mutations in exons 5–9 and insertional or deletional frame shifts has been reported throughout different stages of oesophageal tumourigenesis and in 50% of epithelial cancers (Neshat *et al.*, 1994, Hamelin *et al.*, 1994). Abnormalities of differentiation, however, also characterize the earlier stages of Barrett's tumourigenesis (Jankowski *et al.*, 1995) and in particular the tyrosine kinase gene product. RT-PCR confirmed that expression of c-erbB3 mRNA (Jankowski *et al.*, 1995) in histological biopsies from Barrett's mucosa is moderately strong and that expression in dysplastic specimens and invasive neoplasia decreases steadily, becoming absent in the most poorly differentiated lesions. Application of RT-PCR analysis indicates that oesophageal mucosa from dysplastic mucosa expresses less c-erbB3 compared with non-dysplastic metaplasia. We applied this technique of RT-PCR to analyse oesophageal cytology. Brushings were taken in 1 cm sweeps from metaplastic epithelium during endoscopy and c-erbB3 was found to be reduced or absent in cytology specimens corresponding with histological features of intestinal metaplasia and dysplasia, respectively (Jankowski *et al.*, 1995) (see Chapter 11).

25.2.3 Surveillance of cancer using peripheral blood lymphocytes

Other methods of molecular analyses can be used to assess inherited genomic mutations in cancer family syndromes by screening peripheral blood white blood cells. This type of analysis can be used for surveillance of many cancer family syndromes such as the MSH2 gene in hereditary non polyposis colorectal cancer and the APC gene in familial adenomatous polyposis. The PCR techniques of 'random errors of replication' and single-stranded conformational polymorphisms can be used for screening purposes on peripheral blood cells (see Chapter 13).

25.3 Prognostic assessment

25.3.1 *Staging of colorectal cancer*

The Dukes staging and degree of tubular differentiation have been noted to be independent prognostic variables in colorectal cancer (Jass, Strudley and Faludy, 1984). Oncogenes associated with deregulated proliferation such as c-ras, p53 and multiple allele specific analysis (MASA) have also been used to make prognostic assessment in primary tumours and neighbouring lymph nodes respectively (Bell *et al.*, 1993; Hayashi *et al.*, 1993). In oesophageal cancer the EGFR and c-erbB2 genes are amplified (Al-Kasspooles *et al.*, 1993; Blount *et al.*, 1994) and it has been indicated that these are independent prognostic variables along with depth of invasion and distant metastases (Nakamura *et al.*, 1994).

25.4 Therapeutic intervention

Surgery is still the best curative therapy but for inoperable gastrointestinal tumours 5-fluorouracil (5-FU) and vincristine chemotherapy may be beneficial in a few cases. Molecular pharmacology can specifically modulate the aberrant biological processes in cancer, for example it has been shown that both retinoic acid and tamoxifen may both improve and decrease the incidence of metastases in breast cancer by increasing cell adhesion as a result of upregulated E-cadherin expression (Caceres *et al.*, 1994). Other aspects of molecular pharmacology include antibody-directed enzyme prodrug therapy (ADEPT) or viral-directed enzyme prodrug therapy (VDEPT).

In general terms, gene therapy can be used to alter the genome of the germ cells or somatic cells and this has been advocated for the management of some congenital or inherited diseases (Ledley, 1993) but its role in cancer therapy is much less clear (Sikora, 1994). Many of the reasons for this are because there are amplifications of oncogenes, multiple genetic defects and dominant-negative gene defects in most cancers. In cases where single genes are mutated and their expression is lost, such as occurs with the DCC gene in colorectal tumourigenesis, gene augmentation strategies are possible. Gene replacement techniques would be necessary, however, for the complete replacement of truncated or dysfunctional proteins, such as occurs with the oncosuppressor p53 gene and the cell–cell adhesion molecule E-cadherin, respectively. Furthermore, getting genes into cells *in vivo* requires more efficient transfection techniques than used *in vitro*, such as calcium phosphate transfection, liposomal encapsulation and electroporation (Sikora, 1994; Cohen, 1989). Viral vectors are more efficient in this regard but both small retroviral vectors and larger adenoviruses may have limitations such as lack of tissue tropism or immunogenicity. As a result, gene therapy applications that do not require homologous recombination to suppress endogenous genes or extremely high expression levels seem more feasible with current technology. In this regard immunotherapy relies on the introduction of immunostimulatory genes into cancer cells and boosting the systemic immune reactions against the unmodified cells (Helene, 1994; Hock *et al.*, 1994), thereby enhancing the host's natural immune mechanisms against cancer cells. This can be achieved either by the use of viral vectors such as MMLV retroviruses, adenoviruses or nonviral receptor-mediated endocytosis.

25.5 Summary and future prospects

The amplification of oncogenes such as c-*ras*, viral genomes such as Epstein–Barr and other biomarkers of differentiation in fine needle aspirates or brush cytology is already being utilized routinely to aid in early diagnosis and prognostic assessment of some epithelial tumours. The issue of gene therapy is discussed in detail in another section (see Chapter 27).

In addition, improved experimental model systems will allow the study of these mutant genes with their wild-type counterparts. One enormous advance is the Cre-loxP transgenic system, where a gene of interest can be flanked by viral sequences allowing it to remain dormant in embryonic stem cells (Barinaga, 1994). Once the animal has the normal adult phenotype the gene can be activated by a tissue-specific promotor allowing a perfect reproduction of the human disease in the host animal.

On a global scale, the human genome project still represents the most likely route by which many diseases will have genetic mutations identified. At present the main thrust is to search segments of DNA for linkages between known DNA markers and a genetic disorder from which the associated gene can be isolated and characterized. This method will particularly benefit single-gene defects but an indirect benefit has been the improved gene bank facilities of human, mice and eukaryotic DNA.

In the clinical arena, applications are feasible such as the identification of illegitimate transcription of 'foreign' genetic transcripts in peripheral lymphocytes circulating past diseased cells. At present this phenomenon has been identified in blood samples from patients with hypertrophic cardiomyopathy. This technique clearly has several applications to the screening and diagnosis of premalignancy and many sporadic human cancers.

Perhaps the topic that is the paradigm of molecular medicine is that of gene therapy and therapeutic manipulation strategies. At present there are many potential problems in applying these techniques to the multigene aberrations that occur in most human cancers *in vivo*. It seems clear, however, that this approach will allow a wider therapeutic strategy to be devised for many cancers in the future.

References

Al-Kasspooles, M., Moore, J. H., Orringer, M. B. and Beer, D. G. (1993) Amplification and over-expression of the EGFR and c-erbB2 genes in human esophageal adenocarcinoma. *Int. J. Cancer* **54**, 213–9.

Barinaga, M. (1994) Knockout mice: round two. *Science* **265**, 26–8.

Bell, S. M., Scott, N., Cross, D. *et al.* (1993) Prognostic value of p53 overexpression and c-Ki-ras gene mutations in colorectal cancer. *Gastroenterol.* **104**, 57–64.

Blount, P. L., Galipeau, P. C., Sanchez, C. A. *et al.* (1994) 17p allelic losses in diploid cells of patients with Barrett's esophagus who develop aneuploidy. *Cancer Res.* **54**, 2292–5.

Caceres, J. F., Stamm, S., Helfman, D. M. and Krainer, A. R. (1994) Regulation of alternative splicing in vivo by overrepression of antagonist splicing factors. *Science* **265**, 1706–9.

Cawkwell, L., Bell, S. M., Lewis, F. A. *et al.* (1993) Rapid detection of allele loss in colorectal tumours using microsatellites and fluorescent DNA technology. *Br. J. Cancer* **67**, 1262–7.

Cohen, J. S. (1989) Oligodeoxynucleotides: antisense inhibitors of gene expression. Macmillan, London.

Fearon, E. and Vogelstein, B. (1990) A genetic model for colorectal tumorigenesis. *Cell* **61**, 759–67.

Hamelin, R., Flejou, J. F., Muzeau, F. *et al.* (1994) TP53 gene mutations and p53 protein immunoreactivity in malignant and premalignant Barrett's esophagus. *Gastroenterol.* **107**, 1012–8.

Hayashi, N., Arakawa, H., Nagase, H. *et al.* (1993) Genetic diagnosis identifies occult lymph-node metastases undetectable by the histopathological method. *Cancer Res.* **54**, 3853–6.

Helene, C. (1994) The anti-gene strategy: control of gene expression by triplex-forming oligonucleotides. *Anticancer Drug Design* **6**, 569–84.

Hock, H., Dorsch, M., Richter, G. *et al.* (1994) Tumor-cell-targeted cytokine gene transfer in experimental models for cancer therapy. *Natural Immun.* **13**, 85–92.

Jankowski, J., Hopwood, D., Hurst, H. and Wright, N. A. (1995) Molecular surveillance of Barrett's esophagus and putative strategies for genetic therapy. *Dis. Esophagus* **8**, 113–18.

Jankowski, J. and McGarrity, T. (1993) Screening of alimentary cancer: outstanding problems and a strategy for research. *Eur. J. Cancer Prevent.* **2**, 211–14.

Jankowski, J., Newham, P., Kandemir, O. *et al.* (1994) Differential expression of E-cadherin in normal, metaplastic and dysplastic oesophageal mucosa: a putative biomarker. *Int. J. Oncol.* **4**, 441–8.

Jass, J. R., Strudley, I. and Faludy, J. (1984) Histochemistry in epithelial metaplasia and dysplasia in human stomach and colorectum. In: *Gastrointestinal carcinogenesis* (ed. J. M. Polak, S. R. Bloom, N. A. Wright and A. G. Butler), pp. 109–30. *Scand. J. Gastroenterol.*, Universitetsforlaget, Oslo.

Ledley, F. D. (1993) Prenatal application of somatic gene therapy. *Ob. Gyn. Clinics North Am.* **20**, 611–20.

Levine, D. S., Haggitt, R. C., Blount, P. L. *et al.* (1994) An endoscopic biopsy protocol can differentiate high-grade dysplasia from early adenocarcinoma in Barrett's esophagus. *Gastroenterol.* **105**, 40–50.

Meltzer, S. J., Mane, S. M. and Wood, T. (1990) Activation of Ki-ras in human gastrointestinal dysplasias determined by direct sequencing of polymerase chain reaction products. *Cancer Res.* **50**, 3627–30.

Nakamura, T., Nekarda, H., Hoelscher, A. H. *et al.* (1994) Prognostic value of DNA ploidy and c-erbB2 oncoprotein in adenocarcinoma of Barrett's esophagus. *Cancer* **73**, 1785–94.

Neshat, K., Sanchez, C. A., Galipeau, P. C. *et al.* (1994) p53 mutations in Barrett's adenocarcinoma and high grade dysplasia. *Gastroenterol.* **106**, 1589–95.

Sidransky, D., Tokino, T., Hamilton, S. R. *et al.* (1992) Identification of ras oncogene mutations in the stool of patients with curable colorectal tumors. *Science* **256**, 102–5.

Sikora, K. (1994) Genes, deams and cancer. *Br. Med. J.* **308**, 1217–21.

26

Clinical applications of molecular biology to haematological malignancies

SUSAN RIDGE

26.1 Introduction

Numerous clonal cytogenetic lesions, such as the loss or gain of specific chromosomes and chromosomal translocations, have been detected in haematological tumours. This has provided useful information for diagnosis and prognosis. More recently, cytogenetic methods have been complemented by the use of molecular techniques such as Southern blotting and the polymerase chain reaction (PCR). These methods have provided more sensitive and accurate ways of detecting the presence of the specific aberrations including gene translocations and point mutations in oncogenes such as *ras* and *p53*. Molecular techniques have also provided extremely sensitive methods for monitoring minimal residual disease following treatment.

26.2 Gene translocations

There are numerous examples of clonal pathological gene rearrangements in leukaemias and lymphomas (Rabbitts, 1994). Examples include the t(9;22), which results in the fusion of the *ABL* gene on chromosome 9 with the *BCR* gene on chromosome 22 and translocations involving the *MLL* gene on chromosome 11q23. PCR can be used for the detection of these and other translocations. The most common starting template for these reactions is cDNA (RT-PCR), as in most cases the breakpoints are spread over too large an area for PCR to be carried out using genomic DNA as a template.

26.2.1 Translocations involving MLL

Translocations of *MLL* are detectable in a number of different types of leukaemia, but are especially common in infants with acute lymphoblastic leukaemia (ALL) (reviewed by Ridge and Wiedemann, 1994). Although the t(4;11), t(11;19) and t(9;11) are the most commonly detected reciprocal translocations, many other variants involving other partner genes are also found. As the breakpoints on chromosome 11q23 are clustered within a small genomic region, translocations can be detected using Southern blotting, if a sufficient sample of DNA is available. This method has been shown to detect the presence of translocations not detected by conventional cytogenetics (Chen *et al.*, 1993). Southern blotting was also used in a study that demonstrated that the *MLL* rearrangements present in infants with leukaemia can occur *in utero* (Ford *et al.*, 1993).

To detect molecularly which particular reciprocal translocation is present, however, it is necessary to use an RT-PCR procedure using a number of different primers that can detect, for example, the t(4;11), t(11;19), t(9;11) and t(6;11). In one example, Repp *et al.* (1994) used primers for *MLL* and primers for the four different reciprocal genes *AF4*, *ENL*, *AF9* and *AF6* in a multiplex RT-PCR protocol. It was possible to identify which translocation was present from the size of the resulting PCR product.

The accurate identification of patients with *MLL* rearrangements is important, as it has prognostic implications (Chen *et al.*, 1993). In infant ALL, it has been demonstrated that patients with a germline *MLL* gene have an 80% event-free survival at 46 months compared with a 15% event-free survival at 46 months for those with a rearranged *MLL* gene. It may therefore be appropriate to treat the latter patients with more intensive therapy.

The subsequent monitoring of patients after therapy for minimal residual disease can also provide useful information. In a recent study seven patients in complete remission after treatment for null ALL with a t(4;11), were studied using RT-PCR for the presence of minimal residual disease (Janssen *et al.*, 1994). In no case were the MLL-AF4 fusion transcripts detected, suggesting that the leukaemic cells had largely been eradicated. Continued follow up of these patients may help predict the onset of relapse.

26.2.2 Translocations of BCR/ABL

The Philadelphia chromosome resulting from the translocation t(9;22) is visible cytogenetically in most cases of chronic myeloid leukaemia (CML) and in some cases of acute leukaemia. The translocation results in the fusion between the *ABL* gene on chromosome 9 and the *BCR* gene on chromosome 22. With the use of molecular techniques to detect these rearrangements, the frequency of Philadelphia positive ALL is now known to be higher than originally thought (Westbrook *et al.*, 1992). As the presence of this translocation is also associated with a poor prognosis in adult and childhood ALL, it is important to be able to identify its presence, as this can allow selection of patients for more intensive therapy, such as allogeneic bone marrow transplantation (BMT).

A number of groups have used RT-PCR for the detection of the t(9;22), in the study of minimal residual disease following allogeneic BMT as treatment for CML. Guerrassio *et al.* (1992) demonstrated that although *BCR/ABL* fusion transcripts can be commonly detected early after transplant, the presence of these transcripts was not associated with a higher relapse risk and most of the patients eventually became PCR negative. In another study, it has been suggested that the presence of *BCR/ABL* transcripts at a later stage (more than 6 months) post BMT identifies a group of patients at increased risk of relapse (Hughes *et al.*, 1991). This information can prove useful in the selection of patients for further clinical intervention (see below).

26.3 Point mutations

Unlike chromosomal translocations, which are detectable using cytogenetic methods, the detection of point mutations in oncogenes such as *ras* and *p53* was not possible before the advent of molecular techniques, especially PCR. These studies have provided valuable information on the molecular hetero-geneity present in patients with phenotypically similar diseases.

26.3.1 Mutations in the ras genes

Point mutations in *ras* genes have been detected in a wide range of haemo-poietic malignancies, with the highest frequency detected in patients with the preleukaemic syndrome of myelodysplasia (MDS) (reviewed by Carter, Ridge and Padua 1992). Some groups have reported that the presence of a *ras* mutation in MDS patients is a poor prognostic indicator associated with a reduction in survival and/or an increased chance of progression to overt leukaemia.

The detailed study of patients with *ras* mutations has shown that in some cases the mutations have occurred at an early stage of the disease in a multipotential cell and are present in a number of different lineages in the bone marrow and peripheral blood. In other patients, the mutations appear to have occurred later in the evolution of the disease. Indeed, in some patients more than one *ras* mutation has been detected, segregating in different clones (Bashey *et al.*, 1992). This offers clear evidence that these mutations have occurred subsequent to the initiating leukaemic lesion.

26.3.2 Mutations in the p53 gene

Mutations in the tumour suppressor gene, *p53*, which is located on chromo-some 17p, have been detected in patients with a variety of haematological malignancies (reviewed by Imamura, Miyoshi and Koeffler, 1994). The fre-quency of these mutations is low in patients with MDS, acute myeloid leukaemia (AML), common ALL, chronic lymphocyte leukaemia and chronic phase CML. A much higher frequency of mutations is present in blast crisis CML and high grade lymphomas. The presence of a *p53* mutation is associated with disease progression. Examples of this are in the evolution from chronic phase CML to blast crisis CML and in the progression of MDS to AML. *p53* mutations are also more common in patients with relapsed B or T cell ALL.

26.4 Physiological gene rearrangements

In addition to the detection of pathological gene rearrangements, it has also been possible to use the physiological rearrangement of the antigen receptor genes as a marker of clonality in lymphoid cells. Both the immunoglobulin and T cell receptor genes undergo physiological rearrangement in B or T cells respectively, and this results in the generation of the hugh diversity of antigen receptors required by the immune system. The detection of these gene re-arrangements by PCR can be used to distinguish between a monoclonal and polyclonal population of cells (reviewed by McCarthy and Wiedemann, 1995). In the presence of a clonal population of cells, derived from a single cell with a specific antigen receptor rearrangement, a single PCR band should be de-tected. In a polyclonal population of cells, amplification of the many antigen receptor rearrangements that are present will be visualized as a smear, when analysed using gel electrophoresis. This method has been used primarily for the analysis of minimal residual disease following therapy.

26.5 Clinical intervention

Although there is a large accumulation of molecular data from the studies of patients at diagnosis, remission and relapse, there are very few examples, so far, of this information being used for clinical intervention.

One recently published example described the use of RT-PCR, along with conventional cytogenetics and haematological evaluation, to identify patients at risk of relapse after BMT for CML (van Rhee *et al.*, 1994). Patients with molecular, cytogenetic or haematological indicators of relapse were treated with transfusions of leukocytes from the original bone marrow donor as it is believed that these lymphocytes have an anti-leukaemic effect. In those patients (7/7) with molecular or cytogenetic indicators of relapse molecular remission with no BCR/ABL transcripts was achieved after leukocyte trans-fusions. Only 3/7 (43%) of those with haematological relapse responded. Thus the ability to carry out therapeutic intervention at an early stage of relapse may be clinically beneficial.

26.6 Conclusions

Some examples of the application of molecular techniques in the study of haematological malignancies have been described. These studies have helped increase our understanding of what lesions are present and have provided useful information regarding diagnosis and prognosis. In addition to the techniques described, advances in the use of fluorescent *in situ* hybridization (FISH) for detecting genetic aberrations in interphase and metaphase cells, has proved clinically informative (reviewed by Price, 1993). It is possible to use the FISH detection of clone-specific markers such as trisomy of chromosome 8, in combination with immunophenotype analysis to study which cell lineages are involved in disease (Price *et al.*, 1992).

The molecular identification of the genes involved in the generation of the leukaemic phenotype will provide valuable information about the normal and abnormal functions of these genes. It is hoped that this information will ultimately allow the design of novel more specific therapeutic approaches.

Acknowledgements

I would like to thank Drs Cathy Price and Graham Carter for their helpful comments. The author is funded by the Leukaemia Research Fund of Great Britain.

References

Bashey, A., Gill, R., Levi, S. *et al.* (1992) Mutational activation of the N-ras oncogene assessed in primary clonogenic culture of acute myeloid leukemia; implications for the role of N-ras mutation in AML pathogenesis. *Blood* **79**, 981–9.

Carter, G., Ridge, S. and Padua, R. A. (1992) Genetic lesions in Preleukemia. *Crit. Rev. Oncogenesis* **3**, 339–64.

Chen, C. S., Sorenson, P. H. B., Domer, P. H. *et al.* (1993) Molecular rearrangements on chromosome 11q23 predominate in infant acute lymphoblastic leukemia and are associated with specific biologic variables and poor outcome. *Blood* **81**, 2380–3.

Ford, A. M., Ridge, S. A., Cabrera, M. E. *et al.* (1993) *In utero* rearrangements in the trithorax-related oncogene in infant leukaemias. *Nature* **363**, 358–60.

Guerrassio, A., Martinelli, G., Saglio, G. *et al.* (1992) Minimal residual disease status in transplanted chronic myelogenous leukemia patients: low incidence of polymerase chain reaction positive cases among 48 long disease free subjects who received unmanipulated allogeneic bone marrow transplants. *Leukemia* **6**, 507–17.

Hughes, T. P., Morgan, G. J., Martiat, P. and Goldman, J. M. (1991) Detection of residual leukemia after bone marrow transplantation for chronic myeloid leukemia: role of polymerase chain reaction in predicting relapse. *Blood* **77**, 874–8.

Imamura, J., Miyoshi, I. and Koeffler, H. P. (1994) p53 in hematologic malignancies. *Blood* **84**, 2412–21.

Janssen, J. W. G., Ludwig, W. D., Borkhardt, A. *et al.* (1994) Pre-pre-B acute lymphoblastic leukemia: high frequency of alternatively spliced ALL1-AF4 transcripts and absence of minimal residual disease during complete remission. *Blood* **84**, 3835–42.

McCarthy, K. M. and Wiedemann, L. M. (1995) The PCR-based detection of gene rearrangements. In *PCR applications in pathology* (ed. D. S. Latchman). Oxford University Press, Oxford (in press).

Price, C. M. (1993) Fluorescence in situ hybridisation. *Blood Reviews* **7**, 127–34.

Price, C. M., Kanfer, E. J., Colman, S. M. *et al.* (1992) Simultaneous genotypic and immunophenotypic analysis of interphase cells using dual color fluorescence: a demonstration of lineage involvement in polycythaemia vera. *Blood* **80**, 1033–8.

Rabbitts. T. H. (1994) Chromosomal translocations in human cancer. *Nature* **372**, 143–9.

Repp, R., Borkhardt, A., Haupt, E. *et al.* (1994) Detection of four different 11q23 chromosomal abnormalities by multiplex-PCR and fluorescence-based automatic DNA fragment analysis. *Leukemia* **9**, 210–13.

Ridge, S. A. and Wiedemann, L. M. (1994) Chromosome 11q23 abnormalities in leukaemia. *Leukemia and Lymphoma* **14**, 11–17.

van Rhee, F., Lin, F., Cullis, J. O. *et al.* (1994) Relapse of chronic myeloid leukemia after allogeneic bone marrow transplant: the case for giving donor leukocyte transfusions before the onset of hematological relapse. *Blood* **83**, 3377–83.

Westbrook, C. A., Hooberman, A. L., Spino, C. *et al.* (1992) Clinical significance of the BCR/ABL fusion gene in adult acute lymphoblastic leukemia; A cancer and leukemia group B study (8762). *Blood* **80**, 2983–90.

27

Genetic therapies

KAROL SIKORA

27.1 Introduction

We have witnessed an amazing era in biology over the last 50 years. We now have understanding of a wide range of physiological processes at a molecular level. Along with this new comprehension has come a remarkable set of tools to measure, evaluate and even to intervene in the course of many diseases. At a therapeutic level there are two main interelated thrusts. The first, the production of complex molecular therapies using the recombinant DNA technology, was developed in the mid 1970s and has given us highly purified pharmaceuticals that have already proven useful in many clinical syndromes. The second rapidly advancing area is the emerging discipline of gene therapy – the insertion or manipulation of DNA *in vivo*.

27.2 Genetic pharmacy

The advent of gene cloning rapidly led to a large number of hitherto elusive compounds becoming available in large quantities in a controlled manner. This made serious study of efficacy possible. The interferons are a good example. Discovered some 25 years ago, interferon was found to have partially beneficial effects against chronic viral infections such as hepatitis and cytomegalovirus infection and a range of cancers. However, batch to batch variation, a high level of potentially toxic contaminant cytokines and the high cost of purification led to very confusing results. With cloned interferon alpha, clinical studies in diseases such as cancer rapidly established their true role with remarkable consistency. We now routinely use recombinant interferon in the treatment of hairy cell leukaemia and chronic myeloid leukaemia, cutaneous T cell lymphoma, melanoma and renal cell cancer. Interleukin 2 is another example of a recombinant cytokine that is in our formularies for its anti-cancer effects but its true role is unclear. Eighteen different cytokines are now in clinical trial for different cancer types, although it is too early to say how effective they will turn out to be. They may also have a role in inflammatory and immunologically based diseases.

Recombinant vaccines have changed the face of immunization programmes, reducing complication risks and have allowed a new generation of far more potent products to enter routine use. Skilful design of interactive epitopes will almost certainly further enhance their efficacy. Unlimited quantities of novel targeting molecules may well be developed by learning the rules of *in vivo* three-dimensional molecular interaction. These could be used to carry drugs selectively to certain tissues or areas of disease as in cancer targeting (Table 27.1).

27.3 Gene therapy

The widespread application of gene therapy techniques to many diseases is already breaking down the traditional boundaries into which modern high technology medicine has been split. The drive to develop novel vector systems and techniques to knock out specific genes and to selectively control transcription applies equally to cardiology, nephrology, paediatrics and oncology. The common problem is to get access to the genome in a controlled manner, ideally by homologous recombination throughout a tissue or a whole patient. Many of the problems that currently appear insurmountable will be solved over the next decade. There are already over 100 protocols accepted for clinical trial – the majority in the United States. Most are for cancer where the risk – benefit ratio is the lowest.

There are many diseases being targeted for genetic intervention (Table 27.2). Initially simple, single gene inherited disorders seemed the best candidates. However, even establishing a protein factory in a patient with a non-functional protein variant, e.g. factor VIII in haemophilia has proved a complex task. Furthermore, the risks of insertional mutagenesis with the possibility of cancer are significant. With other congenital disorders irreversible damage may have already been incurred. A good example is the CNS problems in children with the Lesch–Nyhan syndrome. Cystic fibrosis is a more realistic target in the short term. Here the mutant gene – the cystic fibrosis transmembrane regulator – is cloned and the relevant tissue accessible to manipulation by direct bronchoscopy.

By far the majority of research protocols are for cancer treatment. Several strategies are being examined and are outlined below.

Table 27.1. *Gene therapy – diseases*

- Life threatening
- Gene responsible cloned
- Precise regulation of gene not essential
- Delivery system available
- Measurable surrogate end points
- Potentially reversible condition

Table 27.2. *Gene therapy – current targets*

– Immunodeficiencies
– Cystic fibrosis
– Haemoglobinopathies
– Metabolic disorders
Gaucher's disease
Lesch–Nyhan syndrome
hypercholesterolaemia
urea cycle abnormalities
– Arthritis
– HIV infection
– CNS diseases
– Muscle disorders
– Cardiovascular diseases
– Cancer

27.3.1 Genetic tagging

A marker gene is used to follow the course of a tumour through treatment. Decisions on the number of cycles of chemotherapy can be based on the logical assessment of minimal residual disease. In autologous bone marrow transplantation, genetic marking of the marrow prior to its reinfusion will allow the determination of origin of any subsequent recurrence.

27.3.2 Tumour vaccines

One of the reasons why tumours do not elicit a strong immune response is the inability of T helper lymphocytes to recognize them as foreign. By inserting an expression vector containing cytokine genes it may be possible for genetically modified vaccines to attract cytolytic T cells directly.

27.3.3 Selective drug activation

Tumour- or tissue-specific promoters can be coupled upstream of prodrug activating enzymes and used as suicide vectors. Some of the commonly used anticancer drugs can thus be given considerably increased selectivity.

27.3.4 Correction of genetic lesions

The use of antisense constructs to mutated oncogenes or the correct version of tumour suppressor genes can suppress the malignant phenotype *in vitro*. Despite the lack of good systemic vector systems for gene delivery, clinical studies mimicking these approaches are underway.

The remarkable advances in molecular genetics will change the way we practice medicine. We are clearly rapidly leaving an age of therapeutic empiricism.

Appendices

APPENDIX 1. Selected bibliography

Molecular biology monographs

Sambrook, J., Fritsch, E. F. and Maniatis, T. (1989) *Molecular cloning: a laboratory manual*. Cold Spring Harbour Laboratory Press, Cold Spring Harbor, New York.

Goeddel, D. (1989) *Gene expression technology*. Raven Press, New York.

Ausabel, F. M. (1990) *Current protocols in molecular biology*. John Wiley & Son Inc., New York.

Boothwell, A., Yancopoulos, G. D. and Alt, F. W. (1990) *Methods for cloning and analysis of eukaryotic genes*. Jones and Barlett Publishers, Boston.

Innis, M. A., Gelfand, D. H., Sninsky, J. J. and White, T. J. (1990) *PCR protocols*. Academic Press, San Diego.

Trent, R. J. (1993) *Molecular medicine*. Churchill Livingstone, Edinburgh.

Wilkinson, D. G. (1992) *In situ hybridization; a practical approach*. IRL/Oxford University Press, Oxford.

Harlow, E. and Lane, D. (1988) *Antibodies; a laboratory manual*. Cold Spring Harbor Laboratory, New York.

Cuello, A. L. (1993) *Immunocytochemistry II*. John Wiley & Sons, Chichester.

Molecular biology periodicals

Biotechniques, *Biotechnology*, *Cell Current Opinion* series, *EMB*, *Gene*, *Nucleic Acid Research*, *Nature*, *Nature Medicine*, *Proceedings of the National Academy of Sciences*, *Science*, *Trends in Biotechnology* and *Trends in Genetics (TIGS)*.

Other relevant periodicals

British Journal of Cancer, *International Journal of Cancer*, *Journal of Bacteriology*, *Journal of Virology*, *Journal of Biological Chemistry*, *Journal of Cell Biology*, *Oncogene*, *Cancer Research* and *Trends in Biochemical Sciences (TIBS)*.

APPENDIX 2. Commercial suppliers

For molecular genetic materials

Biotechnology Research Enterprises, Clontech, Invitrogen, New England Biolabs, Pharmacia, Promega, Stratagene, United States Biochemicals, Sigma, Techne, 5prime–3prime Inc.

For consumables

Boehringer Mannheim Biochem, Dupont Nen Research products, Bethesda Research Laboratories (Gibco/BRL), Bio-Rad laboratories, Sartorius, Whatman.

Radiochemicals

Amersham and Du Pont.

APPENDIX 3. Equipment

Facilities

Dark room, cold room, wet-ice machine and dry-ice storage, fume hood, autoclave, gas and double-distilled water supply, radiochemical disposal, incineration room.

Equipment

1. Centrifuges: microcentrifuge, desk top centrifuge for use with radioactive materials, refrigerated centrifuge and ultracentrifuge.
2. Heating appliances: shaking water bath (variable temp, 37 °C, 65 °C), incubator, 200 °C oven, microwave oven and dri-block Eppendorf tube heater plate.
3. Electrical appliances: high (> 2000 V) and low (< 200 V) voltage power packs.
4. Electrophoresis apparatus: horizontal electrophoresis slab gel apparatus and vertical polyacrylamide gel running apparatus.
5. Mixers and shakers: vortex rotamixer, shaking tray and heater/magnetic stirrer and stirrer bars/fleas.
6. Coolers; fridge, freezer and -80 °C freezer/liquid nitrogen supply.
7. Tissue homogenizer.
8. UV and visible light spectrophotometer (and plastic and crystal cuvettes).
9. pH meter (2–12 range and accurate to two decimal points), double junction gel-filled electrode.
10. Thermal cycler PCR machine.
11. Vacuum drier, spin vac or vacuum pump and lyophilizer.
12. 10, 20, 200, 1000 and 5000 μl pressure displacement pipettes or, even better, volume displacement pipetters.
13. Bunsen burner.
14. Accurate balance (0.01 g–100 g range).
15. UV crosslinker.
16. Densitometer or image analysis system.

Equipment required for radioactive work

1. UV shielded goggles.
2. Radiation shielded goggles.
3. Radioactive emission detectors: Gamma counter, Geiger–Mueller counter, liquid scintillation counter (542A/544A type for very high β emitters ^{32}P and high γ rays

^{125}I and bremsstrahlung) (E type for high energy β (^{32}P) and EL type for soft energy β (^{35}S).

4. Autoradiography cassettes/intensifying screens (important for β emitters).
5. MP-4 (Polaroid) camera with ultraviolet transilluminator.
6. Hybridization oven (optional).

APPENDIX 4. Consumables

1. Glassware: beakers, biurets, duran bottles and glass pipettes (narrow and agar tips), glass funnels, glass cell scrapers.
2. Pipette tips (10 μl, 200 μl, 1 ml and 5 ml), microfuge tubes (1.55 and 0.6 ml volumes).
3. Gloves.
4. Absorbent paper (towels) and Whatman filter paper.
5. Bleach.
6. Alcohols, acids and alkalis.
7. 3.5 cm and 9 cm bacterial plates.
8. Sterile reagent 0.2 and 0.45 μm filters (115 and 250 ml).
9. Nitrocellulose Hybond C (pore size 0.1 μm) and Nylon Hybond N+ membranes.
10. Sterile toughened plastic bags (radioactive and general use).
11. Falcon 2059 (10 ml) and 2070 (50 ml) tubes.
12. Centrifuge polypropylene tubes 10 ml, 50 ml and 250 ml.
13. Cling film/Saran wrap.
14. Stretch film (Parafilm).
15. Silver foil.
16. Filter paper Grade 1 (large pore, 18.5 or 24 cm diameter) and Grade 3 (medium pore).
17. Box Sterilin 20 ml bottles.
18. Polaroid 667 (ASA 3000) for UV films.
19. Autoclave tape.
20. Minisart NML 0.2 μm syringe filter holders (Sartorius).
21. Tissue culture flasks, 25, 75 and 250 cm^3.
22. Kodak medium and large gel film (with or without non-scratch coating).
23. Nalgene 0.22 μm and 0.45 μm syringe filters.
24. Weighing boats.
25. Permanent marker pens.
28. Dako paraffin pen.
24. Liquid media:
    LB (Luria Bertani) broth for bacteria: an undefined medium, tryptone 10 g/l, yeast extract 5 g/l and NaCl 10 g/l.
    Dulbecco's modified Eagles medium for mammalian cells.
    Fetal calf serum (FCS).

APPENDIX 5. Solutions

Acids

Conc hydrochloric acid = 11.6 molar
Conc nitric acid = 15.7 molar
Conc sulphuric acid = 18.3 molar
Conc formic acid = 23.6 molar
Conc acetic acid = 17.4 molar

Alkalis

1 M sodium hydroxide: 4 g NaOH made up to 100 ml with distilled water and auto-claved.

Antibiotics

Add antibiotic at or below 50 °C to media from filter-sterilized stock solutions.

Ampicillin × 200 stock: 10 mg/ml in 50% alcohol (50 μg/ml in solid agar plates but 25 μg/ml in broth).
Tetracycline × 500 stock: 12.5 mg/ml in 75% alcohol (25 μg/ml for solid agar but 12.5 μg/ml for broth). Light sensitive therefore wrap in silver foil.
Kanamycin × 700 stock: 5 mg/ml in water (50 μg/ml in plates and broth).

Buffers (selected list only)

Molecular buffers

0.5 M EDTA (pH 7, 7.5, 8)

18.6 g disodium salt of EDTA (ethanolamine diamine tetraacetic acid) in 80 ml, pH with very strong NaOH solution or NaOH pellets, add water to 100 ml. (NB this compound will not dissolve until the pH is adjusted to near 7).

3 M Sodium acetate (pH 5.2)

24.6 g sodium acetate (anhydrous) in 80 ml water. Adjust pH to 5.2 with glacial acetic acid and make up to 100 ml with water.

Phosphate buffered saline (PBS)

For 1 l: 8 g NaCl, 0.34 g $KH_2 PO_4$, 2.21 g $Na_2 HPO_4$, adjust pH to 7.3 and autoclave.

1 M Tris/HCl (pH 7.4, 8 or 9.5)

12.1 g Tris base in 80 ml water. Adjust pH as required using conc. HCl and make up to 100 ml.

10 mM Tris/HCl (pH 8) containing 1 mM EDTA

0.12 g Tris base and 20.3 mg of disodium EDTA in 80 ml, adjust pH then add water to 100 ml.

Electrophoresis buffers

TAE × 50

121 g Tris base in 200 ml water, 28 ml glacial acetic acid and 100 ml 0.5 M EDTA (pH = 8.0). Adjust final pH to 8.3 and make up to 500 ml.

TBE × 5

54 g Tris base, 27.5 g boric acid and 20 ml 0.5 M EDTA (pH = 8), make up to 1 l.

Protein analysis solutions

Peptide extraction buffer

1 M hydrochloric acid, 5% (v/v) formic acid, 1% (w/v) sodium chloride, 1% trifluro-acetic acid (TFA), 1 mg per ml ascorbic acid if oxidation of the peptide is likely, i.e. if rich in methionine.

Peptide SDS gel-loading buffer

100 mM Tris/HCl (pH 6.8), 4% (w/v) SDS (sodium dodecyl sulphate) (electrophoresis grade), 0.2% (w/v) Bromophenol Blue, 20% (v/v) glycerol, 200 mM dithiothreitol (add this just prior to use).

RNA analysis solutions

Phenol preparation

1. Take 100 ml water-saturated redistilled phenol (phenol must be distilled), add 100 ml 10 × TES buffer (below), mix well and leave for 1 h at room temp., remove aqueous phase, repeat 1–2 times.
2. Add 100 ml 1 × TES buffer, mix well, stand at room temp. then remove aqueous phase and discard, repeat 1–2 times.
3. Add 100 ml 1 × TES buffer and then 100 ml chloroform and store at 4 °C in the dark (with time the phenol will oxidize and change to an orange colour).

10 × TES buffer

30 g Tris base, 50 ml 0.5 M EDTA pH 8 (autoclaved), 2.5 g SDS, make up to 250 ml with autoclaved distilled water and filter-sterilize through a 0.2 μm membrane filter.

Deionized formamide

If formamide is yellow add 5 g Dowex XG8 mixed-bed resin for each 100 ml sample, stir using a magnetic stirrer for 1 h and filter twice through Whatman No 1 and store in batches at −70 °C (mixed-bed resin consists of Resin H+ and Resin OH− added to ions such as NaCl → Resin Na$^+$ and resin Cl$^-$ and water, i.e. neutralization yields water).

MOPS × 10 stock

41.8 g MOPs, 3.72 g EDTA (di-sodium), 4.10 g sodium acetate, make volume to 900 ml with distilled water, adjust pH to 7 with NaOH crystals (or > 5 M NaOH solution), increase volume to 1 l. (This solution is light-sensitive; straw-coloured MOPs will work well but darker buffer does not.)

50 × Denhardt's

1% (w/v) BSA, 1% (w/v) Ficoll, 1% (w/v) PVP (polyvinyl pyrollidone).

mRNA lysis buffer

400 μl 5 M NaCl, 2 ml 1 M Tris/HCl (pH 7.5), 1.5 μl 1 M MgCl$_2$, 1 ml 20% (w/v) SDS, 100 μl Proteinase K (20 mg/ml), 6.5 ml water. (This is the amount required for 10 ml/10^8 cells.)

RNA running buffer × 5

37.28 g KCl, 1.21 g Tris base, make up to 180 ml with distilled water, adjust pH to 7.5 with HCl, then make up to 200 ml.

RNA sample buffer

200 μl × 10 MOPs (see above), 1 ml formamide (preferably deionized), 356 μl formaldehyde (38% v/v stock).

RNA dye solution

FBX

15% (w/v) Ficoll 400, 0.25–2% (w/v) Bromophenol Blue, 0.25% (w/v) xylene cyanol in distilled water.

GBX

30% (v/v) glycerol, 0.25% (w/v) Bromophenol Blue, 0.25% (w/v) xylene cyanol (store at 4 °C).

RNA extraction/homogenization buffer

4 M (47.3 g) guanidinium isothiocyanate (or thiocyanate), 10 ml of 50 mM sodium citrate pH 7 (DEPC autoclaved), × 20 SSC 8.3 ml, 2.5 ml of 20% (w/v) stock sodium lauryl sarcosyl.

RNA cushion

5.7 M CsCl in 0.01 M EDTA, pH 7.5 (make up in 100 ml batches by dissolving 96 g CsCl in 75 ml distilled water, add 20 ml 0.5 M EDTA pH 7.5, and add DEPC, stand overnight, autoclave, then adjust volume to 100 ml with DEPC (diethylpyrocarbonate) treated water).

SSC × 20 stock

174 g sodium chloride, 88.2 g trisodium citrate dihydrate in 800 ml distilled water, adjust pH to 7 with 5 M NaOH pellets and make up to 1 l.

SSPE × 20

3.6 M NaCl, 0.2 M sodium phosphate, 0.02 M EDTA, pH 7.7.

DNA extraction

STE

15% (w/v) sucrose, 25 mM Tris/HCL pH 7.5, 10 mM EDTA, filter sterilize and store at 4 °C.

TELT

50 mM Tris/HCl pH 8, 62.5 mM EDTA pH 8, 2.5 M lithium chloride, 4% (v/v) Triton X-100.

Transformation buffer (FSB – used for cells for storage)

10 mM potassium acetate (pH 7.5), 45 mM $MnCl_2$, 10 mM $CaCl_2$, 100 mM KCl, 3 mM hexamine cobalt chloride, 10% (v/v) glycerol.

Index

glutamine 382–3
glutaraldehyde
    fixation 62, 126, 173
    in klh-peptide coupling 101
glyceraldehyde-3-phosphate dehydrogenase
        (GAPDH) 168, 281
glycerol 56, 355
glyoxal 186
growth
    curves 42–4, 157
    rate measurements 151–9
growth hormone gene 361–2
guanidinium–caesium chloride (CsCl) method
    DNA extraction 196
    RNA extraction 180–2

H3 histone 159
haemangiopericytoma 214
haematological malignancies 205–6, 426–30
    clinical intervention 429
    gene translocations 426–8
    point mutations 428
    see also leukaemia
haemocytometer 85, 86
haemopoietic cell lines 66
Hanker's method 139
HAT medium 103
health hazards 398–400
HeLa cells 47, 48
hepatocytes, cultured 70, 72, 93
hereditary non-polyposis colorectal cancer
        (HNPCC) 211, 232, 421
heteroduplex analysis 235
heterozygosity 222
    loss of, *see* loss of heterozygosity
histones, proliferation specific 159
history, of molecular genetics 6–9
Hoechst 33258 test 395–6
homeobox (*HOX*) genes
    in human disease 415–16
    identification of novel 291, 293, 295, 296
homogenization 180, 181, 402–3
horseradish peroxidase
    in immunocytochemistry 118
    in *in situ* hybridization 129
    probe labelling/detection 203
host range, modification of 398
HOT (hydroxylamine and osmium tetroxide;
        CCM) technique 235, 238–41, 247
    automation 246
    carbodiimide alternative 236
    protocol 240–1
HPLC, reverse-phase 379
HT-29 colorectal carcinoma cells 48
HT medium 103
HTLV-I 377
human genome project 423
Huntington's disease (HD) 409–10
hyaluronidase 53, 206
hybridization 12
    comparative genome (CGH) 233–5
    in situ, *see in situ* hybridization
    low stringency 291–3

non-specific 131–2
    in Northern blotting 188–90
    in Southern blotting 200
    subtractive 290–1
hybridomas
    cloning by limiting dilution 107–8
    formation 103–5
    screening 105–7
hydrogen peroxide (H$_2$O$_2$) 137, 138
hydroxylamine and osmium tetroxide (HOT)
        analysis, *see* HOT technique
hygromycin B resistance 352

image analysis 122
immortalization, cell 38, 45–6, 74
immune complexes, purification 177
immunization 99–102
immunoblotting 117
immunocytochemistry 113–22
    antibodies 116
    assessment of gene expression 333
    bromodeoxyuridine 154
    combined with *in situ* hybridization 134, 135
    combined with proliferation markers 158–9
    controls 121, 405
    cultured cell lines 64–5
    equipment 21
    immunostaining methods 117–19, 136–41
    multiple immunostaining 121–2
    proliferation markers 154–5
    protocols 134–41
    quantification 122
    testing antibody specificity 117
    tissue fixation/processing 114–16, 135–6
immunodiffusion 117
immunoelectrophoresis 117
immunofluorescence 117–18
    double 121
    indirect 117, 136–7, 139–41
immunoglobulin genes 429
immunoperoxidase method, indirect 118,
        137–8, 139–41
immunoprecipitation (IP), proteins 175–7
immunostaining
    immunocytochemistry 117–19, 136–41
    multiple 121–2
    Western blots 173–5
immunotherapy 423
IMR-32 neuroblastoma cells 48
in situ end labelling (ISEL) 160
in situ hybridization (ISH) 122–34
    assessment of gene expression 333
    automation 132
    combined with immunocytochemistry 134,
        135
    combined with PCR (*in situ* PCR) 133–4
    combined with proliferation studies 159
    controls and troubleshooting 131–2
    ease of application 132
    equipment 21
    fluorescence, *see* fluorescence *in situ*
        hybridization
    future developments 132–4

by microinjection 358
protoplast fusion 358
retroviral vectors, *see* retroviral vectors
protocols 362–71
stable 352
transient 333, 334–8, 351–2
protocol 335–7
troubleshooting 338
transfer RNA (tRNA) 12, 13, 14
function 14–15
transformation
bacterial 352–3, 354, 362–3
cell 45–6, 74–8
E. coli with recombinant DNA 267–9, 309
transformed cells 46
anchorage independence 49
characterization 47
growth in culture 43
transforming growth factor alpha (TGFα) 168,
271–2
translation 14–15
translocations
in haematological malignancies 426–8
in solid tumours 214
Transwells 69, 77
trinucleotide repeats, unstable 229–31, 410
protocol 230–1
triplex-forming oligonucleotides 378–9
tritium (³H) 22, 23
probe labelling 124
troubleshooting 402–5
Trypan Blue 85, 86
trypsin
antigen 'unmasking' 114
in cell line generation 41, 54
electroporation 369
inhibitor (PMSTI) 169
tissue disaggregation 40, 52
tumour cells 46, 74–5
cytotoxicity assays 87–92, 93, 95
genetic changes 210–11, 218
invasion and metastasis 76–8
tumour suppressor genes 231, 234
tumourigenesis, multistep 210–18
tumourigenicity 74, 75–6
tumours 420–3
allele loss and instability 231–3
cytogenetic analysis 206–7, 233
diagnostic assessment 420–1
gene therapy 433–4
genetic changes 205–18, 233
genetic tagging 434
prognostic assessment 422
restriction fragment length polymorphisms
208, 209
therapeutic intervention 422–3
tissue processing 402
293 cells 361

ultracentrifugation 180, 181
unlabelled antibody method, *see* peroxidase
anti-peroxidase method

uridine, tritiated ([³H]uridine) 87
UV light 24–5

vaccines
recombinant 432–3
tumour 434
van der Waals forces 17
variable number tandem repeats (VNTRs)
224–5
Vectabond 126
vectors
adenoviral 354, 360–1, 370–1, 412
cloning 27, 256
cDNA libraries 273, 285–6
DNA sequencing 304–5
ligation 266–7
preparation 265–6
expression 27
reporter 27
retroviral, *see* retroviral vectors
screening 315
transcription 27
Vent 31, 195, 284, 296
Vero cells 48
viability
assays 84–7
subcultured cells 44
videography, time lapse 156–7
vimentin 63–4, 65
vinca alkaloids 157
viral-directed enzyme prodrug therapy
(VDEPT) 422
viral vectors 399, 422–3
see also adenoviral vectors; retroviral vectors
viruses, contaminating 394–5

Waardenburg's syndrome 416
water, type 1 405
wax sections
immunocytochemistry 136, 138
in situ hybridization 126
Western blotting 117, 167, 168, 169–75
assessment of gene expression 333
blotting procedure 172–3
controls 405
immunostaining 173–5
protein extraction 169–70
SDS-PAGE electrophoresis 170–2
troubleshooting 175
whole mount preparations 115–16

X-gal (BCIG) 28, 267, 268
xylene cyanol 24, 186, 297

yeasts
contamination by 393, 405
plasmids 28–9

Zamboni's solution 115, 135–6
ZR-75-1 mammary carcinoma cells 48